Lecture Notes in Mathematics 1977

Editors:
J.-M. Morel, Cachan
F. Takens, Groningen
B. Teissier, Paris

FONDAZIONE
CIME
ROBERTO CONTI

CENTRO INTERNAZIONALE MATEMATICO ESTIVO
INTERNATIONAL MATHEMATICAL SUMMER CENTER

C.I.M.E. means Centro Internazionale Matematico Estivo, that is, International Mathematical Summer Center. Conceived in the early fifties, it was born in 1954 and made welcome by the world mathematical community where it remains in good health and spirit. Many mathematicians from all over the world have been involved in a way or another in C.I.M.E.'s activities during the past years.

So they already know what the C.I.M.E. is all about. For the benefit of future potential users and co-operators the main purposes and the functioning of the Centre may be summarized as follows: every year, during the summer, Sessions (three or four as a rule) on different themes from pure and applied mathematics are offered by application to mathematicians from all countries. Each session is generally based on three or four main courses (24–30 hours over a period of 6-8 working days) held from specialists of international renown, plus a certain number of seminars.

A C.I.M.E. Session, therefore, is neither a Symposium, nor just a School, but maybe a blend of both. The aim is that of bringing to the attention of younger researchers the origins, later developments, and perspectives of some branch of live mathematics.

The topics of the courses are generally of international resonance and the participation of the courses cover the expertise of different countries and continents. Such combination, gave an excellent opportunity to young participants to be acquainted with the most advance research in the topics of the courses and the possibility of an interchange with the world famous specialists. The full immersion atmosphere of the courses and the daily exchange among participants are a first building brick in the edifice of international collaboration in mathematical research.

C.I.M.E. Director
Pietro ZECCA
Dipartimento di Energetica "S. Stecco"
Università di Firenze
Via S. Marta, 3
50139 Florence
Italy
e-mail: zecca@unifi.it

C.I.M.E. Secretary
Elvira MASCOLO
Dipartimento di Matematica
Università di Firenze
viale G.B. Morgagni 67/A
50134 Florence
Italy
e-mail: mascolo@math.unifi.it

For more information see CIME's homepage: http://www.cime.unifi.it

CIME activity is carried out with the collaboration and financial support of:

– INdAM (Istituto Nazionale di Alta Mathematica)

Matthew J. Gursky · Ermanno Lanconelli
Andrea Malchiodi · Gabriella Tarantello
Xu-Jia Wang · Paul C. Yang

Geometric Analysis and PDEs

Lectures given at the
C.I.M.E. Summer School
held in Cetraro, Italy
June 11–16, 2007

Editors:
Antonio Ambrosetti
Sun-Yung Alice Chang
Andrea Malchiodi

 Springer

FONDAZIONE
CIME
ROBERTO CONTI

Editors
Alice Chang
Princeton University
Fine Hall
Washington Road
Princeton, NJ 08544-1000
USA
chang@math.princeton.edu

Andrea Malchiodi
SISSA
Via Beirut 2-4
34014 Trieste
Italy
malchiod@sissa.it

Antonio Ambrosetti
SISSA
Via Beirut 2-4
34014 Trieste
Italy
ambr@sissa.it

Authors: see List of Contributors

ISSN 0075-8434   e-ISSN: 1617-9692
ISBN: 978-3-642-01673-8   e-ISBN: 978-3-642-01674-5
DOI: 10.1007/978-3-642-01674-5
Springer Dordrecht Heidelberg London New York

Library of Congress Control Number: 2009926187

Mathematics Subject Classification (2000): 35J60, 35A30, 35J20, 35H20, 53C17, 35Q40

Cover design: SPi Publisher Services

Printed on acid-free paper

Springer is part of Springer Science+Business Media (www.springer.com)

Preface

This volume contains the notes of the lectures delivered at the CIME course *Geometric Analysis and PDEs* during the week of June 11–16 2007 in Cetraro (Cosenza). The school consisted in six courses held by M. Gursky (*PDEs in Conformal Geometry*), E. Lanconelli (*Heat kernels in sub-Riemannian settings*), A. Malchiodi (*Concentration of solutions for some singularly perturbed Neumann problems*), G. Tarantello (*On some elliptic problems in the study of selfdual Chern-Simons vortices*), X.J. Wang (*The k-Hessian Equation*) and P. Yang (*Minimal Surfaces in CR Geometry*).

Geometric PDEs are a field of research which is currently very active, as it makes it possible to treat classical problems in geometry and has had a dramatic impact on the comprehension of three- and four-dimensional manifolds in the last several years. On one hand the geometric structure of these PDEs might cause general difficulties due to the presence of some invariance (translations, dilations, choice of gauge, etc.), which results in a lack of compactness of the functional embeddings for the spaces of functions associated with the problems. On the other hand, a geometric intuition or result might contribute enormously to the search for natural quantities to keep track of, and to prove regularity or a priori estimates on solutions. This two-fold aspect of the study makes it both challenging and complex, and requires the use of several refined techniques to overcome the major difficulties encountered. The applications of this subject are many while for the CIME course we had to select only a few, trying however to cover some of the most relevant ones, with interest ranging from the pure side (analysis/geometry) to the more applied one (physics/biology). Here is a brief summary of the topics covered in the courses of this school.

M. Gursky treated a class of elliptic equations from conformal geometry: the general aim is to deform conformally (through a dilation which depends on the point) the metric of a given manifold so that the new one possesses special properties. Classical examples are the uniformization problem of two-dimensional surfaces and the Yamabe problem in dimension greater or equal to three, where one requires the Gauss or the scalar curvature to become

constant. After recalling some basic facts on these problems, which can be
reduced to semilinear elliptic PDEs, Gursky turned to their fully nonlinear counterparts. These concern the prescription of the symmetric forms in
the eigenvalues of the *Schouten tensor* (a combination of the Ricci tensor
and the scalar curvature), and turn out to be elliptic under suitable conditions on their domain of definition (admissible functions). The solvability
of these equations has concrete applications in geometry, since for example they might guarantee pinching conditions on the Ricci tensor, together
with its geometric/topological consequences. After recalling some regularity
estimates by Guan and Wang, existence was shown using blow-up analysis techniques. Finally, the functional determinant of conformally invariant
operators in dimension four was discussed: the latter turns out to have a universal decomposition into three terms which respectively involve the scalar
curvature, the Q-curvature and the Weyl tensor. Some conditions on the
coefficients of these three terms guarantee coercivity of the functional, and in
these cases existence of extremal metrics was obtained using a minimization
technique.

E. Lanconelli covered some topics on existence and sharp estimates on
heat kernels of subelliptic operators. Typically, in a domain or a manifold
Ω of dimension n, k vector fields $X_1, \ldots, X_k$ are given (with $2 \leq k < n$)
which satisfy the *Hörmander condition*, namely their Lie brackets span all
of the tangent spaces to Ω. One considers then linear operators L (or their
parabolic counterpart) whose principal part is given by $\sum_{i=1}^{k} X_i^2$. During the
lectures, existence and regularity (Hörmander) theory for such operators was
recalled, and in particular the role of the *Carnot-Caratheodory distance*, mea-
sured through curves whose velocities belong to the linear span of the X_i's.
This distance is not homogeneous (at small scales), and it is very useful to
describe the degeneracy of the operators in the above form. One of the main
motivations for this study is the problem of prescribing the *Levi curvature*
of boundaries of domains in $\mathbb{C}^n$, which for graphs amounts to solving a fully
nonlinear degenerate equation, whose linearization is of the form previously
described. Gaussian bounds for heat kernels were then given, first for constant
coefficient operators modeled on Carnot groups, and then for general operators using the method of the parametrix. Finally, applications to Harnack
type inequalities were derived in terms of the heat kernel bounds.

The course by A. Malchiodi on singularly perturbed Neumann problems
dealt with elliptic nonlinear equations where a small parameter (the singular
perturbation) is present in front of the principal term (the Laplacian). The
study is motivated by considering a class of reaction-diffusion systems (in particular the Gierer-Meinhardt model) and the (focusing) nonlinear Schrödinger
equation. First a finite-dimensional reduction technique, which incorporates
the variational structure of the problem, was presented: by means of this
method, existence of solutions concentrating at points of the boundary of
the domain was studied. Here the geometry of the boundary is significant,
as concentration occurs at critical points of the mean curvature. After this,

existence of solutions concentrating at the whole boundary was proved: the phenomenon here is rather different, since the latter family has a diverging Morse index when the singular perturbation parameter tends to zero. Initially, accurate approximate solutions were constructed (depending on the second fundamental form of the domain), and then the invertibility of the linearized equation was shown (primarily using Fourier analysis), which made it possible to prove existence using local inversion arguments.

G. Tarantello's course focused on self-dual vortices in Chern-Simons theory. The physical phenomenon of superconductivity is described by a system of coupled gauge-field equations whose components stand for the wave function and the electromagnetic potential. A relatively well understood model is the abelian-Higgs (corresponding in a non-relativistic context to the Ginzburg-Landau) variant, for which much has been accomplished even away from the self-dual regime. A more sophisticated alternative is the Chern-Simons model, which compared to the previous one has the advantage of predicting the fact that gauge vortices carry electrical charge in addition to magnetic flux, although its mathematical description is at the moment less complete. After describing the main features of these models, Tarantello presented the approach of Taubes to the selfdual regime for the abelian-Higgs case, which reduces the system to a semilinear elliptic problem with exponential nonlinearities. This method partially extends to the Chern-Simons case, where some natural requirements on solutions can be proved, like their asymptotic behavior at infinity, integrability properties and the decay of their derivatives. The structure of C-S vortices is however more rich (and, as we remarked, far from being completely characterized) in comparison to the abelian-Higgs ones: in addition to the *topological solutions*, which have a well defined winding number at infinity, there are also non-topological solutions, which display different asymptotic behavior.

X.J. Wang treated k-Hessian equations, a class of fully nonlinear equations related to the problem of prescribing the Gauss curvature of a hypersurface, and to the Monge-Ampére equation, which is of interest in complex geometry. First the class of *admissible functions* was defined, where the equations are elliptic, and then existence for Dirichlet boundary value problems was obtained by means of a priori estimates and a continuation argument. Next, interior gradient estimates were derived, which imply Harnack inequalities, plus Sobolev-type inequalities for admissible functions which vanish on the boundary of the domain: the embedding which follows from the latter inequality possesses compactness properties analogous to the classical ones, and makes it possible to derive L^∞ estimates for solutions of equations with sufficiently integrable right-hand sides. These estimates make it possible to treat equations with nonlinear (subcritical or critical) reaction terms, where min-max methods can be applied. After this, the notion of k-admissibility was extended to non smooth functions using the concept of hessian measure, and applied to the existence of weak solutions and to potential-theoretical results. Finally, parabolic equations and several examples were treated.

The course by P. Yang concerned minimal surfaces in three-dimensional CR manifolds, which possess a subriemannian structure modeled on the Heisenberg group $\mathbb{H}^1$. In this setting the volume form of M is naturally defined in terms of the contact form θ as $\theta \wedge d\theta$: the *p-area* and the *p-mean curvature* (p standing for *pseudo*) of a two-dimensional surface were defined looking at the first and second variation of the volume form. p-minimal (regular) graphs in $\mathbb{H}^1$ were then considered, showing that they are ruled surfaces. A study of the *singular set* (where the tangent plane to the surface coincides with the contact plane, the kernel of θ) was then performed, and it was shown that it consists either of isolated points or of smooth curves: applications were given to the classification of entire p-minimal graphs. Existence of weak solutions (minimizers) to boundary value problems (of Plateau type) was considered, showing the uniqueness, comparison principles and geometric properties of the solutions. The regularity issue was then discussed, which is a rather delicate one since the global C^2 regularity of minimizers might fail in general.

Some of the students who attended the school were also supported by SISSA, GNAMPA and by MiUR under the PRIN 2006 *Variational methods and nonlinear differential equations*.

Finally, we would like to express our warm gratitude to the CIME Foundation, to the CIME Director Prof. P. Zecca, to the CIME Board Secretary Prof. E. Mascolo and to the CIME staff for their invaluable help and support, and for making the environment in Cetraro so stimulating and enjoyable.

Alice S.Y. Chang
Antonio Ambrosetti
Andrea Malchiodi

List of Contributors

Matthew J. Gursky
Department of Mathematics, University of Notre Dame, Notre Dame,
IN 46556, USA, mgursky@nd.edu

Ermanno Lanconelli
Dipartimento di Matematica, Universita' degli Studi di Bologna, P.zza di
Porta S. Donato, 5 40127 - Bologna, Italy, lanconel@dm.unibo.it

Andrea Malchiodi
SISSA, Sector of Mathematical Analysis, Via Beirut 2-4, 34014 Trieste,
Italy, malchiod@sissa.it

Gabriella Tarantello
Universita' di Roma Tor Vergata-Dipartimento di Matematica, via della
ricerca scientifica, 00133 Rome, Italy, tarantel@mat.uniroma2.it

Xu-Jia Wang
Mathematical Sciences Institute, Australian National University, Canberra,
ACT 0200, Australia, wang@maths.anu.edu.au

Paul Yang
Department of Mathematics, Princeton University, Fine Hall, Washington
Road, Princeton, NJ 08544, USA, yang@math.princeton.edu

Contents

PDEs in Conformal Geometry

Matthew J. Gursky

1 Introduction

In these lectures I will discuss two kinds of problems from conformal geometry, with the goal of showing an important connection between them in four dimensions.

The first problem is a fully nonlinear version of the Yamabe problem, known as the σ_k-Yamabe problem. This problem is, in general, not variational (or at least there is not a natural variational interpretation), and the underlying equation is second order but possibly not elliptic. Moreover, in contrast to the Yamabe problem, there is very little known (except for some examples and counterexamples) when the underlying manifold is negatively curved.

The second problem we will discuss involves the study of a fourth order semilinear equation, and arose in the context of a natural variational problem from spectral theory. Despite their differences–higher order semilinear versus second order fully nonlinear, variational versus non-variational–both equations are invariant under the action of the conformal group, and we have to address the phenomenon of "bubbling." Therefore, in the first few sections of the notes we will present the necessary background material, including a careful explanation of the idea of a "standard bubble".

After covering the introductory material, we give a description of the σ_k-Yamabe problem, culminating in a sketch of the solution in the four-dimensional case. Modulo some technical regularity estimates, the proof is reduced to a global geometric result (Theorem 5.7) that is easy to understand.

M.J. Gursky
Department of Mathematics, University of Notre Dame, Notre Dame, IN 46556
e-mail: mgursky@nd.edu
Research supported in part by NSF Grant DMS-0500538.

S.-Y.A. Chang et al. (eds.), *Geometric Analysis and PDEs*,
Lecture Notes in Mathematics 1977, DOI: 10.1007/978-3-642-01674-5_1,

In the last section of the notes we discuss the functional determinant of a four-manifold, a variational problem which is based on a beautiful calculation of Branson-Ørsted. We end with a sketch of the existence of extremals for the determinant functional for manifolds of positive scalar curvature. Here, the missing technical ingredient is a sharp functional inequality due to Adams (Theorem 65), but the proof is again reduced to Theorem 5.7. Therefore, we see the underlying unity of the two problems in a very concrete way.

In closing, I wish to express my gratitude to the Fondazione C.I.M.E. for their invitation and their support. The success of the meeting *Geometric Analysis and PDEs* was a result of the considerable efforts of the local organizers (especially Andrea Malchiodi), the scientific contributions of the participants, and the hospitality of our hosts in Cetraro.

2 Some Background from Riemannian Geometry

In this section we review some of the basic notions from Riemannian geometry, including the basic differential operators (gradient, Hessian, etc.) and curvatures (scalar, Ricci, etc.) This is not so much an introduction to the subject–which would be impossible in so short a space–but rather a summary of definitions and formulas.

2.1 Some Differential Operators

1. The Hessian

Let (M^n, g) be an n-dimensional Riemannian manifold, and let ∇ denote the Riemannian connection.

Definition 2.1. The *Hessian* of $f : M^n \to \mathbf{R}$ is defined by

$$\nabla^2 f(X, Y) = \nabla_X df(Y). \tag{1}$$

It is easy to see the Hessian is symmetric, bilinear form on the tangent space of M^n at each point. In a local coordinate system $\{x^i\}$, the *Christoffel symbols* are defined by

$$\nabla_{\frac{\partial}{\partial x^i}} \frac{\partial}{\partial x^j} = \sum_k \Gamma_{ij}^k \frac{\partial}{\partial x^k}.$$

Using (1), in local coordinates we have

$$(\nabla^2 f)_{ij} = \nabla_i \nabla_j f$$
$$= \frac{\partial^2 f}{\partial x^i \partial x^j} - \sum_k \Gamma_{ij}^k \frac{\partial f}{\partial x^k}.$$

2. The Laplacian and Gradient

Definition 2.2. The *Laplacian* is the trace of the Hessian: Let $\{e_1, \ldots, e_n\}$ be an orthonormal basis of the tangent space at a point; then

$$\Delta f = \sum_i \nabla^2 f(e_i, e_i). \qquad (2)$$

In local coordinates $\{x^i\}$,

$$\Delta f = g^{ij} \left(\frac{\partial^2 f}{\partial x^i \partial x^j} - \sum_k \Gamma_{ij}^k \frac{\partial f}{\partial x^k} \right),$$

where $g^{ij} = (g^{-1})_{ij}$. Another useful formula is

$$\Delta f = \frac{1}{\sqrt{g}} \frac{\partial}{\partial x^i} \left(g^{ij} \sqrt{g} \frac{\partial f}{\partial x^j} \right),$$

where $g = det(g_{ij})$.

The *gradient vector field* of f, denoted ∇f, is the vector field dual to the 1-form df; i.e., for each vector field X,

$$g(\nabla f, X) = df(X).$$

In local coordinates $\{x^i\}$,

$$\nabla_j f = \sum_i g^{ij} \frac{\partial f}{\partial x^i}.$$

3. The Curvature Tensor

For vector fields X, Y, Z, the Riemannian curvature tensor of (M, g) is defined by

$$R(X, Y)Z = \nabla_{[X,Y]} Z - [\nabla_X, \nabla_Y] Z,$$

where $[\cdot, \cdot]$ is the Lie bracket. With respect to a local coordinate system $\{x^i\}$, the curvature tensor is given by

$$R \left(\frac{\partial}{\partial x^k}, \frac{\partial}{\partial x^l} \right) \frac{\partial}{\partial x^j} = \sum_i R^i_{jkl} \frac{\partial}{\partial x^i}.$$

Let $\Pi \subset T_p M^n$ be a tangent plane with orthonormal basis $\{E_1, E_2\}$. The *sectional curvature* of Π is the number

$$K(\Pi) = \langle R(E_1, E_2) E_1, E_2 \rangle.$$

($K(\Pi)$ does not depend on the choice of ON-basis.)

Example 1. For $\mathbf{R}^n$ with the Euclidean metric, all sectional curvatures are zero.

Example 2. Let $\mathbf{S}^n = \{\mathbf{x} \in \mathbf{R}^{n+1} \mid \|\mathbf{x}\| = 1\}$ with the metric it inherits as a subspace of $\mathbf{R}^{n+1}$. Then all sectional curvatures are $+1$.

Example 3. Let $\mathbf{H}^n = \{\mathbf{x} \in \mathbf{R}^n \mid \|\mathbf{x}\| < 1\}$, endowed with the metric

$$g = 4 \sum_i \frac{(dx^i)^2}{(1 - \|\mathbf{x}\|^2)^2}.$$

Then all sectional curvatures are -1.

The preceding examples are referred to as *spaces of constant curvature*, or *space forms*. A theorem of Hopf says that any complete, simply connected manifold of constant curvature is isometric to one of these examples (perhaps after scaling). Thus, curvature determines the local geometry of a manifold.

Another way of thinking about curvature is that it measures the failure of derivatives to commute:

Lemma 2.3. *In local coordinates,*

$$\nabla_i \nabla_j \nabla_k f - \nabla_j \nabla_i \nabla_k f = \sum_m R^m_{kij} \nabla_m f.$$

So third derivatives do not commute unless $R = 0$, i.e., the manifold is *flat*.

4. Ricci and Scalar Curvatures

Definition 2.4. The *Ricci curvature tensor* is the bilinear form $Ric : T_p M \times T_p M \to \mathbf{R}$ defined by

$$Ric(X, Y) = \sum_i \langle R(X, e_i) Y, e_i \rangle,$$

where $\{e_1, \ldots, e_n\}$ is an orthonormal basis of $T_p M$.

In local coordinates, the components of Ricci are given by

$$R_{ij} = \sum_m R_{ijm}^m.$$

For spaces of constant curvature, the Ricci tensor is just a constant multiple of the metric:

$$\mathbf{S}^n : \quad Ric = (n-1)g,$$
$$\mathbf{R}^n : \quad Ric = 0,$$
$$\mathbf{H}^n : \quad Ric = -(n-1)g.$$

The Ricci tensor is symmetric: $Ric(X,Y) = Ric(Y,X)$. Therefore, at each point $p \in M$ we can diagonalize Ric with respect to an orthonormal basis of T_pM:

$$Ric = \begin{pmatrix} \rho_1 & & & \\ & \rho_2 & & \\ & & \ddots & \\ & & & \rho_n \end{pmatrix}$$

where $(\rho_1, \ldots, \rho_n)$ are the eigenvalues of Ric. To say that (M^n, g) has positive (negative) Ricci curvature means that all the eigenvalues of Ric are positive (negative).

In two dimensions, the Ricci curvature is determined by the Gauss curvature K:

$$Ric = Kg.$$

Definition 2.5. The *scalar curvature* is the trace of the Ricci curvature:

$$R = \sum_i Ric(e_i, e_i),$$

where $\{e_1, \ldots, e_n\}$ is an orthonormal basis.

If $\{\rho_1, \ldots, \rho_n\}$ are the eigenvalues of the Ricci curvature at a point $p \in M$, then the scalar curvature is given by

$$R = \rho_1 + \cdots + \rho_n.$$

For the spaces of constant curvature, the scalar curvature is a constant function:

$$\mathbf{S}^n : \quad R = n(n-1),$$
$$\mathbf{R}^n : \quad R = 0,$$
$$\mathbf{H}^n : \quad R = -n(n-1).$$

Furthermore, in two dimensions the scalar curvature is twice the Gauss curvature:

$$R = 2K.$$

3 Some Background from Elliptic Theory

In this section we summarize some important results from functional analysis and the theory of partial differential equations.

1. Sobolev Spaces

These are important for discussing some of the PDE topics in these lectures. Let (M, g) be a compact Riemannian manifold. For $1 \leq k < \infty$ and $1 \leq p \leq \infty$, introduce the norms

$$\|u\|_{k,p}^p = \sum_{0 \leq j \leq k} \int |\nabla^j u|^p \, dV,$$

where $\nabla^j u$ denotes the iterated j^{th}-covariant derivative.

Example. For $k = 1$, $p = 2$,

$$\|u\|_{1,2}^2 = \int u^2 \, dV + \int |du|^2 \, dV.$$

The Sobolev space $W^{k,p}(M)$ is the completion of $C^\infty(M)$ in the norm $\|\cdot\|_{k,p}$.

Theorem 3.1. (Sobolev Embedding Theorems; see [GT83])
(*i*) *If*

$$\frac{1}{r} = \frac{1}{m} - \frac{k}{n},$$

then $W^{k,m}(M)$ is continuously embedded in $L^r(M)$:

$$\|u\|_r \leq C\|u\|_{k,m}.$$

(*ii*) *Suppose $0 < \alpha < 1$ and*

$$\frac{1}{m} \leq \frac{k - \alpha}{n}.$$

Then $W^{k,m}$ is continuously embedded in C^α.

(iii) *(Rellich-Kondrakov)* *If*

$$\frac{1}{r} > \frac{1}{m} - \frac{k}{n},$$

then the embedding $W^{k,m} \hookrightarrow L^r$ is compact: i.e., a sequence which is bounded in $W^{k,m}$ has a subsequence which converges in L^r.

2. Linear Operators

Consider the linear differential operator L:

$$Lu = a^{ij}(x)\partial_i\partial_j u + b^k(x)\partial_k u + c(x)u,$$

where the coefficients a^{ij}, b^k, c are defined in a domain $\Omega \subset \mathbf{R}^n$.

Definition 3.2. The operator L is *elliptic* in Ω if $\{a^{ij}(x)\}$ is positive definite at each point $x \in \Omega$. If there is a constant $\lambda > 0$ such that

$$a^{ij}(x)\xi_i\xi_j \geq \lambda|\xi|^2$$

for all $\xi \in \mathbf{R}^n$ and $x \in \Omega$, then L is *strictly elliptic* in Ω. If, in addition, there is another constant $\Lambda > 0$ such that

$$\Lambda|\xi|^2 \geq a^{ij}(x)\xi_i\xi_j \geq \lambda|\xi|^2,$$

then we say that L is *uniformly elliptic* in Ω.

We can formulate a similar definition for operators defined on a Riemannian manifold; e.g., by introducing local coordinates. Of course, the laplacian $L = \Delta$ is an example of a linear, uniformly elliptic operator.

3. Weak Solutions

We say that $u \in W^{1,2}(M)$ is a *weak* solution of the equation

$$\Delta u = f(x) \tag{3}$$

if for each $\varphi \in W^{1,2}$,

$$\int -\langle \nabla u, \nabla\varphi\rangle \, dV = \int f\varphi \, dV. \tag{4}$$

Of course, a smooth solution of (3) satisfies (4) by virtue of Green's Theorem. Weak solutions of elliptic equations like (3) in fact satisfy much better estimates, as we shall see.

4. Elliptic Regularity

Theorem 3.3. (See [GT83]) *Suppose* $u \in W^{1,2}$ *is a weak solution of*

$$\Delta u = f$$

on M.
(i) If $f \in L^m$, *then*

$$\|u\|_{2,m} \leq C(\|f\|_m + \|u\|_m). \tag{5}$$

(ii) (Schauder estimates) If $f \in C^{\ell,\alpha}$ *then*

$$\|u\|_{C^{\ell+2,\alpha}} \leq C(\|f\|_{C^{\ell,\alpha}} + \|u\|_{C^{\ell,\alpha}}). \tag{6}$$

How are such estimates used?

• To prove the regularity of weak solutions.

Weak solutions are often easier to find, for example, by variational methods.

• To prove *a priori* estimates of solutions, that is, estimates which are necessarily satisfied by any solution of a given equation.

Often *a priori* estimates can be combined with a topological argument to establish existence.

Example. To illustrate some of these results we consider an equation that will be an important model for much of the subsequent material.

Theorem 3.4. *Suppose* $u \geq 0$ *is a (weak) solution of*

$$\Delta u + c(x)u = K(x)u^p, \tag{7}$$

where c, K *are smooth functions, and*

$$1 \leq p < \frac{n+2}{n-2}.$$

If

$$\int u^{\frac{2n}{(n-2)}} \, dV \leq B, \tag{8}$$

then u *satisfies*

$$\sup_M u \leq C(p, B).$$

In fact, we can estimate u with respect to any Hölder norm, all in terms of p and B. The main point here is that the assumption $p < \frac{n+2}{n-2}$ is *crucial*.

Proof. Using the preceding elliptic regularity theorem, we know that u satisfies

$$
\begin{aligned}
\|u\|_{2,m} &\leq C\big(\|\Delta u\|_m + \|u\|_m\big) \\
&\leq C\big(\|u^p\|_m + \|u\|_m\big) \\
&\leq C\big(\|u\|_{mp}^p + \|u\|_m\big).
\end{aligned}
\tag{9}
$$

Denote

$$
m_0 = \frac{2n}{n-2},
$$

and choose m so that

$$
mp = m_0.
$$

It follows from (9) that

$$
\|u\|_{2,m} \leq C(p, B).
$$

We now use the Sobolev embedding theorem, which says

$$
\|u\|_r \leq C\|u\|_{2,m}
$$

where

$$
\frac{1}{r} = \frac{1}{m} - \frac{2}{n} = \frac{n-2m}{mn},
$$

or,

$$
r = \frac{mn}{n-2m} = \frac{(\frac{m_0}{p})n}{n - 2(\frac{m_0}{p})}.
$$

So, we've passed from one Lebesgue-space estimate to another. Have things improved?

The answer is yes, as long as

$$
\frac{(\frac{m_0}{p})n}{n - 2(\frac{m_0}{p})} > m_0.
$$

Solving this inequality, we see that it will hold provided p satisfies

$$
p < \frac{n+2}{n-2}.
$$

In this case, we iterate this process an arbitrary number of times, and conclude that

$$\|u\|_{2,m} \leq C(m, p, B) \quad \forall m \gg 1.$$

Once m is large enough, though, we can once more appeal to the Sobolev embedding theorem, part (ii), and conclude that u is Hölder continuous–in particular, u is bounded as claimed.

Remarks.

1. For higher order regularity we turn to the Schauder estimates, since we actually proved that u is Hölder continuous. Iterating the Schauder estimates, we can prove the Hölder continuity of derivatives of all orders.

2. As we mentioned above, and will soon see by explicit example, the preceding result is false if $p = (n+2)/(n-2)$. However, it can be "localized": that is, if

$$\int_{B(x_0,r)} u^{\frac{2n}{(n-2)}} \, dV \leq \epsilon_0$$

for some $\epsilon_0 > 0$ small enough, then

$$\sup_{B(x_0,r/2)} u \leq C(r).$$

3. A Corollary of Theorem 3.4 is that weak solutions of (7) are regular, for all $1 \leq p \leq (n+2)/(n-2)$.

4 Background from Conformal Geometry

In this, the final section of the introductory material, we present some basic ideas from conformal geometry.

Definition 4.1. Let (M^n, g) be a Riemannian manifold. A metric h is *pointwise conformal* to g (or just *conformal*) if there is a function f such that

$$h = e^f g.$$

The function e^f is referred to as the *conformal factor*. We used the exponential function to emphasize the fact that we need to multiply by a positive function (since h must be positive definite). However, in some cases it will be more convenient to write the conformal factor differently.

We can introduce an equivalence relation on the set of metrics: $h \sim g$ iff h is pointwise conformal to g. The equivalence class of a metric g is called its *conformal class*, and will be denoted by $[g]$. Note that

$$[g] = \{e^f g \mid f \in C^\infty(M)\}.$$

Definition 4.2. Let (M, g) and (N, h) be two Riemannian manifolds. A diffeomorphism $\varphi : M \to N$ is called *conformal* if

$$\varphi^* h = e^f g.$$

We say that (M, g) and (N, h) are *conformally equivalent*. Note h and g are pointwise conformal if and only if the identity map is conformal.

Example 1. Let $\delta_\lambda(x) = \lambda^{-1} x$ be the dilation map on $\mathbf{R}^n$, where $\lambda > 0$. Then δ_λ is easily seen to be conformal; in fact,

$$\delta_\lambda^* ds^2 = \lambda^{-2} ds^2,$$

where ds^2 is the Euclidean metric.

Example 2. Let $P = (0, \ldots, 0, 1)$ be the north pole of $\mathbf{S}^n \subset \mathbf{R}^{n+1}$. Let $\sigma : \mathbf{S}^n \setminus \{P\} \to \mathbf{R}^{n+1}$ denote stereographic projection, defined by

$$\sigma(\zeta^1, \ldots, \zeta^n, \xi) = \left(\frac{\zeta^1}{1 - \xi}, \ldots, \frac{\zeta^n}{1 - \xi} \right).$$

Then $\sigma : (\mathbf{S}^n \setminus \{P\}, g_0) \to (\mathbf{R}^{n+1}, ds^2)$ is conformal, where g_0 is the standard metric on $\mathbf{S}^n$.

Since the composition of conformal maps is again conformal, we can use σ to construct conformal maps of $\mathbf{S}^n$ to itself: for $\lambda > 0$, let

$$\varphi_\lambda = \sigma^{-1} \circ \delta_\lambda \circ \sigma : \mathbf{S}^n \to \mathbf{S}^n.$$

Then

$$\varphi_\lambda^* g_0 = \Psi_\lambda^2 g_0,$$

where

$$\Psi_\lambda(\zeta, \xi) = \frac{2\lambda}{(1 + \xi) + \lambda^2 (1 - \xi)}.$$

Note

$$(\zeta, \xi) = (0, 1) \Rightarrow \Psi_\lambda \to \infty \quad \text{as } \lambda \to \infty,$$
$$(\zeta, \xi) \neq (0, 1) \Rightarrow \Psi_\lambda \to 0 \quad \text{as } \lambda \to \infty.$$

The set of conformal maps of a given Riemannian manifold is a Lie group; the construction above shows that the conformal group of the sphere is *non-compact*. This fact distinguishes the sphere:

Theorem 4.3. (Lelong-Ferrand) *A compact Riemmanian manifold with non-compact conformal group is conformally equivalent to the sphere with its standard metric.*

This fact is the source of many of the analytic difficulties we will encounter in the PDEs we are about to describe.

1. Curvature and Conformal Changes of Metric

Let $h = e^{-2u}g$ be conformal metrics, and let $Ric(h), R(h)$ denote the Ricci and scalar curvatures of h, and $Ric(g), R(g)$ denote the Ricci and scalar curvatures of g. Then

$$
\begin{aligned}
Ric(h) = {}& Ric(g) + (n-2)\nabla^2 u + \Delta u g \\
& + (n-2)du \otimes du - (n-2)|\nabla u|^2 g, \\
R(h) = {}& e^{2u}\{ R(g) + 2(n-1)\Delta u \\
& - (n-1)(n-2)|\nabla u|^2 \},
\end{aligned}
$$

where $\nabla^2 u$ and Δu denote the Hessian and laplacian of u with respect to g.

2. The Uniformization Theorem and Yamabe Problem.

Let (M^2, g) be a closed (no boundary), compact, two-dimensional Riemannian manifold. Let K denotes its Gauss curvature.

Theorem 4.4. (The Uniformization Theorem) *There is a conformal metric* $h = e^{-2u}g$ *with constant Gauss curvature.*

See ([Ber03], p. 254) for some historical background on the result. Let $K_h = const.$ denote the Gauss curvature of the metric h; then the sign of K_h is determined by the Gauss-Bonnet formula:

$$
\begin{aligned}
2\pi\chi(M^2) &= \int K_h \, dA_h \\
&= K_h \cdot Area(h).
\end{aligned}
$$

Note the geometric/topological significance of the Uniformization Theorem: Since h has constant curvature, by the Hopf theorem the universal cover $\tilde{M}$ is isometric to either $\mathbf{S}^2, \mathbf{R}^2$, or $\mathbf{H}^2$, each case being determined by the sign of the Euler characteristic.

Now let (M^n, g) be a closed, compact, Riemannian manifold of dimension $n \geq 3$. In higher dimensions there are obstructions to being even *locally* conformal to a constant curvature metric. This leads to

Question: How do we generalize the Uniformization Theorem to higher dimensions?

A major theme of these lectures is the various ways one might answer this question (there are yet others). The first attempt we will discuss is the *Yamabe Problem*: Find a conformal metric $h = e^{-2u}g$ whose *scalar curvature* is constant.

By the formulas above, solving the Yamabe problem is equivalent to solving the semilinear PDE

$$2(n-1)\Delta u - (n-1)(n-2)|\nabla u|^2 + R(g) = \mu e^{-2u}$$

for some constant μ. This formula can be simplified if we write $h = v^{4/(n-2)}g$, where $v > 0$. Then v should satisfy

$$-\frac{4(n-1)}{(n-2)}\Delta v + R(g)v = \lambda v^{\frac{n+2}{n-2}}. \tag{10}$$

Notice the exponent! This equation is of the form

$$\Delta v + c(x)v = K(x)v^p,$$

where $p = (n+2)/(n-2)$. This is the critical case of the equation we considered in Theorem 3.4.

3. The Case of the Sphere

Recall the conformal maps of the sphere described above, $\varphi_\lambda : \mathbf{S}^n \to \mathbf{S}^n$. Then $h = \varphi_\lambda^* g_0 = \Psi_\lambda^2 g_0$ has the same scalar curvature as the standard metric. Therefore, writing

$$h = v_\lambda^{4/(n-2)} g_0,$$

where

$$v_\lambda = \Psi_\lambda^{\frac{(n-2)}{2}},$$

we have a family $\{v_\lambda\}$ of solutions to

$$-\frac{4(n-1)}{(n-2)}\Delta v_\lambda + n(n-1)v_\lambda = n(n-1)v_\lambda^{\frac{n+2}{n-2}}.$$

As we observed above, if P is the North pole, then

$$v_\lambda(P) \to \infty \quad \text{as } \lambda \to \infty,$$

whereas if $x \neq P$, then

$$v_\lambda(x) \to 0 \quad \text{as } \lambda \to \infty.$$

To summarize, there is good news and bad. The good news is that there are many solutions of the Yamabe equation. The bad news is that it will be impossible to prove *a priori* estimates for solutions of (10). Of course, the non-compactness of the set of solutions arises precisely because of the influence of the conformal group. Thus, on manifolds other than the sphere, one would expect that the set of solutions is compact. Put another way, ideally we would like to show that non-compactness implies the underlying manifold is $(\mathbf{S}^n, g_0)$.

2. The Yamabe Problem: A variational Approach.

There is an approach to solving the Yamabe problem by the methods of the calculus of variations. Define the functional $\mathcal{Y} : W^{1,2} \to \mathbf{R}$ by

$$\mathcal{Y}[v] = \frac{\int \left(\frac{4(n-1)}{(n-2)} |\nabla v|^2 + R(g)v^2 \right) dV}{\left(\int v^{\frac{2n}{(n-2)}} \, dV \right)^{(n-2)/n}}. \tag{11}$$

Using the formulas above, one can check that

$$\mathcal{Y}[v] = Vol(h)^{-(n-2)/n} \int R(h) \, dV(h),$$

where $h = v^{4/(n-2)} g$. The quantity on the right-hand side is called the *total scalar curvature* of h.

Lemma 4.5. *A function $v \in W^{1,2}$ is a critical point of $\mathcal{Y}$ iff v is a weak solution of the Yamabe equation.*

By critical point, we mean that

$$\frac{d}{dt}\mathcal{Y}(v + t\varphi)\big|_{t=0} = 0$$

for all $\varphi \in W^{1,2}$.

Recall that weak solutions of (10) are regular. Also, by the Sobolev embedding theorem the number

$$Y(M^n, [g]) = \inf_{v \in W^{1,2}} \mathcal{Y}(v) \tag{12}$$

is $> -\infty$. This number is called the *Yamabe invariant* of the conformal class of g.

Some historical notes: H. Yamabe claimed to have proved the existence of a minimizer of $\mathcal{Y}$, for all manifolds (M^n, g). However, N. Trudinger found a serious gap in his proof, which he was able to fix provided $Y(M^n, [g])$ was sufficiently small (for example, if $Y(M^n, [g]) \leq 0$). Subsequently, T. Aubin proved that for all n-dimensional manifolds

$$Y(M^n, [g]) \leq Y(\mathbf{S}^n, [g_0]),\tag{13}$$

and that whenever this inequality was strict, a minimizing sequence converges (weakly) to a (smooth) solution of the Yamabe equation. Aubin also showed that a strict inequality holds in (13) if (M^n, g) was of dimension $n \geq 6$ and not locally conformal to a flat metric.

Finally, the remaining cases were solved by Schoen: that is, he showed that when M^n has dimension $3, 4$, or 5, or if M is locally conformal to a flat metric, then the inequality (13) is strict unless (M^n, g) is conformally equivalent to (S^n, g_0). An excellent survey of the Yamabe problem can be found in [LP87].

5 A Fully Nonlinear Yamabe Problem

In this section we begin our discussion of the σ_k-*Yamabe problem*, a more recent attempt to generalize the Uniformization Theorem to higher dimensions. To do so, we need to introduce another notion of curvature:

Definition 5.1. The Schouten tensor of (M, g) is

$$A = \frac{1}{(n-2)} \left(Ric - \frac{1}{2(n-1)} R \cdot g \right).\tag{14}$$

Example. For spaces of constant curvature ± 1 (e.g. the sphere or hyperbolic space), the Schouten tensor is

$$A = \operatorname{diag}\left\{ \pm \frac{1}{2}, \ldots, \pm \frac{1}{2} \right\}.$$

From the perspective of conformal geometry, the Schouten is actually more natural than the Ricci tensor (but this takes some time to explain). Here's one indication: Suppose $\widehat{g} = e^{-2u} g$. Then the Schouten tensor of $\widehat{g}$ is given by

$$\widehat{A} = A + \nabla^2 u + du \otimes du - \frac{1}{2} |du|^2 g.\tag{15}$$

A complicated formula; but just think of it as saying

$$\widehat{A} = \nabla^2 u + \cdots$$

where $\cdots$ indicates lower order terms. Contrast this with the more complicated formulas for the Ricci tensor, which also involves the Laplace operator.

The equations we will consider involve symmetric functions of the eigenvalues of A. Let $\lambda_1, \ldots, \lambda_n$ denote the eigenvalues of A; suppose we choose a local basis which diagonalizes A:

$$A = \begin{pmatrix} \lambda_1 & & & \\ & \lambda_2 & & \\ & & \ddots & \\ & & & \lambda_n \end{pmatrix}.$$

Then define

$$\sigma_k(A) = \sum_{i_1 < \cdots < i_k} \lambda_{i_1} \cdots \lambda_{i_k}, \tag{16}$$

i.e., σ_k is the kth elementary symmetric polynomial in n variables. Note that

$$\sigma_1(A) = \mathrm{trace}(A) = \frac{R}{2(n-1)},$$

just a multiple of the scalar curvature. In general, the quantity $\sigma_k(A)$ is called the kth-scalar curvature, or σ_k-curvature, of the manifold.

Now, we can rephrase the Yamabe problem in the following way: Given (M^n, g) find a conformal metric $\widehat{g} = e^{-2u}g$ with constant σ_1-curvature. This naturally leads to the σ_k-Yamabe problem: Given (M^n, g), find a conformal metric $\widehat{g} = e^{-2u}g$ such that the σ_k-curvature is constant. By the formula above, this is equivalent to solving the PDE

$$\sigma_k\left(A + \nabla^2 u + du \otimes du - \frac{1}{2}|du|^2 g\right) = \mu e^{-2ku} \tag{17}$$

for some constant μ. Note the exponential weight on the right comes from the fact that we are computing the eigenvalues of $\widehat{A}$ w.r.t. $\widehat{g}$.

These equations are closely related to the Hessian equations covered in Prof. Xu-Jia Wang's C.I.M.E. course. The differences will come from (1) The conformal invariance, and (2) The lower order (gradient) terms.

The σ_k-Yamabe problem was first formulated by J. Viaclovsky in his thesis [Via00]. Viaclovsky is also the author of a recent survey article on the subject, [Via06].

5.1 Ellipticity

Recall from Professor Wang's lectures that the *Hessian equations*

$$\sigma_k(\nabla^2 u) = f(x)$$

are elliptic provided u is *admissible*, or k-convex. That is, if

$$\sigma_j(\nabla^2 u) > 0, \quad 1 \le j \le k.$$

In particular, a necessary condition is that $f(x) > 0$. We will need to impose a similar ellipticity condition:

Definition 5.2. A metric g is *admissible* (or *k-admissible*) if the Schouten tensor satisfies

$$\sigma_j(A) > 0, \quad 1 \le j \le k$$

at each point of M^n.

What is the *geometric* meaning of admissibility? One can think of it as a kind of "positivity" condition on the Schouten tensor. When $k = n$, it means the Schouten tensor is positive definite; when $k = 1$, it means the trace (i.e., the scalar curvature) is positive. Here is a more precise result, due to Guan-Viaclovsky-Wang [GVW03]:

Theorem 5.3. *If* (M^n, g) *is* k-*admissible then*

$$Ric \ge \frac{2k - n}{2n(k - 1)} R \cdot g.$$

In particular, if $k > n/2$ then admissibility means positive <u>Ricci</u> curvature. We can also define *negative admissibility*, which just means that $(-A)$ is k-convex.

As in the usual Yamabe problem, there is a non-compact family of solutions to the σ_k-Yamabe problem on S^n:

$$g_\lambda = \varphi_\lambda^* g_0 = \Psi_\lambda^2 g_0.$$

In particular, this gives an obstruction to proving *a priori* estimates (as it does for the Yamabe problem). Thus, we are faced with some of the same technical difficulties. However, there are some important technical differences between the σ_k- and classical Yamabe problems. For example, equation (17) does not have an easy variational description (though there are some important geometric cases where it does).

A more mysterious contrast arises when studying manifolds of negative curvature. If (M^n, g) has negative scalar curvature, the Yamabe problem is very easy to solve–indeed, the solution is unique. But for negative admissible metrics there are at this time no general existence results for the σ_k-Yamabe problem. In fact, Sheng-Trudinger-Wang showed by example that the local estimates of Guan-Wang are *false* for solutions in the negative cone (see [STW05]).

Finally, we remark that the condition of admissibility can be very restrictive: for example, the manifold $X^3 = S^2 \times S^1$ does not admit a k-admissible metric for $k = 2$ or 3. Of course, one can consider the Yamabe problem for *any* conformal class on X^3.

5.2 From Lower to Higher Order Estimates

Our goal is to explain the main issues involved in solving the σ_k-Yamabe problem, and sketch the proof of a particular case. As we shall see, the central problem is establishing *a priori* estimates. Owing to a fundamental result of Evans, Krylov ([Eva82], [Kry83]), plus the classical Schauder estimates, we only need to worry about estimating derivatives up to order <u>two</u>. That is,

$$|u| + |\nabla u| + |\nabla^2 u| \le C_2$$
$$\Downarrow$$
$$|u| + |\nabla u| + \cdots + |\nabla^k u| \le C(k, C_2).$$

Of course, even C^2-estimates will fail without further assumptions, again because of the sphere. However, let's look closer: Let $\varphi_\lambda : S^n \to S^n$ be the 1-parameter family of conformal maps, and write

$$g_\lambda = \varphi_\lambda^* g_0 = e^{-2u_\lambda} g_0.$$

Note that as $\lambda \to \infty$, the conformal factor grows like

$$\max e^{-2u_\lambda} \sim \lambda^2,$$

while the gradient and Hessian of u grow like

$$\max |\nabla u_\lambda|^2 \sim \lambda^2, \quad \max |\nabla^2 u_\lambda| \sim \lambda^2.$$

In particular, for this family we have

$$|\nabla u|^2 + |\nabla^2 u| \approx \max e^{-2u_\lambda}.$$

So the optimal estimate one could hope for would be

$$\max \left(2^{nd} \text{ derivatives of } u \right) \le C \max e^{-2u}. \tag{18}$$

It turns out that such an estimate always holds:

Theorem 5.4. (See Guan-Wang, [GW03]) *Assume $u \in C^4$ is an admissible solution of the σ_k-Yamabe equation on $B(1)$. Then*

$$\max_{B(1/2)} \left[|\nabla u|^2 + |\nabla^2 u| \right] \leq C(1 + \max_{B(1)} e^{-2u}).$$

In view of this result, and the Evans and Krylov results, we see that

$$\min_M u \geq C \Rightarrow \|u\|_{C^{k,\alpha}(M)} \leq C(k).$$

Therefore, if we can somehow rule out "bubbling", we obtain estimates of all orders. Once estimates are known, there are various topological methods to prove the existence of solutions. This shows the geometric nature of the problem: i.e., we need to detect the global geometry of the manifold in order to get estimates, hence existence.

5.3 An Existence Result: Four Dimensions

To finish our discussion of the σ_k-Yamabe problem, we want to sketch its solution in four dimensions. This case is special because, in 4-d, the integral

$$\int \sigma_2(A) \, dV$$

is conformally invariant. That is, if $\widehat{g} = e^{-2u}g$, then

$$\int \sigma_2(\widehat{A}) \, d\widehat{V} = \int \sigma_2(A) \, dV.$$

You can check this by hand using the formulas above along with the fact that

$$d\widehat{V} = e^{-4u}dV.$$

Eventually, you will find that

$$\sigma_2(\widehat{A}) \, d\widehat{V} = \sigma_2(A) \, dV + \text{(divergence terms)}.$$

We will provide some details for the case $k = 2$; this was first treated by Chang-Gursky-Yang [CGY02b], and later by Gursky-Viaclovsky [GV04]. For $k = 3$ or 4, the scheme of the proof is essentially the same. However, the proof presented here is a simplified version of the original one, since we will use the local estimates of Guan-Wang (which appeared several years after [CGY02b]). As we emphasized above, existence eventually boils down to estimates: this is what we will prove.

To begin, let us write the equation in the case $k = 2$:

$$\sigma_2^{1/2}\left(A + \nabla^2 u + du \otimes du - \frac{1}{2}|du|^2 g\right) = f(x)e^{-2u}. \tag{19}$$

where $f \in C^\infty$. Using the definition of σ_2, this actually reads:

$$-|\nabla^2 u|^2 + (\Delta u)^2 + c_1 \nabla_i \nabla_j u \nabla_i u \nabla_j u$$
$$+ c_2 \Delta u |\nabla u|^2 + c_3 |\nabla u|^4 + \cdots = f^2(x)e^{-4u}.$$

We will prove:

Theorem 5.5. *Suppose (M^4, g) is (i) admissible, and (ii) not conformally equivalent to the round sphere. If $u \in C^4$ is a solution of (19), then there is a constant $C = C(g, f)$ such that*

$$\min u \geq -C.$$

Consequently,

$$\|u\|_{C^k} \leq C(k).$$

Proof. Suppose to the contrary there is a sequence of solutions $\{u_i\}$ of (19) with $\min u_i \to -\infty$. Let's imagine that there is a point P with

$$\min_M u_i = u_i(P)$$

and by introducing local coordinates we can identify P with the origin in $\mathbb{R}^4$ and think of u_i as being defined in a neighborhood Ω of 0. (In reality, the location of the minimum point will vary, but this doesn't affect the argument in a significant way).

It is time to use conformal invariance. Recall the dilations on Euclidean space are conformal. Define

$$w_i(x) = u(\epsilon_i x) + \log\frac{1}{\epsilon_i},$$

where $\epsilon_i > 0$ is chosen so that

$$w_i(0) = 0.$$

The w_i's are defined on $\frac{1}{\epsilon_i}\Omega$, and satisfy

$$\sigma_2^{1/2}\left(\epsilon_i^2 A + \nabla^2 w_i + dw_i \otimes dw_i - \frac{1}{2}|dw_i|^2 g\right)$$
$$= f(\epsilon_i x)e^{-2w_i}.$$

After applying the local estimates of Guan-Wang, we can take a subsequence $\{w_i\}$ which converges in $C_{loc}^{k,\alpha}$ to a solution of

$$\sigma_2^{1/2}\left(\nabla^2 w + dw \otimes dw - \frac{1}{2}|dw|^2 g\right) = \mu e^{-2w}. \tag{20}$$

with $\mu > 0$.

We now appeal to the following uniqueness result

Lemma 5.6. (See Chang-Gursky-Yang, [CGY02a]) *Up to scaling, the unique solution of (20) is realized by*

$$e^{2w} ds^2 = (\sigma^{-1})^* g_0, \tag{21}$$

where σ is the stereographic projection map, ds^2 the Euclidean metric, and g_0 is the round metric on the sphere.

It is easy to check that each solution given by (21) satisfies

$$\int_{\mathbb{R}^4} \sigma_2(\widetilde{A})\, d\widetilde{V} = 4\pi^2.$$

where $\widetilde{g} = e^{2w} ds^2$. (Remember that $A = \mathrm{diag}\{1/2, \ldots, 1/2\}$, and $\mathrm{Vol}(S^4) = 8\pi^2/3$). Also, since our solution w comes from blowing up a little piece of the original manifold, for each $\widehat{g}_i = e^{-2u_i} g$ we must have

$$\int_{M^4} \sigma_2(\widehat{A}_i)\, d\widehat{V}_i \geq 4\pi^2.$$

The proof now follows from the following global geometric result:

Theorem 5.7. (See Gursky, [Gur99]) *If (M^4, g) has positive scalar curvature, then*

$$\int_{M^4} \sigma_2(A)\, dV \leq 4\pi^2,$$

and equality holds if and only if (M^4, g) is conformally equivalent to the sphere.

It follows that each (M^4, g_i) is conformally equivalent to the round sphere, a contradiction. Therefore, assuming the manifold (M^4, g) is not conformally the sphere, any sequence of solutions remains bounded, as claimed.

Important Remark. The following remark is for the benefit of experts: The proof of the preceding theorem does <u>not</u> use the Positive Mass Theorem! (Or, to be precise, it uses an extremely weak form). Therefore, we are not solving the σ_k-Yamabe problem by somehow reducing it to the classical Yamabe problem.

6 The Functional Determinant

In the final section we will introduce a higher order elliptic problem which has its origins in spectral theory. Although this problem is semilinear and not fully nonlinear, the structure of the Euler equation is related to the σ_2-Yamabe equation in 4-d. Moreover, for 4-manifolds of positive scalar curvature, the same result (Theorem 5.7) plays a crucial role in the existence theory.

Suppose (M^n, g) is a closed Riemannian manifold, and let Δ denote the Laplace-Beltrami operator associated to g. We can label the eigenvalues of $(-\Delta)$ (counting multiplicities) as

$$0 = \lambda_0 < \lambda_1 \leq \lambda_2 \leq \lambda_3 \leq \ldots \tag{22}$$

The *spectral zeta function* of (M^2, g) is defined by

$$\zeta(s) = \sum_{j \geq 1} \lambda_j^{-s}. \tag{23}$$

By Weyl's asymptotic law,

$$\lambda_j \sim j^{2/n}.$$

Consequently, (23) defines an analytic function for $\text{Re}(s) > n/2$. In fact, one can meromorphically continue ζ in such a way that ζ becomes regular at $s = 0$ (see [RS71]). Note that formally–that is, if we take the definition in (23) literally–then

$$\zeta'(0) = -\sum_{j \geq 1} \log \lambda_j$$

$$= -\log \left\{ \prod_{j \geq 1} \lambda_j \right\} \tag{24}$$

$$= -\log \det(-\Delta_g).$$

In view of this *ansatz*, it is natural to define the regularized determinant of $(-\Delta_g)$ as

$$\det(-\Delta_g) = e^{-\zeta'(0)}. \tag{25}$$

6.1 The Case of Surfaces

Clearly, the determinant is not a local quantity. Therefore, it is rather remarkable that Polaykov ([Pol81a], [Pol81b]) was able to compute a closed formula for the *ratio* of the determinants of the laplacians of two conformally related surfaces:

Theorem 6.1. *Let* $(\Sigma, g), (\Sigma, \hat{g} = e^{2w}g)$ *be conformal surfaces. Then*

$$\log \frac{\det(-\Delta_{\hat{g}})}{\det(-\Delta_g)} = -\frac{1}{12\pi} \int_{\Sigma} \left[|\nabla w|^2 + 2Kw \right] dA, \qquad (26)$$

where K is the Gauss curvature and dA the surface measure associated to (Σ^2, g).

Remarks.

1. The formula (26) naturally defines a (relative) action on the space of conformal metrics. That is, once we fix a metric g, we have the functional

$$\hat{g} \in [g] \mapsto \log \frac{\det(-\Delta_{\hat{g}})}{\det(-\Delta_g)}.$$

However, since the determinant is not scale-invariant, we should consider the *normalized functional determinant*

$$S[w] = \int_{\Sigma} \left[|\nabla w|^2 + 2Kw \right] dA - \left(\int_{\Sigma} K \, dA \right) \log \left(\fint_{\Sigma} e^{2w} \, dA \right), \qquad (27)$$

so that

$$S[w] = -12\pi \log \frac{\det(-\Delta_{\hat{g}})}{\det(-\Delta_g)} + 2\pi\chi(\Sigma) \log \mathrm{Area}(\hat{g}),$$

while

$$S[w + c] = S[w].$$

2. A first variation shows that w is a critical point of S if and only if w satisfies

$$\Delta w + ce^{2w} = K, \qquad (28)$$

where c is a constant. Now, if $\hat{g} = e^{2w}g$, then the Gauss curvature $\hat{K}$ of $\hat{g}$ is related to the Gauss curvature of g via

$$\Delta w + \hat{K}e^{2w} = K, \tag{29}$$

this is called the *Gauss curvature equation*. Comparing (28) and (29), we see that w is a critical point of S if and only if $(\Sigma, \hat{g})$ has constant Gauss curvature. In particular, a metric extremizes the functional determinant if and only if it uniformizes; i.e., it is a conformal metric of constant Gauss curvature.

3. In a series of papers ([Osg88b], [Osg88a]) Osgood-Phillips-Sarnak gave a proof of the Uniformization Theorem by showing that each conformal class on a surface admits a metric that extermizes the determinant. Like the Yamabe problem and its fully nonlinear version discussed earlier in the article, the main difficulty is the invariance of the determinant under the action of the conformal group. And like the analysis of the Yamabe problem, the solution involves the study of sharp functional inequalities. A very nice overview of the study of the functional determinant and related material can be found in [Cha].

6.2 Four Dimensions

The key property of the Laplacian that Polyakov exploited in his calculation was its *conformal covariance*:

$$\Delta_{e^{2w}g} = e^{-2w}\Delta_g. \tag{30}$$

More generally, we say that the differential operator $A = A_g : C^\infty(M^n) \to C^\infty(M^n)$ is *conformally covariant* of bi-degree (a, b) if

$$A_{e^{2w}g}\varphi = e^{-bw}A_g(e^{aw}\varphi). \tag{31}$$

In fact, this definition makes perfect sense for operators acting on smooth sections of bundles (spinors, forms, etc.) as well as on functions. Two examples of note are

Example 1. The conformal laplacian of (M^n, g), where $n \geq 3$, is

$$L = -\frac{4(n-1)}{(n-2)}\Delta + R, \tag{32}$$

where R is the scalar curvature. Then L is conformally covariant with

$$a = \frac{n-2}{2}, \ b = \frac{n+2}{2}.$$

Example 2. Let (M^4, g) be a four-dimensional Riemannian manifold. The *Paneitz operator* is

$$P = (\Delta)^2 + \text{div}\{(\frac{2}{3}Rg - 2Ric) \circ d\}. \tag{33}$$

Then P is conformally covariant with

$$a = 4, \ b = 0.$$

An analogue of Polyakov's formula for conformally covariant operators defined on four-manifolds was computed by Branson-Ørsted in [Bra91]. To explain the Branson-Ørsted formula we need to introduce three functionals associated to a Riemannian 4-manifold (M^4, g). Each functional is defined on $W^{2,2}$, the Sobolev space of functions with derivatives up to order two in L^2. The first functional is zeroth order in w:

$$I[w] = 4 \int w|W|^2 \ dV - (\int |W|^2 \ dV) \log \fint e^{4w} \ dV, \tag{34}$$

where W is the Weyl curvature tensor and dV the volume form of g. If $w \in W^{2,2}$, The Moser-Trudinger inequality ([GT83]) implies that

$$e^w \in L^p, \text{all } p \geq 1.$$

Therefore, $I : W^{2,2} \to \mathbb{R}$.

The second functional is analogous to the functional S defined in (27):

$$II[w] = \int wPw \ dV + 4 \int Qw \ dV - (\int Q \ dV) \log \fint e^{4w} \ dV, \tag{35}$$

where P is the Paneitz operator and Q is the Q-curvature:

$$Q = \frac{1}{12}(-\Delta R + R^2 - 3|Ric|^2). \tag{36}$$

Here we see the parallel between the Laplace-Beltrami operator/Gauss curvature of a surface and the Paneitz operator/Q-curvature of a 4-manifold.

The third functional is

$$III[w] = 12 \int (\Delta w + |\nabla w|^2)^2 \ dV - 4 \int (w\Delta R + R|\nabla w|^2) \ dV. \tag{37}$$

The geometric meaning of this functional is apparent if we rewrite it in terms of the scalar curvature $R_{\hat{g}}$ and volume form $d\hat{V}$ of the conformal metric $\hat{g} = e^{2w}g$:

$$III[w] = \frac{1}{3}[\int R_{\hat{g}}^2 \ d\hat{V}] - \int R^2 \ dV. \tag{38}$$

Therefore, III is the L^2-version of the Yamambe functional in (11).

With these definitions, we can give the Branson-Ørsted formula: Suppose $A = A_g$ is a conformally covariant differential operator satisfying certain "naturality" conditions (see [Bra91] for details). Then there are numbers, $\gamma_i = \gamma_i(A), 1 \leq i \leq 3$, such that

$$F_A[w] = \log \frac{\det A_{e^{2w}g}}{\det A_g} = \gamma_1 I[w] + \gamma_2 II[w] + \gamma_3 III[w]. \tag{39}$$

We remark that the Branson-Ørsted formula is normalized; i.e., $F_A[w + c] = F_A[w]$.

Example 1. For the conformal laplacian, Branson-Ørsted calculated

$$\gamma_1(L) = 1, \ \ \gamma_2(L) = -4, \ \ \gamma_3(L) = -\frac{2}{3}. \tag{40}$$

Example 2. Later, in [Bra96], Branson calculated the coefficients for the Paneitz operator:

$$\gamma_1(L) = -\frac{1}{4}, \ \ \gamma_2(L) = -14, \ \ \gamma_3(L) = \frac{8}{3}. \tag{41}$$

Neglecting lower order terms, the log det functional is of the form

$$\log \frac{\det A_{e^{2w}g}}{\det A_g} = \gamma_1 \int (\Delta w)^2 \, dV + \gamma_3 \int [\Delta w + |\nabla w|^2]^2 \, dV$$
$$+ \kappa_A \log \fint e^{4w} \, dV + \cdots , \tag{42}$$

where κ_A is given by

$$\kappa_A = -\gamma_1 \int |W|^2 \, dV - \gamma_2 \int Q \, dV, \tag{43}$$

a conformal invariant. In particular, when γ_2 and γ_3 have the same sign (as they do for the conformal laplacian), the main issue from the variational point of view is the interaction of the highest order terms with the exponential term. However, when the signs of γ_2 and γ_3 differ, then the highest order terms are a non-convex combination of II and III, and the variational structure can be quite complicated.

6.3 The Euler Equation

As we observed above, critical points of the functional determinant on a surface corresponds to metrics of constant Gauss curvature. In four dimensions the geometric meaning of the Euler equation is less straightforward: Suppose $\hat{g} = e^{2w}g$ is a critical point of F_A; then the curvature of $\hat{g}$ satisfies

$$\gamma_1 |W_{\hat{g}}|^2 + \gamma_2 Q_{\hat{g}} + \gamma_3 \Delta_{\hat{g}} R_{\hat{g}} = -\kappa_A \cdot Vol(\hat{g})^{-1}. \tag{44}$$

The geometric significance of this condition is, at first glance, difficult to fathom. However, this equation in some sense includes all the significant curvature conditions studied in four-dimensional conformal geometry, as can be seen by considering special values of the γ_i's:

- Taking $\gamma_1 = \gamma_2 = 0$ and $\gamma_1 = 1$, equation (44) becomes

$$\Delta_{\hat{g}} R_{\hat{g}} = const. = 0, \tag{45}$$

which is equivalent to the Yamabe equation

$$R_{\hat{g}} = const.$$

- Taking $\gamma_1 = 0$ and $\gamma_2 = -12\gamma_3$, equation (44) becomes

$$\sigma_2(A_{\hat{g}}) = const., \tag{46}$$

that is, a critical point is a solution of the σ_2-Yamabe problem.

- Taking $\gamma_1 = \gamma_3 = 0$ and $\gamma_2 = 1$, then

$$Q_{\hat{g}} = const. \tag{47}$$

Thus, critical points are solutions of the Q-curvature problem.

Geometric properties of critical metrics were used in [Gur98] to prove various vanishing theorems, and as a regularization of the σ_2-Yamambe problem in [CGY02b].

Turning to analytic aspects of the Euler equation, it is clear from (42) that it is fourth order in w. A precise formula is

$$\mu e^{4w} = (\frac{1}{2}\gamma_2 + 6\gamma_3)\Delta^2 w + 6\gamma_3 \Delta |\nabla w|^2 - 12\gamma_3 \nabla^i [(\Delta w + |\nabla w|^2)\nabla_i w] \tag{48}$$

$$+ \gamma_2 R_{ij}\nabla_i\nabla_j w + (2\gamma_3 - \frac{1}{3}\gamma_2)R\Delta w + (2\gamma_3 + \frac{1}{6}\gamma_2)\langle \nabla R, \nabla w\rangle \tag{49}$$

$$+ (\gamma_1 |W|^2 + \gamma_2 Q - \gamma_3 \Delta R), \tag{50}$$

where R_{ij} are the components of the Ricci curvature and

$$\mu = -\frac{\kappa_A}{\int e^{4w}}. \tag{51}$$

To simplify this expression we can divide both sides of (49) by $6\gamma_3$, then rewrite the lower order terms to arrive at

$$(1+\alpha)\Delta^2 w = f(x)e^{4w} - \Delta|\nabla w|^2 + 2\nabla^i[(\Delta w + |\nabla w|^2)\nabla_i w] \\ + a^{ij}\nabla_i\nabla_j w + b^k\nabla_k w + c(x), \tag{52}$$

where

$$\alpha = \frac{\gamma_2}{12\gamma_3}. \tag{53}$$

Although writing the equation in this form clearly reveals the divergence structure, for some purposes it is better to expand the terms on the right and write

$$(1+\alpha)\Delta^2 w = f(x)e^{4w} - 2|\nabla^2 w|^2 + 2(\Delta w)^2 + 2\langle\nabla|\nabla w|^2, \nabla w\rangle \\ + 2\Delta w|\nabla w|^2 + \text{(lower order terms)}. \tag{54}$$

In particular, we see that the right-hand side does not involve any third derivatives of the solution.

The regularity of extremal solutions of (49) was proved by Chang-Gursky-Yang in [CGY99]), and for general solutions by Uhlenbeck-Viaclovsky in [UV00]. Similar to the harmonic map equation in two dimensions, the main difficulty is that the right-hand side of (54) is only in L^1 when $w \in W^{2,2}$, ruling out the possibility of using a naive bootstrap argument to prove regularity.

6.4 Existence of Extremals

The most complete existence theory for extremals of the functional determinant was done by Chang-Yang in [CY95]:

Theorem 6.2. *Assume*

$$\gamma_2, \gamma_3 < 0, \tag{55}$$

and

$$\kappa_A < 8\pi^2(-\gamma_2). \tag{56}$$

Then $\sup_{W^{2,2}} F_A$ *is attained by some some* $w \in W^{2,2}$.

Remarks.

1. Recall that

$$\kappa_A = -\gamma_1 \int |W|^2 \, dV - \gamma_2 \int Q \, dV.$$

If $\gamma_1 > 0$, then

$$\kappa_A \leq -\gamma_2 \int Q \, dV.$$

Therefore, assuming $\gamma_2 < 0$, then (56) holds provided

$$\int Q \, dV < 8\pi^2. \tag{57}$$

By the definition of the Q-curvature,

$$Q = 2\sigma_2(A) - \frac{1}{12}\Delta R. \tag{58}$$

Therefore,

$$\int Q \, dV = 2 \int \sigma_2(A) \, dV.$$

In particular, for manifolds of positive scalar curvature, by Theorem 5.7 it follows that

$$\int Q \, dV \leq 8\pi^2, \tag{59}$$

with equality if and only if (M^4, g) is conformal to the round sphere. Thus, combining the existence result of Chang-Yang with the sharp inequality of Theorem 5.7, we conclude

Corollary 6.3. *If (M^4, g) has positive scalar curvature, then an extremal for F_L exists.*

2. It is easy to construct examples of 4-manifold–necessarily with negative scalar curvature–for which

$$\int Q \, dV >> 8\pi^2.$$

Thus, the existence theory for the functional determinant is quite incomplete. This shows another parallel with the σ_k-Yamabe problem (and contrast with the classical Yamabe problem): the case of negative curvature is much more difficult than the positive case.

3. Branson-Chang-Yang proved that on the sphere S^4, the functionals II and III are minimized by the round metric and its images under the conformal group [Bra]. In particular, the round metric is the unique extremal (up to conformal transformation) of F_L. Later, in [Gur97], Gursky showed that the round metric is the unique critical point.

6.5 Sketch of the Proof

In the following we give a sketch of the proof of Theorem 6.2. By Corollary 6.3, this will give the existence of extremals for F_A on any 4-manifold of positive scalar curvature.

To begin, we write the functional as

$$
\begin{aligned}
F_A[w] &= \gamma_1 I[w] + \gamma_2 II[w] + \gamma_3 III[w] \\
&= \gamma_1 \int (\Delta w)^2 + \gamma_2 \int (\Delta w + |\nabla w|^2)^2 + \kappa_A \log \fint e^{4(w-\bar{w})} + (l.o.t.).
\end{aligned}
$$
(60)

Next, divide by γ_2, and denote $\tilde{F} = (1/\gamma_2) F_A$:

$$
\tilde{F}[w] = \int (\Delta w)^2 + \beta \int (\Delta w + |\nabla w|^2)^2 - \left(\frac{\kappa_A}{-\gamma_2}\right) \log \fint e^{4(w-\bar{w})} + (l.o.t.),
$$
(61)

where

$$
\beta = \gamma_3/\gamma_2 > 0.
$$
(62)

Since $\gamma_2 < 0$, we are trying to prove the existence of *minimizers* of $\tilde{F}$.

Let us first consider the easy case, when $\kappa_A \le 0$. Then

$$
-\left(\frac{\kappa_A}{-\gamma_2}\right) \ge 0.
$$

Also, by Jensen's inequality,

$$
\log \fint e^{4(w-\bar{w})} \ge 0.
$$

Therefore,

$$
\tilde{F}[w] \ge \int (\Delta w)^2 + \beta \int (\Delta w + |\nabla w|^2)^2 + (l.o.t.)
$$
(63)

Now suppose $\{w_k\}$ is a minimizing sequence for $\tilde{F}$; from (63) we conclude

$$C \geq \int (\Delta w_k)^2 + (l.o.t.),$$

which implies, for example by the Poincare inequality, that $\{w_k\}$ is bounded in $W^{2,2}$. It follows that a subsequence converges weakly to a minimizer $w \in W^{2,2}$.

For the more difficult case when $\kappa_A > 0$, first observe that by hypothesis, $\kappa_A < 8\pi^2(-\gamma_2)$. Therefore,

$$\frac{\kappa_A}{-\gamma_2} = 8\pi^2(1 - \epsilon) \tag{64}$$

for some $\epsilon > 0$. The significance of the constant $8\pi^2$ is apparent from the following sharp Moser-Trudinger inequality due to Adams:

Proposition 6.4. (See [Ada]) *If (M^4, g) is a smooth, closed 4-manifold, then there is a constant $C_1 = C_1(g)$ such that*

$$\log \fint e^{4(w-\bar{w})} \leq \frac{1}{8\pi^2} \int (\Delta w)^2 + C_1. \tag{65}$$

Using Adams' inequality, we will show that the positive terms in $\tilde{F}$ dominate the logarithmic term. To see why, we argue in the following way: by the arithmetic-geometric mean,

$$2\beta xy \geq -\beta(1 + \delta)x^2 - \beta(\frac{1}{1+\delta})y^2,$$

for any real numbers x, y, as long as $\beta, \delta > 0$. From this inequality it follows that

$$\int (\Delta w)^2 + \beta \int (\Delta w + |\nabla w|^2)^2 \geq \int (1 - \delta\beta)(\Delta w)^2 + \beta(\frac{\delta}{1+\delta})|\nabla w|^4. \tag{66}$$

Therefore, by (64) and (66),

$$\tilde{F}[w] \geq \int (1-\delta\beta)(\Delta w)^2 + \beta(\frac{\delta}{1+\delta}) \int |\nabla w|^4 - 8\pi^2(1-\epsilon) \log \fint e^{4(w-\bar{w})} + (l.o.t.).$$

By Adams' inequality, the logarithmic term above can be estimated by

$$-8\pi^2(1 - \epsilon) \log \fint e^{4(w-\bar{w})} \geq -(1 - \epsilon) \int (\Delta w)^2 - C.$$

Substituting this above, we get

$$\tilde{F}[w] \geq \int (\epsilon - \delta\beta)(\Delta w)^2 + \beta(\frac{\delta}{1+\delta}) \int |\nabla w|^4 + (l.o.t.).$$

By choosing $\delta > 0$ small enough, we conclude

$$\tilde{F}[w] \geq \delta' \int \left[(\Delta w)^2 + |\nabla w|^4 \right] - C.$$

Arguing as we did in the previous case, it follows that $\tilde{F}$ is bounded below, and a minimizing sequence converges (weakly) to a smooth extremal.

Remarks.

1. The lower order terms that we neglected in the proof can actually dominate the expression when $\gamma_3 = 0$, e.g., when studying the Q-curvature problem. In particular, there are known examples of manifolds for which the functional II in not bounded below.

2. When γ_2 and γ_3 have different signs–for example, when $A = P$, the Paneitz operator–the situation is even worse. In fact, F_P is never bounded from below. However, manifolds of constant negative curvature are always local extremals of F_P.

References

[Ada]   Adams, *A sharp inequality of J. Moser for higher order derivatives.*
[Ber03]   Marcel Berger, *A panoramic view of Riemannian geometry*, Springer-Verlag, Berlin, 2003.
[Bra]   Sun-Yung A.; Yang Paul C. Branson, Thomas P.; Chang.
[Bra91]   Bent Branson, Thomas P.; rsted, *Explicit functional determinants in four dimensions*, Proc. Amer. Math. Soc. **113** (1991), 669–682.
[Bra96]   Thomas P. Branson, *An anomaly associated with 4-dimensional quantum gravity*, Comm. Math. Phys. **178** (1996), 301–309.
[CGY99]   Sun-Yung A. Chang, Matthew J. Gursky, and Paul C. Yang, *Regularity of a fourth order nonlinear PDE with critical exponent*, Amer. J. Math. **121** (1999), no. 2, 215–257.
[CGY02a] Sun-Yung A. Chang, Matthew J. Gursky, and Paul Yang, *An a priori estimate for a fully nonlinear equation on four-manifolds*, J. Anal. Math. **87** (2002), 151–186, Dedicated to the memory of Thomas H. Wolff.
[CGY02b] Sun-Yung A. Chang, Matthew J. Gursky, and Paul C. Yang, *An equation of Monge-Ampère type in conformal geometry, and four-manifolds of positive Ricci curvature*, Ann. of Math. (2) **155** (2002), no. 3, 709–787. MR 1 923 964
[Cha]   Sun-Yung Alice Chang, *The moser-trudinger inequality and applications to some problems in conformal geometry*, Nonlinear partial differential equations in differential geometry.
[CY95]   Sun-Yung A. Chang and Paul C. Yang, *Extremal metrics of zeta function determinants on 4-manifolds*, Ann. of Math. (2) **142** (1995), no. 1, 171–212.
[Eva82]   Lawrence C. Evans, *Classical solutions of fully nonlinear, convex, second-order elliptic equations*, Comm. Pure Appl. Math. **35** (1982), no. 3, 333–363. MR 83g:35038
[GT83]   David Gilbarg and Neil S. Trudinger, *Elliptic partial differential equations of second order*, second ed., Springer-Verlag, Berlin, 1983.

[Gur97] Matthew J. Gursky, *Uniqueness of the functional determinant*, Comm. Math. Phys. **189** (1997), no. 3, 655–665.

[Gur98] ———, *The Weyl functional, de Rham cohomology, and Kähler-Einstein metrics*, Ann. of Math. (2) **148** (1998), 315–337.

[Gur99] ———, *The principal eigenvalue of a conformally invariant differential operator, with an application to semilinear elliptic PDE*, Comm. Math. Phys. **207** (1999), no. 1, 131–143.

[GV04] Matthew Gursky and Jeff Viaclovsky, *Volume comparison and the σ_k-Yamabe problem*, Adv. Math. **187** (2004), 447–487.

[GVW03] Pengfei Guan, Jeff Viaclovsky, and Guofang Wang, *Some properties of the Schouten tensor and applications to conformal geometry*, Trans. Amer. Math. Soc. **355** (2003), no. 3, 925–933 (electronic).

[GW03] Pengfei Guan and Guofang Wang, *Local estimates for a class of fully nonlinear equations arising from conformal geometry*, Int. Math. Res. Not. (2003), no. 26, 1413–1432. MR 1 976 045

[Kry83] N. V. Krylov, *Boundedly inhomogeneous elliptic and parabolic equations in a domain*, Izv. Akad. Nauk SSSR Ser. Mat. **47** (1983), no. 1, 75–108. MR 85g:35046

[LP87] John M. Lee and Thomas H. Parker, *The Yamabe problem*, Bull. Amer. Math. Soc. (N.S.) **17** (1987), no. 1, 37–91.

[Osg88a] R.; Sarnak-P. Osgood, B.; Phillips, *Compact isospectral sets of surfaces*, J. Funct. Anal. **80** (1988), 212–234.

[Osg88b] ———, *Extremals of determinants of laplacians*, J. Funct. Anal. **80** (1988), 148–211.

[Pol81a] A. M. Polyakov, *Quantum geometry of bosonic strings*, Phys. Lett. B **103** (1981), 207–210.

[Pol81b] ———, *Quantum geometry of fermionic strings*, Phys. Lett. B **103** (1981), 211–213.

[RS71] D. B. Ray and I. M. Singer, *R-torsion and the Laplacian on riemannian manifolds*, Adv. in Math. **7** (1971), 145–210.

[STW05] Weimin Sheng, Neil S. Trudinger, and Xu-Jia Wang, *The Yamabe problem for higher order curvatures*, preprint, 2005.

[UV00] Karen K. Uhlenbeck and Jeff A. Viaclovsky, *Regularity of weak solutions to critical exponent variational equations*, Math. Res. Lett. **7** (2000), no. 5-6, 651–656.

[Via00] Jeff A. Viaclovsky, *Conformal geometry, contact geometry, and the calculus of variations*, Duke Math. J. **101** (2000), no. 2, 283–316.

[Via06] ———, *Conformal geometry and fully nonlinear equations*, to appear in World Scientific Memorial Volume for S.S. Chern., 2006.

Heat Kernels in Sub-Riemannian Settings

Ermanno Lanconelli

1 Lecture Topics

In this lectures we present a series of results concerning a class of diffusion second order PDE's of heat-type. The results we show have been obtained in collaboration with M.Bramanti, L.Brandolini and F. Uguzzoni (see [9], [10], [24]). The exended version of the main results presented in these notes is contained in [10].

The operators we are dealing with can be written in the general form

$$\sum_{i,j=1}^{N} q_{i,j}(z)\partial^2_{x_i,x_j} + \sum_{j=1}^{N} q_j(z)\partial_{x_j} - \partial_t.$$

The coefficients $q_{i,j} = q_{j,i}, q_j$ are of class C^∞ in the strip

$$S := \{z = (t,x) \; : \; x \in \mathbb{R}^N, \; T_1 < t < T_2\}$$
$$=]T_1, T_2[\times \mathbb{R}^N$$

where $-\infty \leq T_1 < T_2 \leq \infty$. We assume the characteristic form

$$q_{\mathcal{H}}(z,\xi) = \sum_{i,j=1}^{N} q_{i,j}(z)\xi_i\xi_j$$

is non-negative definite, and not identically zero, at any point $z \in S$. Moreover, the operator $\mathcal{H}$ is supposed to be hypoelliptic in S. This means that

E. Lanconelli

Dipartimento di Matematica Universita' degli Studi di Bologna, P.zza di Porta S. Donato, 5 40127 - Bologna - Italy

e-mail: lanconel@dm.unibo.it

S.-Y.A. Chang et al. (eds.), *Geometric Analysis and PDEs*, 35
Lecture Notes in Mathematics 1977, DOI: 10.1007/978-3-642-01674-5_2,
© Springer-Verlag Berlin Heidelberg 2009

every distributional solution to $\mathcal{H}u = f$ in an open set $\Omega \subset S$ is of class C^∞ whenever f is C^∞.

Together with the previous qualitative properties, we assume that $\mathcal{H}$ has a *fundamental solution* Γ satisfying the *Gaussian estimates*

$$\frac{1}{\Lambda} G_{b_0}(z, \zeta) \le \Gamma(z, \zeta) \le \Lambda G_{a_0}(z, \zeta),$$

where, $G_a(z, \zeta) \doteq G_a(t, x; \tau, \xi) = 0$ if $t \le \tau$, and

$$G_a(t, x; \tau, \xi) = \frac{1}{|B(x, \sqrt{t - \tau})|} \exp\left(-a \frac{d^2(x, \xi)}{t}\right)$$

if $t > \tau$. Λ, a_0, b_0 are positive constants.

Hereafter d is a *metric* in $\mathbb{R}^N$ and $|B(x, r)|$ denotes the Lebesgue measure of the d-ball $B(x, r)$. We assume that the metric space $(\mathbb{R}^N, d)$ is of *sub-Riemannian type*, i.e. that it satisfies the following conditions:

- the d-topology is the Euclidean topology
- $\text{diam}_d(\mathbb{R}^N) = \infty$
- $(\mathbb{R}^N, d)$ is a doubling space w.r to the Lebesgue measure, i.e. $0 < |B(0, 2r)| \le c_d|B(0, 2r)|$, for every $x \in \mathbb{R}^N$, and $r > 0$
- $(\mathbb{R}^N, d)$ has the segment property, i.e. for every $x, y \in \mathbb{R}^N$ there exists $\gamma : [0, 1] \to \mathbb{R}^N$, continuous and such that $d(x, y) = d(x, \gamma(t)) + d(\gamma(t), y)$ for every $t \in [0, 1]$

We shall denote by $|\mathcal{H}|$ the constant

$$|\mathcal{H}| := \Lambda + a_0 + b_0 + c_d$$

Under the previous assumptions, we show the following *scale invariant Harnack inequality* which extends to our sub-Riemannian setting the classical Hadamard-Pini parabolic Harnack inequality

Harnack inequality. If $\mathcal{H}u = 0$ and $u \ge 0$ in an open set containing $[t_0 - R^2, t_0] \times B(x_0, R)$, where $(t_0, x_0) =: z_0 \in S$, then

$$\max_{C_R(z_0)} u \le M\, u(z_0).$$

Here $C_R(z_0) := [\tau_0 - \gamma R^2, \tau_0 - \frac{\gamma}{2}R^2] \times \overline{B_d(\xi_0, \gamma R)}$

The constants $M > 0$ and $0 < \gamma < 1$ are independent of R and z_0. They *depend on the operator* $\mathcal{H}$ *only through the constant* $|\mathcal{H}|$.

Our basic example of *diffusion operator* , satisfying the assumption stated before is the following ones

$$\mathcal{H} := \mathcal{L} - \partial_t := \sum_{i,j=1}^q a_{ij}(z)X_iX_j + \sum_{k=1}^q a_k(z)X_k - \partial_t$$

where:

- $X_1, X_2, \ldots, X_q$ are smooth vector fields in the open set $\Omega \subset \mathbb{R}^N$ satisfying the Hörmander condition

$$\text{rank Lie}\{X_i, i = 1, 2, \ldots, q\} = n \text{ at any point of } \Omega.$$

Then $\mathcal{H}$ is hypoelliptic in Ω [17].
- $A(z) = (a_{ij}(z))_{i,j=1}^q$ is a symmetric matrix such that

$$\frac{1}{\lambda}|\xi|^2 \leq \sum_{i,j=1}^q a_{ij}(z)\xi_i\xi_j \leq \lambda|\xi|^2$$

for every $z = (t, x) \in S$, $\xi = (\xi_1, \ldots, \xi_q) \in \mathbb{R}^q$

A natural distance for the operator $\mathcal{H}$ is the *Carnot-Carathéodory* distance d generated by the vector fields $X_1, X_2, \ldots, X_q$. We would like to stress that d is well defined since the system $X = \{X_1, X_2, \ldots, X_q\}$ satisfies the Hörmander rank condition.

In [10] we proved that X can be extended to a system of Hörmander vector fields, defined all over $\mathbb{R}^N$, in such a way that the associated Carnot-Carathéodory distance satisfies all the assumptions stated before.

In [10] the operator $\mathcal{H}$ is also extended to the whole $\mathbb{R}^{N+1}$ in such a way that outside of a compact set in the spatial variable, it becomes the classical Heat operator.

Still denoting by $\mathcal{H}$ the extended operator, under the qualitative assumption $a_{i,j}, a_j \ C^\infty$, we proved that $\mathcal{H}$ has a global fundamental solution Γ such that

$$\frac{1}{\Lambda} G_{b_0}(z, \zeta) \leq \Gamma(z, \zeta) \leq \Lambda G_{a_0}(z, \zeta)$$

and, for every $\alpha = (\alpha_i, \alpha_j.\alpha_k)$ with $|\alpha| \leq 2$,

$$|D^\alpha \Gamma(z, \zeta)| \leq \Lambda |t - \tau|^{-\frac{|\alpha|}{2}} G_{a_0}(z, \zeta).$$

Here we have used the notation $D^\alpha := X_i^{\alpha_i} X_j^{\alpha_j} \partial_t^{\alpha_k}$ and $|\alpha| := \alpha_i + \alpha_j + 2\alpha_k$.

In these inequalities, Λ, a_0, b_0 are positive structural constants: they only depend on

- the doubling constant of d
- the constant λ in

$$\frac{1}{\lambda}|\xi|^2 \leq \sum_{i,j=1}^q a_{ij}(z)\xi_i\xi_j \leq \lambda|\xi|^2$$

- the d-Hölder norms of the coefficients $a_{i,j}$ and a_j.

As a consequence: *all our results extend to the operators*

$$\mathcal{H} := \mathcal{L} - \partial_t := \sum_{i,j=1}^{q} a_{ij}(z)X_iX_j + \sum_{k=1}^{q} a_k(z)X_k - \partial_t$$

with d-Hölder continuous coefficients $a_{i,j}$ and a_j.

We construct the heat kernel (the fundamental solution) by an adaptation to our sub-riemannian setting of the classical Levi parametrix method, as in [4]. For this adaptation we used a large amount of ideas, techniques and results due to Rotschild&Stein [34], Jerison&Sanchez-Calle [19], Fefferman&Sanchez-Calle [14] and to Kusuoka&Stroock [20], [21], [22].

2 A Motivation: Levi Curvatures and Motion by Levi Curvatures

There are several motivations for studying heat-type operators in sub-Riemannian settings. We want to show just one of them, arising in Complex Geometry in studying a notion af curvature related to the *Levi form*.

Let M be a real hypersurface of $\mathbb{C}^{n+1}$ which is the boundary of a domain

$$D := \{z \in \mathbb{C}^{n+1} \,:\, f(z) < 0\},$$

where f is real function of class C^2 on $\mathbb{C}^{n+1}$. The complex tangent space to $M = \partial D$ at a point p is given by

$$T_p^{\mathbb{C}}(\partial D) = \{h \in \mathbb{C}^{n+1} \,:\, \langle h, \bar{\partial}_p f \rangle = 0\}.$$

We explicitly remark remark that $T_p^{\mathbb{C}}(\partial D)$ is a complex vector space of dimension n, hence it can be idetified with a real vector space of real dimension $2n$. Therefore, in passing from M to $T_p^{\mathbb{C}}(\partial D)$ *we loose a real dimension*

The *Levi form* of f at p is given by

$$L_p(f,\zeta) = \langle H_f^T(p)\zeta, \zeta \rangle := \sum_{j,k=1}^{n+1} f_{j,\bar{k}}(p)\zeta_j\bar{\zeta}_k, \quad \zeta \in T_p^{\mathbb{C}}(bD).$$

Here $f_j = \frac{\partial f}{\partial z_j}$, $f_{\bar{k}} = \frac{\partial f}{\partial \bar{z}_k}$, $f_{j,\bar{k}} := \frac{\partial^2 f}{\partial z_j \partial \bar{z}_k}$. Let us now fix an orthonormal basis $B = \{u_1, \dots, u_n\}$ of $T_p^{\mathbb{C}}(bD)$ and consider the matrix

$$L_p(f,B) := \left(\frac{1}{|\partial_p f|} \langle H_f^T(p)u_j, u_k \rangle \right)_{k,j=1,\dots,n}.$$

This is a Hermitian $n \times n$ matrix which we call *Normalized Levi Matrix* of f at p. Its eigenvalues $\lambda_1(p), \dots, \lambda_n(p)$ are real and independent of B and the defining function f of D. So that they only depend on the domain D.

Just proceeding as in the real case, one can define the $m-th$ *Levi curvature* of ∂D at p, $1 \leq m \leq n$, as

$$K_p^m(\partial D) = \frac{\sigma^{(m)}(\lambda_1(p), \ldots, \lambda_n(p))}{\binom{n}{m}},$$

where $\sigma^{(m)}$ denotes the m-th elementary symmetric function. More generally, given a generalized symmetric function s, in the sense of Caffarelli-Nirenberg-Spruck [11], one can define the s-Levi curvature of M at p, as follows:

$$S_p(M) = s(\lambda_1(p), ..., \lambda_n(p)).$$

When M is the graph of a function u and one imposes that its s-Levi curvature is equal to a given function, one obtains a second order fully nonlinear partial differential equation, which can be seen as the pseudoconvex counterpart of the usual fully nonlinear elliptic equations of Hessian type, as studied e.g. in [11]. In linearized form, the equations of this new class can be written as (see [29, equation (34) p. 324])

$$\mathcal{L}u \equiv \sum_{i,j=1}^{2n} a_{ij}\left(Du, D^2u\right) X_i X_j u = K\left(x, u, Du\right) \ \text{in} \mathbb{R}^{2n+1} \tag{1}$$

where:

the X_j's are first order differential operators, with coefficients depending on the gradient of u, which form a real basis for the complex tangent space to the graph of u;

the matrix $\{a_{ij}\}$ depends on the function s;

K is a prescribed function.

It has to be noticed that $\mathcal{L}$ only involves $2n$ derivatives, while it lives in a space of dimension $2n + 1$. Then, $\mathcal{L}$ is never elliptic, on any reasonable class of functions. However, the operator $\mathcal{L}$, when restricted to the set of strictly s-pseudoconvexk functions, becomes "elliptic" along the $2n$ linearly independent directions given by the X_i's, while the missing one can be recovered by a commutation. Precisely,

$$\dim\left(\text{span}\left\{X_j, [X_i, X_j], i, j = 1, ..., 2n\right\}\right) = 2n + 1$$

at any point (see [29, equation (36) p. 324]). This is a Hörmander-type rank condition of step 2.

The parabolic counterpart of (1), i.e. the equation

$$\partial_t u(t, x) = \mathcal{L}u(t, x) \ \text{for} \ t \in \mathbb{R}, \ x \in \mathbb{R}^{2n+1} \tag{2}$$

arises in studying the evolution by s-Levi curvature of a real hypersurface of $\mathbb{C}^{n+1}$ (see [18], [28]).

Before closing this section we would like to quote some remarkable *isoperimetric inequalities* involving the Levi curvatures, recently proved by Martino and Montanari in [27]. These inequalities well enlighten the geometric content of the Levi curvatures.

Isoperimetric inequalities Let D be a bounded smooth domain in C^{n+1}. Let $1 \leq m \leq n$ and assume $K_z^m(\partial D) \geq 0$ at any point $z \in \partial D$. Then

$$\int_{\partial D} (K_z^{(m)}(\partial D))^{-\frac{1}{m}} d\sigma(z) \geq (2n+2)meas(D)$$

If $K_z^m(\partial D) = constant$ w.r. to z, in the previous inequality the equality holds if and only if D is a ball.

3 Heat Kernels and Gaussian Bounds: Past History

Gaussian estimates for the fundamental solution of second order partial differential operators of parabolic type, or, somehow more generally, for the density function of heat diffusion semigroups, have a long history, starting with Aronson's work [1]. The relevance of two-sided Gaussian estimates to get scaling invariant Harnack inequalities for positive solutions was firstly pointed out by Nash in the Appendix of his celebrated paper [31]. However, a complete implementation of the method outlined by Nash was given much later by Fabes and Stroock in [13], also inspired by some ideas of Krylov and Safonov. Since then, the full strength of Gaussian estimates has been enlightened by several authors, showing their deep relationship not only with the scaling invariant Harnack inequality, but also with the ultracontractivity property of heat diffusion semigroups, with inequalities of Nash, Sobolev or Poincaré type, and with the doubling property of the measure of "intrinsic" balls. We directly refer to the recent monograph by Saloff-Coste [35] for a beautiful exposition of this circle of ideas, and for an exhaustive list of references on these subjects. Here we explicitly recall just the results in literature strictly close to the core of our lectures.

For heat-type operators

$$H = \partial_t - \sum_{i=1}^{q} X_i^2 \tag{3}$$

with the X_i's left invariant homogeneous vector fields on a sratified Lie group in $\mathbb{R}^n$, Gaussian bounds have been proved by Varopoulos ([40], [41], see also [42]):

$$\frac{1}{ct^{Q/2}} e^{-c\|y^{-1}\circ x\|^2/t} \leq h(t,x,y) \leq \frac{c}{t^{Q/2}} e^{-\|y^{-1}\circ x\|^2/ct} \tag{4}$$

for any $x, y \in \mathbb{R}^n, t > 0$, where Q is the homogeneous dimension of the group, and $\|\cdot\|$ any homogeneous norm of the group. Two-sided Gaussian estimates and a scaling invariant Harnack inequality have been proved by Saloff-Coste and Stroock for the operator

$$H = \partial_t - \sum_{i,j=1}^{q} X_i \left(a_{ij} X_j \right)$$

where $\{a_{ij}\}$ is a uniformly positive matrix with measurable entries, and the vector fields X_i are left invariant with respect to a connected unimodular Lie group with polynomial growth [36].

In absence of a group structure, Gaussian bounds for operators (3) have been proved, on a compact manifold and for finite time, by Jerison and Sanchez-Calle [19], with an analytic approach (see also the previous partial result in [37]), and, on the whole $\mathbb{R}^{n+1}$, by Kusuoka-Stroock, [21], [22], by using Malliavin stochastic calculus.

Unlike the study of "sum of squares" Hörmander's operators, the investigation of non-divergence operators of Hörmander type has a relatively recent history. Stationary operators like

$$L = \sum_{i,j=1}^{q} a_{ij}(x) X_i X_j \qquad (5)$$

with $X_1, ..., X_q$ system of Hörmander's vector fields have been studied by Xu [43], Bramanti, Brandolini [7], [8], Capogna, Han [12]. Evolution operators the previuos one have been considered by Bonfiglioli, Lanconelli, Uguzzoni [3], [4], [5]. In [7] also more general operators of kind

$$L = \sum_{i,j=1}^{q} a_{ij}(x) X_i X_j + a_0(x) X_0 \qquad (6)$$

with $X_0, X_1, ..., X_q$ system of Hörmander's vector fields have been studied.

In these papers, the matrix $\{a_{ij}\}$ is assumed symmetric and uniformly elliptic, and the entries a_{ij} typically belong to some function space defined in terms of the vector fields X_i and the metric they induce. In particular, these operators do not have smooth coefficients, so they are no longer hypoelliptic. Therefore the mere existence of a fundamental solution is troublesome. For the operators (6) (without lower order terms) with X_i left invariant homogeneous Hörmander's vector fields on a stratified Lie group and under assumptions (H1), (H2), (H3), it has been proved by Bonfiglioli, Lanconelli, Uguzzoni in [3], [4], [5] that the fundamental solution h exists and satisfies Gaussian bounds of the kind (4). Previous results about Harnack inequality for general Hörmander's operators date back to Bony's seminal paper [6], where a first qualitative version of this result is proved. A first scaling invariant Harnack inequality for heat-type Hörmander's operators was proved later by Kusuoka-Stroock [21].

4 Carnot-Carathèodory metric

Let $X_1, X_2, \ldots, X_q$ be a system of real smooth vector fields which are defined in some bounded connecetd domain $\Omega \subseteq \mathbb{R}^n$ and satisfying Hörmander's condition of some step s in Ω. Explicitly, this means that:

$$X_i = \sum_{k=1}^{n} b_{ik}(x) \, \partial_{x_k}$$

with $b_{ik} \in C^\infty(\Omega)$, and the vector space spanned at every point of Ω by: the fields X_i; their commutators $[X_i, X_j] = X_i X_j - X_j X_i$; the commutators of the X_k's with the commutators $[X_i, X_j]$; ... and so on, up to some step s, is the whole $\mathbb{R}^n$.

We say that an absolutely continuous curve $\gamma : [0, T] \to \Omega$ is a sub-unit curve with respect to the system $X_1, X_2, \ldots, X_q$ if

$$\gamma'(t) = \sum_{j=1}^{q} \lambda_j(t) X_j(\gamma(t))$$

for a.e. $t \in [0, T]$, with $\sum_{j=1}^{q} \lambda_j(t)^2 \leq 1$ a.e. In the following, this number T will be denoted by $l(\gamma)$.

For any $x, y \in \Omega$, we define

$$d_\Omega(x, y) = \inf\{l(\gamma) \mid \gamma \text{ is } X\text{-subunit}, \gamma(0) = x, \gamma(l(\gamma)) = y\}.$$

It is well known (Carathèodory-Chow's Theorem) that, if the vector fields satisfy Hörmander's condition, the above set in nonempty, so that $d_\Omega(x, y)$ is finite for every pair of points. Moreover, d_Ω is a distance in Ω, called the Carnot-Carathéodory distance (CC-distance) induced by the vector fields X_i's.

A known result by Fefferman-Phong [15] states that

$$c^{-1} |x - y| \leq d_\Omega(x, y) \leq c |x - y|^{1/s} \tag{7}$$

for every $x, y \in K \subset \Omega$, where s is the step of Hörmander condition. In particular, this means that d_Ω induces the usual topology of $\mathbb{R}^n$. Moreover, Sanchez-Calle [37] and Nagel-Stein-Weinger [30] proved that the CC-distance is *locally* doubling with respect to the Lebesgue measure, i.e., denoting by $B(x, r)$ the d_Ω-ball of center x and radius r:

$$|B(x, 2r)| \leq c|B(x, r)| \tag{8}$$

at least for x ranging in a compact set and r bounded by some r_0.

As we have already recalled , in [10] is proved that the given system of vector fields can be exended to the whole $\mathbb{R}^n$ to a new system of Hörmander vector fields whose related Carnot-Carathèodory distance satisfies a global doubling property, and coincide with the original one in a fixed compact subset of Ω. Moreover, in a neighborhood of infinity, the new system is the Euclidean one.

5 Heat-Type Operators with Constant Coefficients

Let us consider the heat-type operator in $\mathbb{R}^{n+1}$

$$H_A = \partial_t - L_A = \partial_t - \sum_{i,j=1}^q a_{ij} X_i X_j \tag{9}$$

where:

(H1)$X_1, X_2, \ldots, X_q$ is a system of Hörmander vector fields in $\mathbb{R}^n$, of step $\leq s$, coinciding with the Euclidean one in a neighborhood of infinity

(H2)$A = \{a_{ij}\}_{i,j=1}^q$ is a real symmetric positive definite matrix with constant entries, and $\lambda > 0$ a constant such that:

$$\lambda^{-1}|\xi|^2 \leq \sum_{i,j=1}^q a_{ij}\xi_i\xi_j \leq \lambda|\xi|^2$$

for every $\xi \in \mathbb{R}^q$. When condition (H2) is fulfilled, we will briefly say that

$$A \in \mathcal{E}_\lambda.$$

Then, the following thoerem holds.

Theorem 5.1. *For any $T > 0$ there exists $c > 0$ such that, for any $t \in (0, T), x, y \in \mathbb{R}^n$ the following bounds hold:*
1. Upper and lower bounds on h_A:

$$\frac{1}{c|B(x, \sqrt{t})|} e^{-cd(x,y)^2/t} \leq h_A(t, x, y) \leq \frac{c}{|B(x, \sqrt{t})|} e^{-d(x,y)^2/ct} \tag{10}$$

2. Upper bounds on the derivatives of h_A of arbitrary order:

$$\left| X_x^I X_y^J \partial_t^i h_A(t, x, y) \right| \leq \frac{c}{t^{i+\frac{|I|+|J|}{2}} |B(x, \sqrt{t})|} e^{-d(x,y)^2/ct} \tag{11}$$

3. Estimate on the difference of the fundamental solutions of two operators (and their derivatives):

$$\left| X_x^I X_y^J \partial_t^i h_A (t,x,y) - X_x^I X_y^J \partial_t^i h_B (t,x,y) \right| \leq \frac{c \, \|A - B\|}{t^{i + \frac{|I| + |J|}{2}} \left| B \left(x, \sqrt{t}\right) \right|} e^{-d(x,y)^2/ct}$$

(12)

(here I, J are arbitrary multiindices, $A, B \in \mathcal{E}_\lambda$). The constants depend on the matrix A only through the number λ; in (11), (12), the constant also depends on the multiindices. The same estimates hold for $h_A (t,y,x)$.

The proof of this theorem is contained in [10], Part I. We followed an approach which is basically inspired to the work by Jerison and Sanchez-Calle [19], integrated with several devices to overcome the new difficulties. The main of them are the following: first, we had to take into account the dependence on the matrix A, getting estimates depending on A only through the number λ; second, the globality of the estimates ; third, the estimates of the difference of the fundamental solutions of two different operators.

The strategy we followed in order to get the uniform Gaussian bounds for h_A was as follows. First, we proved the upper bound for $t \in (0,1)$ and $\varepsilon < d(x,y) < R$. In this range, the bound is equivalent to:

$$h_A (t,x,y) \leq c e^{-1/ct}$$

(13)

and is proved by means of estimates of Gevray type. This means that the exponential decay of h_A for vanishing t is deduced by a control on the supremum of the time derivative of any order of a solution to $H_A u = 0$. This technique makes the constant c in (13) depend on:

$$\sup_{y \in \mathbb{R}^n} \int_0^T d\tau \int_{\varepsilon < d(x,y) < R} h_A (\tau, x, y) \, dx.$$

So the next problem was to prove a uniform upper bound on this quantity (i.e., depending on A only through λ). This is accomplished exploiting suitable estimates on fractional and singular integrals on spaces of homogeneous type, and uniform subelliptic estimates. Next, we had to prove the upper bound in (10) for $t \in (0,1)$ and $d(x,y) < \varepsilon$. For this we used Rothschild-Stein's technique of "lifting and approximation". This allowed, by a rather involved procedure, to deduce the desired uniform bound from the analogous result proved, in the context of homogeneous groups, by Bonfiglioli-Lanconelli-Uguzzoni [3], and to complete the proof of the upper bound in (10) for $t \in (0,1)$ and $d(x,y) < R$. To prove the same upper bound for any $x, y \in \mathbb{R}^n$ and $t \in (0,T)$, we used a comparison argument, exploiting our *ad hoc* extension of the vector fields.

Then, we proved the lower bound in (10), again by using the analogous uniform lower bound which holds in the case of homogeneous vector fields.

Subsequently, we proved the Gaussian bound (11) on the derivatives of h_A. Like in [19], this bound is deduced by the upper bound on h_A applying a powerful result by Fefferman and Sanchez-Calle [14], which assures the existence of a local change of coordinates which is a good substitute of dilations (which in our context do not generally exist).

Finally, we proved the estimate (12) on the difference of two fundamental solutions $h_A - h_B$. It relies on a suitable use of basic properties of the fundamental solution and on the uniform bound (11) on the derivatives of h_A.

6 Heat-Type Operators with Variable Coefficients. The Levi Parametrix Method

The previuos results on the constant coefficients heat-type operators are crucial to deal with operators of this kind

$$H = \partial_t - \sum_{i,j=1}^{m} a_{i,j}\left(t, x\right) X_i X_j - \sum_{k=1}^{m} a_k\left(t, x\right) X_k - a_0\left(t, x\right). \qquad (14)$$

To exploit the results of the previous section, we make the same assumptions on the vector fields and the structure of the matrix of the coefficients in the principal part. Moreover, the coefficients a_{ij}, a_k, a_0 are supposed smooth and globally defined; the matrix $\{a_{ij}\}_{i,j=1}^{m}$ will be assumed symmetric and uniformly positive definite. Under these assumptions one can prove the existence of a (global) fundamental solution for H, satisfying natural basic properties and sharp Gaussian bounds. To state the results we need to introduce some more definitions, notation and assumptions.

Assumptions on the Vector Fields

We assume that:

$X = (X_1, X_2, ..., X_m)$ $(m = n + q)$ is a fixed system of Hörmander's vector fields defined in the whole $\mathbb{R}^n$, and such that

$$X = (0, 0, ..., 0, \partial_{x_1}, \partial_{x_2}, ..., \partial_{x_n}) \text{ in } \mathbb{R}^n \setminus \Omega_0 \qquad (15)$$

where Ω_0 is a fixed bounded domain.

Function Spaces

We start with the following

Definition 6.1. The intrinsic-derivative along the vector field X_j of a function $v(x)$ at a point $x_0 \in \mathbb{R}^n$, is defined to be

$$X_j v(x_0) = \left. \frac{d}{d\sigma} \right|_{\sigma=0} v(\gamma(\sigma))$$

(if such derivative exists), where γ is the solution to

$$\dot{\gamma}(\sigma) = X_j(\gamma(\sigma)), \qquad \gamma(0) = x_0.$$

We can now introduce the natural function spaces for the operator H.

We will denote by d the Carnot-Carathéodory distance induced by the system $\{X_i\}_{i=1}^m$ in the whole $\mathbb{R}^n$ and by $B(x,r)$ the balls in the metric d. Moreover, let d_P be the corresponding "parabolic-CC-distance":

$$d_P((t,x),(s,y)) = \left(d^4(x,y) + (t-s)^2 \right)^{\frac{1}{4}}$$

Definition 6.2. For any $\alpha \in (0,1]$ and domain $U \subseteq \mathbb{R}^{n+1}$, let:

$$|u|_{C^\alpha(U)} = \sup \left\{ \frac{|u(t,x) - u(s,y)|}{d_P((t,x),(s,y))^\alpha} : (t,x),(s,y) \in U, (t,x) \neq (s,y) \right\}$$

$$\|u\|_{C^\alpha(U)} = |u|_{C^\alpha(U)} + \|u\|_{L^\infty(U)}$$

$$C^\alpha(U) = \left\{ u : U \to \mathbb{R} : \|u\|_{C^\alpha(U)} < \infty \right\}.$$

Also, for any positive integer k, and domain $U \subseteq \mathbb{R}^{n+1}$, let

$$C^{k,\alpha}(U) = \left\{ u : U \to \mathbb{R} : \|u\|_{C^{k,\alpha}(U)} < \infty \right\}$$

with

$$\|u\|_{C^{k,\alpha}(U)} = \sum_{|I|+2h \leq k} \left\| \partial_t^h X^I u \right\|_{C^\alpha(U)}$$

where, for any multiindex $I = (i_1, i_2, ..., i_s)$, with $1 \leq i_j \leq q$, we say that $|I| = s$ and

$$X^I u = X_{i_1} X_{i_2} ... X_{i_s} u.$$

Assumptions on the Coefficients

We assume that the matrix $\{a_{ij}\}_{i,j=1}^m$ is block-diagonal, with last block the $n \times n$ identity. Moreover, the functions $a_{ij} = a_{ji}$, a_k, a_0, are smooth and defined on $\mathbb{R}^{n+1}$ and satisfy, for some $\alpha \in (0,1]$ and for some positive constants λ, K,

$$\lambda^{-1} |w|^2 \leq \sum_{i,j=1}^{q} a_{ij}(t,x) w_i w_j \leq \lambda |w|^2 \; \forall w \in \mathbb{R}^q, (t,x) \in \mathbb{R}^{n+1} \quad (16)$$

$$\|a_{ij}\|_{C^\alpha(\mathbb{R}^{n+1})} + \|a_k\|_{C^\alpha(\mathbb{R}^{n+1})} + \|a_0\|_{C^\alpha(\mathbb{R}^{n+1})} \leq K. \quad (17)$$

To state the existence result for the fundamental solution, it is also convenient to denote the Gaussian kernel $G_1(x,t;\xi,0)$ as follows:

$$\mathbf{E}(x,\xi,t) = |B(x,\sqrt{t})|^{-1} \exp\left(-\frac{d(x,\xi)^2}{t}\right), \quad x,\xi \in \mathbb{R}^n, \; t > 0.$$

Main Results

Theorem 6.3 (Fundamental solution for H). *Let H be as in (14). Under the above assumptions, there exists a function*

$$h : \mathbb{R}^{n+1} \times \mathbb{R}^{n+1} \to \mathbb{R}$$

such that:

i) h *is continuous away from the diagonal of* $\mathbb{R}^{n+1} \times \mathbb{R}^{n+1}$;
ii) $h(z,\zeta)$ *is nonnegative, and vanishes for* $t \leq \tau$;
iii) for every fixed $\zeta \in \mathbb{R}^{n+1}$, *we have*

$$h(\cdot;\zeta) \in C_{loc}^{2,\alpha}(\mathbb{R}^{n+1} \setminus \{\zeta\}), \quad H(h(\cdot;\zeta)) = 0 \text{ in } \mathbb{R}^{n+1} \setminus \{\zeta\};$$

iv) the following estimates hold for every $T > 0, z = (t,x), \zeta = (\tau,\xi) \in \mathbb{R}^{n+1}, 0 < t - \tau \leq T$:

$$\mathbf{c}(T)^{-1} \mathbf{E}(x,\xi,\mathbf{c}^{-1}(t-\tau)) \leq h(z;\zeta) \leq \mathbf{c}(T) \mathbf{E}(x,\xi,\mathbf{c}(t-\tau)),$$
$$|X_j(h(\cdot;\zeta))(z)| \leq \mathbf{c}(T)(t-\tau)^{-1/2} \mathbf{E}(x,\xi,\mathbf{c}(t-\tau)'$$
$$|X_i X_j(h(\cdot;\zeta))(z)| + |\partial_t(h(\cdot;\zeta))(z)| \leq \mathbf{c}(T)(t-\tau)^{-1} \mathbf{E}(x,\xi,\mathbf{c}(t-\tau)).$$

v) for any $f \in C^\alpha(\mathbb{R}^{n+1})$ $g \in C(\mathbb{R}^n)$, *both satisfying suitable growth condition at infinity*, $T \in \mathbb{R}$, *the function*

$$u(t,x) = \int_{\mathbb{R}^n} h(t,x;T,\xi)\, g(\xi)\, d\xi + \int_{[T,t] \times \mathbb{R}^n} h(t,x;\tau,\xi)\, f(\tau,\xi) d\tau\, d\xi$$

is a $C_{loc}^{2,\alpha}$ *solution to the following Cauchy problem*

$$Hu = f \text{ in } (T,\infty) \times \mathbb{R}^n, \quad u(T,\cdot) = g \text{ in } \mathbb{R}^n$$

vi) the following reproduction formula holds

$$h(t, x; \tau, \xi) = \int_{\mathbb{R}^n} h(t, x; s, y) \, h(s, y; \tau, \xi) dy,$$

for $t > s > \tau$ and $x, \xi \in \mathbb{R}^n$.

The Levi Paramerix Method

To construct the fundamental solution for the operator

$$H = \partial_t - \sum_{i,j=1}^m a_{i,j}(t, x) X_i X_j - \sum_{k=1}^m a_k(t, x) X_k - a_0(t, x). \qquad (18)$$

we used the classical Levi paramtrix metod, adapted to our sub-Riemann setting. The "Levi method" is a classical technique that allows to construct the fundamental solution of a variable coefficient differential operator, starting from the fundamental solution of the corresponding operator with constant coefficients. This method was originally developed by E. E. Levi at the beginning of 20th century to study uniformly elliptic equations of order $2n$ (see [25], [26]) and later extended to uniformly parabolic operators (see e.g. [16]).

In the context of hypoelliptic ultraparabolic operators of Kolmogorov-Fokker-Planck type, Polidoro in [33] managed to adapt this method, thanks to the knowledge of an explicit expression for the fundamental solution of the "frozen" operator, which had been constructed in [23]. For operators of type (18), structured on homogeneous and invariant vector fields on Carnot groups, no explicit fundamental solution is available in general. Nevertheless, Bonfiglioli, Lanconelli, Uguzzoni in [4] showed how to adapt the same method, exploiting suitable sharp uniform Gaussian bounds on the fundamental solutions of the frozen operators. Here we follow the same line, first giving an outline of the Levi method.

Let us consider the fundamental solution $h_{\zeta_0}(z, \zeta)$ of the "frozen" operator

$$H_{\zeta_0} := \partial_t - L_{\zeta_0} := \partial_t - \sum_{i,j=1}^m a_{i,j}(\zeta_0) X_i X_j. \qquad (19)$$

The function $z \mapsto h_\zeta(z, \zeta)$ is called a *parametrix* for H. The idea of the Levi method is to look for a fundamental solution $h(z, \zeta)$ for H, which could be written in the form:

$$h(z, \zeta) = h_\zeta(z, \zeta) + \int_\tau^t \int_{\mathbb{R}^n} h_\eta(z, \eta) \Phi(\eta, \zeta) \, d\eta \qquad (20)$$

for a suitable, unknown kernel $\Phi(z, \zeta)$. This seems reasonable because we expect h to be a small perturbation of theparametrix, as the integral equation (20) expresses. The following formal computation suggests how to guess the right form of $\Phi(z, \zeta)$. If we set

$$Z_1(z; \zeta) = -H\left(z \mapsto h_\zeta(z, \zeta)\right)(z), \qquad z \neq \zeta \in \mathbb{R}^{n+1}$$

and apply the operator H to both sides of (20) for $z \neq \zeta$, we find:

$$0 = -Z_1(z; \zeta) + \Phi(z, \zeta) - \int_\tau^t \int_{\mathbb{R}^n} Z_1(z, \eta)\Phi(\eta, \zeta)\, d\eta.$$

This means that Φ solves the integral equation

$$Z_1(z, \zeta) = \Phi(z; \zeta) - \int_\tau^t \int_{\mathbb{R}^n} Z_1(z, \eta)\Phi(\eta, \zeta)\, d\eta$$

which, defining the integral operator T with kernel Z_1, can be rewritten as

$$Z_1 = (I - T)\,\Phi$$

whence, formally,

$$\Phi = \sum_{k=0}^\infty T^k Z_1 \equiv \sum_{k=0}^\infty Z_k.$$

To make the above idea rigorous, one has to reverse the order of the previous steps, first studying the properties of the function Z_1, then of the functionns Z_k, hence the convergence of series $\Phi = \sum_{k=0}^\infty Z_k$, then the properties of

$$J(z, \zeta) = \int_\tau^t \int_{\mathbb{R}^n} h_\eta(z, \eta)\Phi(\eta, \zeta)\, d\eta$$

and finally the ones of

$$h(z, \zeta) = h_\zeta(z, \zeta) + J(z, \zeta).$$

This can be accomplished by using all the results related to the constant coefficients heat-type operators.

7 Hadamard-Pini Invariant Harnack Inequality

Here we finally come to the main goal of our lectures: the proof of the invariant parabolic Harnack inequality, of Hadamard-Pini type, for the non negative solution to the equation $\mathcal{H}u = 0$, where

$$\mathcal{H} = \sum_{i,j=1}^N q_{i,j}(z)\partial_{x_i, x_j}^2 + \sum_{j=1}^N q_j(z)\partial_{x_j} - \partial_t$$

is a diffusion operator satisfying all the assumptions fixed in the first section.

The proof is inspired to the paper by Fabes-Stroock [13], who, in turn, exploited the original ideas by Krylov-Safanov about parabolic operators in nondivergence form. In Fabes-Stroock's paper, Harnack inequality is derived by a fairly short but clever combination of estimates based only on the Gaussian bounds (from above and below) on the Green function for a cylinder. The radius of the cylinder incorporates the essential geometrical information, giving dilation invariance to the Harnack estimate.

About at the same time of Fabes-Stroock paper, the same deep ideas were applied by Kusuoka-Stroock [21] in the context of Hörmander's operators $\partial_t - \sum_{i=1}^{q} X_i^2$. Much more recently, this general strategy has been adapted by Bonfiglioli, Uguzzoni [5] to study nonvariational operators structured on Hörmander's vector fields in Carnot groups.

Here we will follow the same line. The striking feature of this proof is the "axiomatic" nature of its core: it depends only on the suitable Gaussian estimates for the Green function, a maximum principle for H, the fact that constants are solutions to $Hu = 0$ (absence of the zero order term), and some simple geometric properties of the distance D: the doubling property of the Lebesgue measure of the d-balls and the segment property of d. However, a first problem arises since, for our operator H, the existence of the Green function is not granted: the cylinders based on the metric balls could be bad domain for the Dirichlet problem. Nevertheless, Lanconelli and Uguzzoni have recently proved in [24] that given two metric balls $B(\xi, \delta R), B(\xi, R)$, (with $\delta \in (0, 1)$), there always exists a domain $A(\xi, R)$, regular for the (stationary) Dirichlet problem, and such that $B(\xi, \delta R) \subseteq A(\xi, R) \subseteq B(\xi, R)$ (see Lemma 7.2). The Green function for H on $\mathbb{R} \times A(\xi, R)$ must be thought as the natural substitute of the Green function for the cylinder $\mathbb{R} \times B(\xi, R)$.

Green Function on Regular Domains

We start with the following definitions:

Definition 7.1. We shall say that a bounded cylinder

$$D = (T_1, T_2) \times \Omega \subseteq \mathbb{R}^{n+1}$$

is H-regular, if for every continuous function φ on the parabolic boundary

$$\partial_p D = ([T_1, T_2] \times \partial\Omega) \cup (\{T_1\} \times \overline{\Omega}),$$

there exists a unique solution u_φ to

$$u \in C^\infty(D) \cap C(D \cup \partial_p D), \quad Hu = 0 \text{ in } D, \quad u = \varphi \text{ in } \partial_p D. \qquad (21)$$

We shall also say that a bounded open set $\Omega \subseteq \mathbb{R}^n$ is H-regular if, for any $T_1 < T_2$, the cylinder $(T_1, T_2) \times \Omega$ is H-regular.

By the Picone Maximum Principle [32] if D is an H-regular domain, for any fixed $z \in D$, the linear functional

$$T : C(\partial_p D) \to \mathbb{R},$$

$$T : \varphi \longmapsto u_\varphi(z) \text{ with } u_\varphi \text{ as in (21)}$$

is continuous. Therefore there exists a measure μ_z^D (supported on $\partial_p D$) so that

$$u_\varphi(z) = \int_{\partial_p D} \varphi(\zeta) d\mu_z^D(\zeta).$$

The measures $\{\mu_z^D\}_{z \in D}$ are called H-caloric measures.

The following lemma, proved in [24], states that it is always possible to approximate any bounded domain $\Omega \subset \mathbb{R}^n$ by H-regular domains, both from the inside and from the outside.

Lemma 7.2. *Let B be a bounded open set of $\mathbb{R}^n$. Then for every $\delta > 0$ there exist H-regular domains A^δ, A_δ such that*

$$\{x \in B | d(x, \partial B) > \delta\} \subseteq A_\delta \subseteq B \subseteq A^\delta \subseteq \{x \in \mathbb{R}^n | d(x, \overline{B}) < \delta\}.$$

At this point, one can prove the existence and basic properties of the Green function for any regular cylinder $\mathbb{R} \times \Omega$.

Theorem 7.3. *Let $\Omega \subseteq \mathbb{R}^n$ be an H-regular domain. Then there exists a Green function $G = G^\Omega$ on the cylinder $\mathbb{R} \times \Omega$, with the properties listed below.*

(i) G is a continuous function defined on the set $\{(z, \zeta) \in (\mathbb{R} \times \overline{\Omega}) \times (\mathbb{R} \times \Omega) : z \neq \zeta\}$. Moreover, for every fixed $\zeta \in \mathbb{R} \times \Omega$, $G(\cdot; \zeta) \in C^\infty((\mathbb{R} \times \Omega) \setminus \{\zeta\})$, and we have

$$H(G(\cdot; \zeta)) = 0 \text{ in } (\mathbb{R} \times \Omega) \setminus \{\zeta\}, \qquad G(\cdot; \zeta) = 0 \text{ in } \mathbb{R} \times \partial \Omega.$$

(ii) We have $0 \leq G \leq h$. Moreover $G(t, x; \tau, \xi) = 0$ if $t < \tau$.

(iii) For every $\varphi \in C(\overline{\Omega})$ such that $\varphi = 0$ in $\partial \Omega$ and for every fixed $\tau \in \mathbb{R}$, the function

$$u(t, x) = \int_\Omega G(t, x; \tau, \xi) \varphi(\xi) d\xi, \qquad x \in \overline{\Omega}, \ t > \tau$$

belongs to the class $C^\infty((\tau, \infty) \times \Omega) \cap C([\tau, \infty) \times \overline{\Omega})$ and solves

$$Hu = 0 \text{ in } (\tau, \infty) \times \Omega,$$

$$u = 0 \text{ in } [\tau, \infty) \times \partial \Omega,$$

$$u(\tau, \cdot) = \varphi \text{ in } \overline{\Omega}.$$

From this theorem, one easily obtains the following corollary.

Corollary 7.4. *Let* $\Omega \subseteq \mathbb{R}^n$ *be an* H-*regular domain and let* G^Ω *denote the related Green function as in Theorem 7.3. The following* reproduction *property of* G^Ω *holds:*

$$G^\Omega(t,x;\tau,\xi) = \int_\Omega G^\Omega(t,x;s,y)\, G^\Omega(s,y;\tau,\xi)dy,$$

for every $t > s > \tau$ *and* $x, \xi \in \Omega$.

Proof. We fix τ, ξ, s as above and we set $\varphi = G^\Omega(\cdot,s;\tau,\xi)$. Then $\varphi \in C(\overline{\Omega})$, $\varphi = 0$ in $\partial\Omega$, by Theorem 7.3-(i). Therefore we can apply Theorem 7.3-(iii) and obtain that the function

$$u(t,x) = \int_\Omega G^\Omega(t,x;s,y)\,\varphi(y)dy, \qquad x \in \overline{\Omega},\ t > s,$$

satisfies $u \in C^\infty((s,\infty)\times\Omega)\cap C([s,\infty)\times\overline{\Omega})$, $Hu = 0$ in $(s,\infty)\times\Omega$, $u = 0$ in $[s,\infty)\times\partial\Omega$, $u(s,\cdot) = \varphi$ in $\overline{\Omega}$. It is now sufficient to observe that $G^\Omega(\cdot;\tau,\xi)$ has the same properties and to use the Picone maximum principle for H ∎

We now specialize the study of Green functions to cylinders based on regular domains which approximate metric balls.

Fix $\delta_0 \in (0,1)$. By Lemma 7.2, for every $\xi_0 \in \mathbb{R}^n$ and $R > 0$, there exists an H-regular domain $A(\xi_0, R)$ of $\mathbb{R}^n$ such that

$$B(\xi_0, \delta_0 R) \subseteq A(\xi_0, R) \subseteq B(\xi_0, R). \tag{22}$$

Then:

Lemma 7.5. *Let* $R_0 > 0$ *and* $\delta \in (0,\delta_0)$. *There exists a constant* $\rho = \mathbf{c}(\delta,\delta_0,R_0)^{-1} \in (0,1)$, *such that*

$$G^{A(\xi_0,R)}(t,x;\tau,\xi) \geq \mathbf{c}(R_0)^{-1}\,\mathbf{E}(x,\xi,\mathbf{c}^{-1}(t-\tau)),$$

for every $\xi_0 \in \mathbb{R}^n$, $R \in (0,R_0]$, $x \in A(\xi_0,R)$, $\xi \in B(\xi_0,\delta R)$, $t,\tau \in \mathbb{R}$ *satisfying* $d^2(x,\xi) < t - \tau < \rho R^2$.

Proof. Let $\xi_0, R, x, \xi, t, \tau$ be as above. Let us set

$$D = (\tau - 1, t + 1) \times A(\xi_0, R).$$

Observing that $\mu^D_{(t,x)}((t,t+1]\times\partial A(\xi_0,R)) = 0$ we have

$$\Psi^{A(\xi_0,R)}(t,x;\tau,\xi) = \Psi^D_{(\tau,\xi)}(t,x) = \int_{\partial_p D} \Gamma(s,y;\tau,\xi)\,\mu^D_{(t,x)}(s,y)$$

$$\leq \mathbf{c}(R_0)\int_{[\tau,t]\partial A(\xi_0,R)} \mathbf{E}(\xi,y,\mathbf{c}(s-\tau))\,\mu^D_{(t,x)}(s,y)$$

$$\leq \mathbf{c}(R_0)\sup_{0<r<t-\tau} |B(\xi,\sqrt{r})|^{-1}\exp\left(-\frac{(\delta_0-\delta)^2\,R^2}{\mathbf{c}\,r}\right).$$

In the last inequality we have used the fact that $\xi \in B(\xi_0, \delta R)$ and that

$$\mu^D_{(t,x)}(\partial_p D) \equiv 1$$

(recall the operator H is homogeneous). We now exploit the Gaussian bounds of Γ, and obtain

$$G^{A(\xi_0,R)}(t,x;\tau,\xi) = \Gamma(t,x;\tau,\xi) - \Psi^{A(\xi_0,R)}(t,x;\tau,\xi)$$

$$\geq \mathbf{c}_1(R_0)^{-1} |B(\xi, \sqrt{t-\tau})|^{-1} \exp\left(-\mathbf{c}\,\frac{d^2(x,\xi)}{t-\tau}\right) \times$$

$$\times \left[1 - \mathbf{c}_2(R_0) \sup_{0<r<t-\tau} \frac{|B(\xi, R\sqrt{\rho})|}{|B(\xi,\sqrt{r})|} \exp\left(\mathbf{c}\,\frac{d^2(x,\xi)}{t-\tau} - \frac{R^2}{\mathbf{c}(\delta,\delta_0)\,r}\right)\right]$$

$$\geq \mathbf{c}_3(R_0)^{-1} \mathbf{E}(x,\xi,\mathbf{c}^{-1}(t-\tau)) \times$$

$$\times \left[1 - \mathbf{c}_4(R_0) \sup_{0<r<t-\tau} \left(\frac{\rho R^2}{r}\right)^{Q/2} \exp\left(-\frac{R^2}{\mathbf{c}(\delta,\delta_0)\,r}\right)\right].$$

It is now sufficient to observe that the expression between square brackets is greater than $1/2$ if $\rho = \rho(\delta, \delta_0, R_0)$ is small enough, as one can easily recognize by showing that the function

$$h \mapsto (\rho\,h)^{Q/2} \exp(-h/\mathbf{c}(\delta,\delta_0))$$

is monotone decreasing on the interval $[\rho^{-1}, \infty)$. ∎

We now prove the lower Gaussian bound for the Green function.

Theorem 7.6. *Let $R_0 > 0$, $T > 1$ and $\gamma \in (0, \delta_0)$. We have*

$$G^{A(\xi_0,R)}(t,x;\tau,\xi) \geq$$

$$\mathbf{c}(T,\gamma,R_0,\delta_0)^{-1} |B(\xi, \sqrt{t-\tau})|^{-1} \exp\left(-\mathbf{c}(\gamma,R_0,\delta_0)\,\frac{d^2(x,\xi)}{t-\tau}\right),$$

for every $\xi_0 \in \mathbb{R}^n$, $R \in (0, R_0]$, $x, \xi \in B(\xi_0, \gamma R)$, and $0 < t - \tau < T R^2$.

Proof.
We set $\delta = (\gamma + \delta_0)/2$ and choose $\rho = \rho(\delta, \delta_0, R_0)$ as in Lemma 7.5. Let us also fix $\xi_0, R, x, \xi, t, \tau$ as above. Let k be the smallest integer greater than

$$M(\gamma, \delta_0) \max\left\{T/\rho,\ d^2(x,\xi)/(t-\tau)\right\},$$

where the constant $M(\gamma, \delta_0) > 1$ will be chosen later, and let us set

$$\sigma = \frac{1}{4}\sqrt{(t-\tau)/(k+1)}.$$

We claim that there exists a chain of points of $\mathbb{R}^n$ $x = x_0, x_1, \ldots, x_{k+1} = \xi$ such that

$$d(x_j, x_{j+1}) \leq \mathbf{c}(\gamma, \delta_0) \frac{d(x, \xi)}{k+1}, \qquad d(x_j, \xi_0) \leq \frac{\gamma + \delta}{2} R. \qquad (23)$$

Indeed, if $d(x, \xi) \leq R(\delta_0 - \gamma)/8$, we can choose $x_1, \ldots, x_k$ laying on a suitable X-subunit path connecting x and ξ, so that $d(x_j, x_{j+1}) \leq 2d(x, \xi)/(k+1)$ and

$$d(x_j, \xi_0) \leq d(x, x_j) + d(x, \xi_0) \leq 2d(x, \xi) + d(x, \xi_0)$$
$$\leq R(\delta_0 - \gamma)/4 + \gamma R = R(\delta + \gamma)/2.$$

On the other hand, if $d(x, \xi) > R(\delta_0 - \gamma)/8$, then we can choose $x_1, \ldots, x_k$ laying on suitable X-subunit paths connecting x with ξ_0 and ξ_0 with ξ, so that

$$d(x_j, x_{j+1}) \leq 2\gamma R/(k+1) < 16\gamma d(x, \xi)(\delta_0 - \gamma)^{-1}(k+1)^{-1}$$

and $d(x_j, \xi_0) \leq \gamma R$. Observing that, by the definition of k and σ, we have $\sigma \leq M(\gamma, \delta_0)^{-1/2} R$, from (23) it follows that we can choose $M(\gamma, \delta_0)$ such that

$$B(x_j, \sigma) \subseteq B(\xi_0, \delta R). \qquad (24)$$

Moreover, up to a new choice of $M(\gamma, \delta_0)$, we also have

$$d(y_j, y_{j+1}) < \sqrt{\frac{t - \tau}{k+1}} \qquad \text{for every } y_j \in B(x_j, \sigma), \ y_{j+1} \in B(x_{j+1}, \sigma). \quad (25)$$

Indeed, from (23) and the definition of k it follows that

$$d(y_j, y_{j+1}) \leq 2\sigma + d(x_j, x_{j+1}) \leq \frac{1}{2}\sqrt{\frac{t - \tau}{k+1}} + \mathbf{c}(\gamma, \delta_0)\frac{d(x, \xi)}{k+1}$$
$$\leq \sqrt{\frac{t - \tau}{k+1}}\left(\frac{1}{2} + \mathbf{c}(\gamma, \delta_0)\, M(\gamma, \delta_0)^{-1/2}\right).$$

Let now $t = t_0, t_1, \ldots, t_{k+1} = \tau$ be such that $t_j - t_{j+1} = (t - \tau)/(k+1)$ for $j = 0, \ldots, k$. Using Corollary 7.4 repeatedly, we obtain (we set $G = G^{A(\xi_0, R)}$, $y_0 = x$, $y_{k+1} = \xi$)

$$G(t, x; \tau, \xi) = \int_{(A(\xi_0, R))^k} G(t, x; y_1, t_1)\, G(y_1, t_1; y_2, t_2) \cdots G(y_k, t_k; \tau, \xi)\, y_1 \cdots y_k$$

$$\geq \int_{\prod_{j=1}^{k} B(x_j, \sigma)} \prod_{j=0}^{k} G(y_j, t_j; y_{j+1}, t_{j+1})\, y_1 \cdots y_k,$$

by (24). Moreover, from (24), (25) and the definition of k, it follows that $y_{j+1} \in B(\xi_0, \delta R)$,

$$d^2(y_j, y_{j+1}) < (t - \tau)/(k+1) = t_j - t_{j+1} < TR^2/(k+1) < \rho R^2.$$

Therefore, we can apply Lemma 7.5 and obtain

$$G(t, x; \tau, \xi) \geq \mathbf{c}(R_0)^{-(k+1)} \int_{\prod_{j=1}^{k} B(x_j, \sigma)} \prod_{j=0}^{k} \mathbf{E}(y_j, y_{j+1}, \mathbf{c}^{-1}(t_j - t_{j+1})) dy_1 \cdots dy_k$$

$$\geq \mathbf{c}(R_0)^{-1} \exp(-\mathbf{c}(R_0) k) \frac{1}{|B(x, \mathbf{c}\sigma)|} \prod_{j=1}^{k} \frac{|B(x_j, \sigma)|}{|B(x_j, \mathbf{c}\sigma)|}$$

$$\geq \mathbf{c}(R_0)^{-1} |B(x, \mathbf{c}\sigma)|^{-1} \exp(-\mathbf{c}(R_0) k)$$

$$\geq \mathbf{c}(R_0)^{-1} |B(x, \sqrt{t - \tau})|^{-1} \exp(-\mathbf{c}(R_0) k),$$

by the definition of σ. Now, if $d^2(x, \xi) \geq T(t - \tau)/\rho$, from the definition of k it follows that $k < \mathbf{c}(\gamma, \delta_0) d^2(x, \xi)/(t - \tau)$ and then

$$G(t, x; \tau, \xi) \geq \mathbf{c}(R_0)^{-1} |B(x, \sqrt{t - \tau})|^{-1} \exp\left(-\mathbf{c}(\gamma, R_0, \delta_0) \frac{d^2(x, \xi)}{t - \tau}\right).$$

On the other hand, if $d^2(x, \xi) < T(t - \tau)/\rho$, the definition of k gives $k < \mathbf{c}(\gamma, \delta_0) T/\rho$ and then

$$G(t, x; \tau, \xi) \geq \mathbf{c}(T, \gamma, R_0, \delta_0)^{-1} |B(x, \sqrt{t - \tau})|^{-1}.$$

This completes the proof.   ∎

Harnack Inequality

We start by giving the following oscillation lemma

Lemma 7.7. *Let $R_0 > 0$ and $\gamma \in (0, \delta_0)$. There exists a constant $\mu \in (0, 1)$, $\mu = \mathbf{c}(\gamma, R_0, \delta_0)$, such that for every $(\tau_0, \xi_0) \in \mathbb{R}^{n+1}$, $R \in (0, R_0]$ and every*

$$u \in C^{\infty}((\tau_0 - R^2, \tau_0) \times B(\xi_0, R)) \cap C([\tau_0 - R^2, \tau_0] \times \overline{B(\xi_0, R)})$$

satisfying $Hu = 0$ in $(\tau_0 - R^2, \tau_0) \times B(\xi_0, R)$, we have

$$\operatorname{osc}_{[\tau_0 - \gamma^2 R^2, \tau_0] \times \overline{B(\xi_0, \gamma R)}} u \leq \mu \operatorname{osc}_{[\tau_0 - R^2, \tau_0] \times \overline{B(\xi_0, R)}} u. \tag{26}$$

Proof. We set

$$D = (\tau_0 - R^2, \tau_0) \times A(\xi_0, R)$$

and

$$S = \left\{ x \in B(\xi_0, \gamma R) \mid u(x, \tau_0 - R^2) \geq (M + m)/2 \right\},$$

where $M = \max_{\overline{D}} u$, $m = \min_{\overline{D}} u$. We also define $w = u - m$ and (for $x \in \overline{A(\xi_0, R)}$, $t > \tau_0 - R^2$)

$$v(t,x) = \int_{A(\xi_0,R)} G^{A(\xi_0,R)}(t,x;\tau_0 - R^2, y)\, w(\tau_0 - R^2, y)\, \varphi(y) dy,$$

where $\varphi \in C_0(A(\xi_0, R))$ is a cut-off function such that $0 \le \varphi \le 1$ and $\varphi \equiv 1$ in $B(\xi_0, \gamma R)$. By means of Theorem 7.3-(iii), v is a solution to $Hv = 0$ in D, $v = 0$ in $[\tau_0 - R^2, \tau_0] \times \partial A(\xi_0, R)$, $v(\cdot, \tau_0 - R^2) = w(\tau_0 - R^2, \cdot)\, \varphi$ in $\overline{A(\xi_0, R)}$. Moreover $Hw = 0$ since we are supposing that $a_0 = 0$. Therefore, by the weak maximum principle for H, $v \le w$ in D. As a consequence, using the estimate in Theorem 7.6, for every

$$(t,x) \in D_\gamma = (\tau_0 - \gamma^2 R^2, \tau_0) \times B(\xi_0, \gamma R)$$

we get

$$w(t,x) \ge v(t,x) \ge \int_S G^{A(\xi_0,R)}(t,x;\tau_0 - R^2, y)\left(\frac{M+m}{2} - m\right) dy$$

$$\ge \int_S \frac{\mathbf{c}}{|B(x,\sqrt{t-\tau+R^2})|} e^{-d(x,y)^2/\mathbf{c}(t-\tau+R^2)} \frac{M-m}{2} dy$$

$$\ge \mathbf{c}(\gamma, R_0, \delta_0)^{-1} |B(x,R)|^{-1} \frac{M-m}{2} |S|$$

$$\ge \mathbf{c}(\gamma, R_0, \delta_0)^{-1} |B(\xi_0,R)|^{-1} \frac{M-m}{2} |S|.$$

Now, if $|S| \ge \frac{1}{2}|B(\xi_0, \gamma R)|$, we infer that

$$\min_{\overline{D}_\gamma} u - m \ge \mathbf{c}(\gamma, R_0, \delta_0)^{-1}(M - m)$$

and then

$$\mathrm{osc}_{\overline{D}_\gamma}\, u \le M - \min_{\overline{D}_\gamma} u \le M - m - \mathbf{c}(\gamma, R_0, \delta_0)^{-1}(M - m)$$

$$= \left(1 - \mathbf{c}(\gamma, R_0, \delta_0)^{-1}\right) \mathrm{osc}_{\overline{D}}\, u.$$

Recalling that $A(\xi_0, R) \subseteq B(\xi_0, R)$, we have proved (26) when $|S| \ge \frac{1}{2}|B(\xi_0, \gamma R)|$. On the other hand, if $|S| < \frac{1}{2}|B(\xi_0, \gamma R)|$, the argument above can be applied to $\tilde{u} := -u$, since (with the natural notation) $|\tilde{S}| > \frac{1}{2}|B(\xi_0, \gamma R)|$. As a consequence, we get (26) for $\tilde{u}$ and the proof is completed, since $\mathrm{osc}\, \tilde{u} = \mathrm{osc}\, u$. ∎

We also have the following crucial lemma.

Lemma 7.8. *Let $R_0 > 0$, $h \in (0,1)$ and $\gamma \in (0, \delta_0)$. There exists a positive constant $\beta = \mathbf{c}(h, \gamma, R_0, \delta_0)$ such that*

$$\sup_{\sigma > 0, \, s \in [\tau_0 - R^2, \tau_0 - hR^2]} \sigma \, |\{y \in B(\xi_0, \gamma R) \, | \, u(s,y) \geq \sigma\}| \leq \beta \, |B(\xi_0, R)| \, u(\xi_0, \tau_0).$$

for every $(\xi_0, \tau_0) \in \mathbb{R}^{n+1}$, $R \in (0, R_0]$ and every

$$u \in C^\infty(B(\xi_0, R) \times (\tau_0 - R^2, \tau_0)) \cap C(\overline{B(\xi_0, R)} \times [\tau_0 - R^2, \tau_0])$$

satisfying $Hu = 0$, $u \geq 0$ in $B(\xi_0, R) \times (\tau_0 - R^2, \tau_0)$.

Proof. Let us fix $s \in [\tau_0 - R^2, \tau_0 - hR^2]$; set $A = A(\xi_0, R)$ and

$$w(t,x) = \int_A G^A(t,x;s,y) \, u(s,y) \, \varphi(y) \, dy, \qquad x \in \overline{A}, \ t > s,$$

where $\varphi \in C_0(A)$ is a cut-off function such that $0 \leq \varphi \leq 1$ and $\varphi \equiv 1$ in $B(\xi_0, \gamma R)$. By means of Theorem 7.3-(iii), w is a solution to $Hw = 0$ in $(s, \infty) \times A$, $w = 0$ in $[s, \infty) \times \partial A$, $w(s, \cdot) = u(s, \cdot) \varphi$ in $\overline{A}$. Therefore, by the weak maximum principle for H, $w \leq u$ in $[s, \tau_0] \times \overline{A}$. As a consequence, using the estimate in Theorem 7.6, we obtain

$$u(\tau_0, \xi_0) \geq w(\tau_0, \xi_0) \geq \int_A \frac{\mathbf{c}u(s,y)\varphi(y)}{|B(\xi_0, \sqrt{\tau_0 - s})|} e^{-d(\xi_0, y)^2/\mathbf{c}(\tau_0 - s)} dy$$

since $\tau_0 - s \geq hR^2$ and $d(\xi_0, y) \leq \mathbf{c}R$

$$\geq \frac{\mathbf{c}(h, \gamma, R_0, \delta_0)}{|B(\xi_0, R)|} \int_{B(\xi_0, \gamma R)} u(s,y) \, dy$$

$$\geq \frac{\mathbf{c}(h, \gamma, R_0, \delta_0)}{|B(\xi_0, R)|} \sigma \, |\{y \in B(\xi_0, \gamma R) \, | \, u(s,y) \geq \sigma\}|$$

for any σ. This ends the proof. ∎

Next theorem follows from the previous two Lemmas, with the same technique used in [13].

Theorem 7.9. *Let $R_0 > 0$, $0 < h_1 < h_2 < 1$ and $\gamma \in (0,1)$. There exists a positive constant $M = \mathbf{c}(h_1, h_2, \gamma, R_0)$ such that for every $(\xi_0, \tau_0) \in \mathbb{R}^{n+1}$, $R \in (0, R_0]$ and every*

$$u \in C^\infty((\tau_0 - R^2, \tau_0) \times B(\xi_0, R)) \cap C([\tau_0 - R^2, \tau_0] \times \overline{B(\xi_0, R)})$$

satisfying $Hu = 0$, $u \geq 0$ in $(\tau_0 - R^2, \tau_0) \times B(\xi_0, R)$, we have

$$\max_{[\tau_0 - h_2 R^2, \tau_0 - h_1 R^2] \times \overline{B(\xi_0, \gamma R)}} u \leq M \, u(\xi_0, \tau_0). \tag{27}$$

Proof of Theorem 7.9. Recall that our previous estimates depend on a number $\delta_0 \in (0,1)$ which can be arbitrarily chosen (see (22)). Then, for any fixed $\gamma \in (0,1)$, pick $\delta_0 = (1+\gamma)/2$. We will apply all our previous results with this particular choice of δ_0.

Let $\mu = \mu(\frac{\delta_0}{2}, R_0, \delta_0) \in (0,1)$, $\beta = \beta(h_1, \frac{\gamma+\delta_0}{2}, R_0, \delta_0) > 0$ be as in Lemma 7.7 and Lemma 7.8 respectively. We define $r : (0,\infty) \to (0,\infty)$,

$$r(\sigma) = 2 \left(\frac{4\,\beta \mathbf{c}_0}{\sigma(1-\mu)} \right)^{1/Q}$$

and we set

$$K = (1+\mu^{-1})/2$$

and

$$M = r^{-1}\left(\delta_0 \,(1-h_2)\,(\delta_0-\gamma)\,(1-K^{-1/Q})/4 \right).$$

We now argue by contradiction and suppose that there exist ξ_0, τ_0, R and u satisfying the hypotheses of the theorem, for which (27) is not true (with the above choice of M). We first observe that $u(\tau_0, \xi_0) \neq 0$, since otherwise (27) would follow from Lemma (7.8). Let now $v = u/u(\tau_0, \xi_0)$. Since v is bounded, in order to get a contradiction and thus prove the theorem, it is sufficient to show that there exists a sequence of points $\{(s_j, y_j)\}_{j \in \mathbb{N}}$ in $[\tau_0 - R^2, \tau_0] \times \overline{B(\xi_0, R)}$ such that

$$v(s_j, y_j) \geq K^j M, \qquad (s_j, y_j) \in [\tau_0 - R^2, \tau_0] \times \overline{B(\xi_0, R)}.$$

Indeed, recalling that $K > 1$, this would give $v(s_j, y_j) \to \infty$. To construct this sequence, we will prove by induction the existence of points $(s_j, y_j) \in [\tau_0 - R^2, \tau_0] \times \overline{B(\xi_0, R)}$ such that:

$$v(s_j, y_j) \geq K^j M \tag{28}$$
$$(s_0, y_0) \in [\tau_0 - h_2 R^2, \tau_0 - h_1 R^2] \times \overline{B(\xi_0, \gamma R)}$$
$$(s_j, y_j) \in [s_{j-1} - \rho_{j-1}^2, s_{j-1}] \times \overline{B(y_{j-1}, \rho_{j-1})} \text{ if } j \geq 1$$
$$\text{with } \rho_j = 2\delta_0^{-1} r(K^j M) R$$

The existence of $(s_0, y_0) \in [\tau_0 - h_2 R^2, \tau_0 - h_1 R^2] \times \overline{B(\xi_0, \gamma R)}$ such that $v(s_0, y_0) \geq M$ follows from the assumption that u does not satisfy (27). We now suppose that, for a fixed $q \in \mathbb{N}$, $(s_0, y_0), \ldots, (s_q, y_q)$ have been defined and satisfy (28) for every $j \in \{0, \ldots, q\}$. We have to prove that we can find (s_{q+1}, y_{q+1}) satisfying (28) for $j = q + 1$. We claim that

$$\overline{B(y_q, \rho_q)} \subseteq B(\xi_0, (\gamma + \delta_0)R/2). \tag{29}$$

Indeed, if $d(y, y_q) \leq \rho_q$, then recalling the definition of M and using (28) for $j \in \{0, \ldots, q\}$, we obtain

$$
\begin{aligned}
d(y, \xi_0) &\leq d(\xi_0, y_0) + \sum_{j=1}^{q} d(y_{j-1}, y_j) + d(y_q, y) \\
&\leq \gamma R + 2\delta_0^{-1} R \sum_{i=0}^{q} r(K^i M) < \gamma R + 2\delta_0^{-1} r(M) R \sum_{i=0}^{\infty} K^{-i/Q} \\
&= (\gamma + (1 - h_2)(\delta_0 - \gamma)/2) R < (\gamma + \delta_0) R/2.
\end{aligned}
$$

Moreover, with a similar computation we can prove that

$$
[s_q - \rho_q^2, s_q] \subseteq (\tau_0 - R^2, \tau_0 - h_1 R^2]. \tag{30}
$$

Indeed, $s_q \leq s_{q-1} \leq \cdots \leq s_0 \leq \tau_0 - h_1 R^2$ and

$$
\begin{aligned}
s_q - \rho_q^2 &= s_0 + \textstyle\sum_{j=1}^{q}(s_j - s_{j-1}) - \rho_q^2 \\
&> \tau_0 - h_2 R^2 - 4\delta_0^{-2}(r(M))^2 R^2 \textstyle\sum_{i=0}^{\infty} K^{-2i/Q} \\
&> \tau_0 - \left(h_2 + (1 - h_2)(1 - K^{-1/Q})(1 + K^{-1/Q})^{-1}\right) R^2 > \tau_0 - R^2.
\end{aligned}
$$

We now apply Lemma 7.8 (with $\sigma = (1 - \mu)K^q M/2$) to v and we obtain (recalling (30) and the definition of r)

$$
\begin{aligned}
&|\{y \in B(\xi_0, (\gamma + \delta_0) R/2) \mid v(y, s_q) \geq (1 - \mu)K^q M/2\}| \\
&\qquad \leq \frac{2\beta |B(\xi_0, R)|}{(1 - \mu)K^q M} = \tfrac{1}{2} \mathbf{c}_0^{-1} \left(\frac{r(K^q M)}{2}\right)^Q |B(\xi_0, R)| \\
&\qquad < \mathbf{c}_0^{-1} \left(\frac{r(K^q M)}{2}\right)^Q |B(y_q, 2R)| \leq |\{B(y_q, r(K^q M)R)\}|.
\end{aligned}
$$

In the last inequality we have used the fact that $r(K^q M) \leq 2$, which follows from the definition of M (see the proof of (29)). As a consequence, since also (29) holds, there exists

$$
\overline{y} \in B(y_q, r(K^q M)R)
$$

such that

$$
v(s_q, \overline{y}) < (1 - \mu)K^q M/2.
$$

Therefore, recalling that we are supposing that (28) holds for $j = q$, we have

$$
\begin{aligned}
(1 + \mu)K^q M/2 &= K^q M - (1 - \mu)K^q M/2 < v(s_q, y_q) - v(s_q, \overline{y}) \\
&\leq \operatorname*{osc}_{\{s_q\} \times \overline{B(y_q, r(K^q M)R)}} v \leq \mu \operatorname*{osc}_{[s_q - \rho_q^2, s_q] \times \overline{B(y_q, \rho_q)}} v
\end{aligned}
$$

by means of Lemma 7.7, (29) and (30) (note that $\rho_q \leq R_0$ by the definition of M). Since $v \geq 0$, it follows that there exists

$$
(s_{q+1}, y_{q+1}) \in [s_q - \rho_q^2, s_q] \times \overline{B}(y_q, \rho_q)
$$

such that
$$v(s_{q+1}, y_{q+1}) > \mu^{-1}(1 + \mu)K^q M/2 = K^{q+1}M.$$
This completes the proof of Theorem 7.9. ∎

References

1. D. G. Aronson: Bounds for the fundamental solution of a parabolic equation. Bull. Amer. Math. Soc. 73 1967 890–896.
2. E. Bedford, B. Gaveau: Hypersurfaces with bounded Levi form. Indiana Univ. Math. J. 27 (1978), no. 5, 867–873.
3. A. Bonfiglioli, E. Lanconelli, F. Uguzzoni: Uniform Gaussian estimates of the fundamental solutions for heat operators on Carnot groups, Adv. Differential Equations, 7 (2002), 1153–1192.
4. A. Bonfiglioli, E. Lanconelli, F. Uguzzoni: Fundamental solutions for non-divergence form operators on stratified groups. Trans. Amer. Math. Soc. 356 (2004), no. 7, 2709–2737.
5. A. Bonfiglioli, F. Uguzzoni: Harnack inequality for non-divergence form operators on stratified groups, to appear in Trans. Amer. Math. Soc.
6. J.-M. Bony: Principe du maximum, inégalite de Harnack et unicité du problème de Cauchy pour les opérateurs elliptiques dégénérés. Ann. Inst. Fourier (Grenoble) 19 1969 fasc. 1, 277–304 xii.
7. M. Bramanti, L. Brandolini: L^p-estimates for uniformly hypoelliptic operators with discontinuous coefficients on homogeneous groups. Rend. Sem. Mat. dell'Univ. e del Politec. di Torino. Vol. 58, 4 (2000), 389–433.
8. M. Bramanti, L. Brandolini: L^p-estimates for nonvariational hypoelliptic operators with VMO coefficients. Trans. Amer. Math. Soc. 352 (2000), no. 2, 781–822.
9. M. Bramanti, L. Brandolini, E. Lanconelli, F. Uguzzoni: Heat kernels for non-divergence operators of Hörmander type. C.R. Acad. Sci. Paris, Mathematique, 343 (2006) 463–466.
10. M. Bramanti, L. Brandolini, E. Lanconelli, F. Uguzzoni: Non-divergence equations structured on Hörmander vector fields: heat kernels and Harnack inequalities, preprint.
11. L. Caffarelli, L. Nirenberg, J. Spruck: The Dirichlet problem for nonlinear second-order elliptic equations. III. Functions of the eigenvalues of the Hessian. Acta Math. 155 (1985), no. 3-4, 261–301.
12. L. Capogna, Q. Han: Pointwise Schauder estimates for second order linear equations in Carnot groups. Harmonic analysis at Mount Holyoke (South Hadley, MA, 2001), 45–69, Contemp. Math., 320, Amer. Math. Soc., Providence, RI, 2003.
13. E. B. Fabes, D. W. Stroock, A new proof of Moser's parabolic Harnack inequality using the old ideas of Nash, Arch. Rational Mech. Anal. 96 (1986), 327–338.
14. C. Fefferman, A. Sánchez-Calle: Fundamental solutions for second order subelliptic operators. Ann. of Math. (2) 124 (1986), no. 2, 247–272.
15. C. Fefferman, D. H. Phong: Subelliptic eigenvalue problems. Conference on harmonic analysis in honor of Antoni Zygmund, Vol. I, II (Chicago, Ill., 1981), 590–606, Wadsworth Math. Ser., Wadsworth, Belmont, CA, 1983.
16. A. Friedman: Partial differential equations of parabolic type. Prentice-Hall, Inc., Englewood Cliffs, N.J. 1964.
17. L. Hörmander: Hypoelliptic second order differential equations. Acta Math. 119 (1967) 147–171.
18. G. Huisken, W. Klingenberg: Flow of real hypersurfaces by the trace of the Levi form. Math. Res. Lett. 6 (1999), no. 5-6, 645–661.
19. D. S. Jerison, A. Sánchez-Calle: Estimates for the heat kernel for a sum of squares of vector fields. Indiana Univ. Math. J. 35 (1986), no. 4, 835–854.

20. S. Kusuoka, D. Stroock: Applications of the Malliavin calculus. II. J. Fac. Sci. Univ. Tokyo Sect. IA Math. 32 (1985), no. 1, 1–76.
21. S. Kusuoka, D. Stroock: Applications of the Malliavin calculus. III. J. Fac. Sci. Univ. Tokyo Sect. IA Math. 34 (1987), no. 2, 391–442.
22. S. Kusuoka, D. Stroock: Long time estimates for the heat kernel associated with a uniformly subelliptic symmetric second order operator. Ann. of Math. (2) 127 (1988), no. 1, 165–189.
23. E. Lanconelli, S. Polidoro: On a class of hypoelliptic evolution operators. Partial differential equations, II (Turin, 1993). Rend. Sem. Mat. Univ. Politec. Torino 52 (1994), no. 1, 29–63.
24. E. Lanconelli, F. Uguzzoni: Potential Analysis for a Class of Diffusion Equations: a Gaussian Bounds Approach, Preprint.
25. E. E. Levi: Sulle equazioni lineari totalmente ellittiche alle derivate parziali. Rend. Circ. Mat. Palermo 24, 275–317 (1907) (Also with corrections in: Eugenio Elia Levi, Opere, vol. II, 28–84. Roma, Edizioni Cremonese 1960).
26. E. E. Levi: I problemi dei valori al contorno per le equazioni lineari totalmente ellittiche alle derivate parziali, Memorie Mat. Fis. Soc. Ital. Scienza (detta dei XL) (3) 16, 3–113 (1909). (Also with corrections in: Eugenio Elia Levi, Opere, vol. II, 207–343. Roma, Edizioni Cremonese 1960).
27. V. Martino, A. Montanari : Integral formulas for a class of curvature PDE's and applications to isoperimetric inequalities and to symmetry problems, Preprint.
28. A. Montanari: Real hypersurfaces evolving by Levi curvature: smooth regularity of solutions to the parabolic Levi equation. Comm. Partial Differential Equations 26 (2001), no. 9-10, 1633–1664.
29. A. Montanari, E. Lanconelli: Pseudoconvex fully nonlinear partial differential operators: strong comparison theorems. J. Differential Equations 202 (2004), no. 2, 306–331.
30. A. Nagel, E. M. Stein, S. Wainger: Balls and metrics defined by vector fields I: Basic properties. Acta Mathematica, 155 (1985), 130–147.
31. J. Nash: Continuity of solutions of parabolic and elliptic equations. Amer. J. Math. 80 (1958) 931–954.
32. M. Picone: Maggiorazione degli integrali di equazioni lineari ellittico-paraboliche alle derivate parziali del secondo ordine. Rend. Accad. Naz. Lincei 5, n.6 (1927) 138–143.
33. S. Polidoro: On a class of ultraparabolic operators of Kolmogorov-Fokker-Planck type. Le Matematiche (Catania) 49 (1994), no. 1, 53–105.
34. L. P. Rothschild, E. M. Stein: Hypoelliptic differential operators and nilpotent groups. Acta Math., 137 (1976), 247–320.
35. L. Saloff-Coste, Aspects of Sobolev-type inequalities. London Mathematical Society Lecture Note Series, 289. Cambridge University Press, Cambridge, 2002.
36. L. Saloff-Coste, D. W. Stroock: Opérateurs uniformément sous-elliptiques sur les groupes de Lie. J. Funct. Anal. 98 (1991), no. 1, 97–121.
37. A. Sanchez-Calle: Fundamental solutions and geometry of sum of squares of vector fields. Inv. Math., 78 (1984), 143–160.
38. Z. Slodkowski, G. Tomassini: Weak solutions for the Levi equation and envelope of holomorphy. J. Funct. Anal. 101 (1991), no. 2, 392–407.
39. G. Tomassini: Geometric properties of solutions of the Levi-equation. Ann. Mat. Pura Appl. (4) 152 (1988), 331–344.
40. N. Th. Varopoulos: Théorie du potentiel sur les groupes nilpotents. C. R. Acad. Sci. Paris Sér. I Math. 301 (1985), no. 5, 143–144.
41. N. Th. Varopoulos, Analysis on nilpotent groups. J. Funct. Anal. 66 (1986), no. 3, 406–431.
42. N. Th. Varopoulos, L. Saloff-Coste, T. Coulhon: Analysis and geometry on groups. Cambridge Tracts in Mathematics, 100. Cambridge University Press, Cambridge, 1992.
43. C. J. Xu: Regularity for quasilinear second-order subelliptic equations. Comm. Pure Appl. Math. 45 (1992), no. 1, 77–96.

Concentration of Solutions for Some Singularly Perturbed Neumann Problems

Andrea Malchiodi

1 Introduction

The purpose of these notes is to present some techniques for constructing solutions to a class of singularly perturbed problems with a precise asymptotic behavior when the perturbation parameter ε tends to zero. We first treat the case of concentration at points, and then the case of concentration at manifolds.

One of the main motivations for the study of these equations arises from reaction-diffusion systems, concerning in particular the so-called *Turing's instability*. More precisely, see [18], [53], it is known that scalar diffusion equations like

$$\begin{cases} u_t = \Delta u + f(u) & \text{in } \Omega; \\ \frac{\partial u}{\partial \nu} = 0 & \text{on } \partial\Omega, \end{cases}$$

for convex domains Ω admit only constant (linearly) stable steady state solutions. On the other hand, as noticed in [62], reaction-diffusion *systems* with different diffusivities might lead to non-homogeneous stable steady states.

A well-know example is the following one (Gierer-Meinhardt)

$$\begin{cases} \mathcal{U}_t = d_1\Delta\mathcal{U} - \mathcal{U} + \frac{\mathcal{U}^p}{\mathcal{V}^q} & \text{in } \Omega \times (0,+\infty), \\ \mathcal{V}_t = d_2\Delta\mathcal{V} - \mathcal{V} + \frac{\mathcal{U}^r}{\mathcal{V}^s} & \text{in } \Omega \times (0,+\infty), \\ \frac{\partial\mathcal{U}}{\partial\nu} = \frac{\partial\mathcal{V}}{\partial\nu} = 0 & \text{on } \partial\Omega \times (0,+\infty), \end{cases} \qquad (GM)$$

introduced in [31] to describe some biological experiment. The functions $\mathcal{U}$ and $\mathcal{V}$ represent the densities of some chemical substances, the numbers p,q,r,s are non-negative and such that $0 < \frac{p-1}{q} < \frac{r}{s+1}$, and it is assumed

A. Malchiodi
SISSA, Sector of Mathematical Analysis, Via Beirut 2-4, 34014 Trieste, Italy
e-mail: malchiod@sissa.it

S.-Y.A. Chang et al. (eds.), *Geometric Analysis and PDEs*,
Lecture Notes in Mathematics 1977, DOI: 10.1007/978-3-642-01674-5_3,
© Springer-Verlag Berlin Heidelberg 2009

that the diffusivities d_1 and d_2 satisfy $d_1 \ll 1 \ll d_2$. In the stationary case of (GM), as explained in [55], [58], when $d_2 \to +\infty$ the function $\mathcal{V}$ is close to a constant (being nearly harmonic and with zero normal derivative at the boundary), and therefore the equation satisfied by $\mathcal{U}$ *converges* to the following one

$$
\begin{cases}
-\varepsilon^2 \Delta u + u = u^p & \text{in } \Omega, \\
\frac{\partial u}{\partial \nu} = 0 & \text{on } \partial\Omega, \\
u > 0 & \text{in } \Omega,
\end{cases}
\qquad (\tilde{P}_\varepsilon)
$$

with $\varepsilon^2 = d_1$. One is interested in general in the profile of solutions, their location and possibly in their stability properties.

The advantage of problem $(\tilde{P}_\varepsilon)$ compared to (the stationary case of) (GM) is that, apart from being a scalar equation, its structure is variational and solutions can be found as critical points of the following Euler-Lagrange functional

$$
\tilde{I}_\varepsilon(u) = \frac{1}{2} \int_\Omega \left(\varepsilon^2 |\nabla u|^2 + u^2 \right) - \frac{1}{p+1} \int_\Omega |u|^{p+1}; \qquad u \in H^1(\Omega).
$$

The typical concentration behavior of solutions u_ε to $(\tilde{P}_\varepsilon)$ is via a scaling of the variables in the form $u_\varepsilon(x) \sim u_0 \left(\frac{x-Q}{\varepsilon} \right)$, where Q is some point of $\overline{\Omega}$, and where u_0 is a solution of the problem

$$
-\Delta u_0 + u_0 = u_0^p \qquad \text{in } \mathbb{R}^n \quad (\text{or in } \mathbb{R}_+^n = \{(x_1, \ldots, x_n) \in \mathbb{R}^n \ : \ x_n > 0\}),
$$
$$
(1)
$$

the domain depending on whether Q lies in the interior of Ω or at the boundary; in the latter case Neumann conditions are imposed.

When $p < \frac{n+2}{n-2}$ (and indeed only if this inequality is satisfied), problem (1) admits positive radial solutions which decay to zero at infinity, see [14], [15], [61]. Some details on the existence theory for (1) are reported below.

Solutions of $(\tilde{P}_\varepsilon)$ which inherit this profile are called *spike-layers*, since they are highly concentrated near some point of $\overline{\Omega}$. There is an extensive literature regarding this type of solutions, beginning from the papers [44], [56], [57]. Indeed their structure is very rich, and we refer for example to the (far from complete) list of papers [21], [27], [33], [34], [35], [36], [42], [43], [63], [64].

In these notes we are going to prove the following two theorems, to give an idea of the methods employed in this area of research. The first concerns the case of local extrema of the mean boundary curvature, while the second deals with non-degenerate critical points. We will present a general perturbative argument to deal with such concentration phenomena, and then apply it to study $(\tilde{P}_\varepsilon)$.

Theorem 1 *Suppose $\Omega \subseteq \mathbb{R}^n$, $n \geq 2$, is a smooth bounded domain, and that $1 < p < \frac{n+2}{n-2}$ ($1 < p < +\infty$ if $n = 2$). Suppose $X_0 \in \partial\Omega$ is a local* *strict maximum or minimum of the mean curvature H of $\partial\Omega$. Then for $\varepsilon > 0$ sufficiently small problem $(\tilde{P}_\varepsilon)$ admits a solution concentrating at X_0.*

Theorem 2 *Suppose $\Omega \subseteq \mathbb{R}^n$, $n \geq 2$, is a smooth bounded domain, and that $1 < p < \frac{n+2}{n-2}$ $(1 < p < +\infty$ if $n = 2)$. Suppose $X_0 \in \partial\Omega$ is a non-degenerate critical point of the mean curvature H of $\partial\Omega$. Then for $\varepsilon > 0$ sufficiently small problem $(\tilde{P}_\varepsilon)$ admits a solution concentrating at X_0.*

We were intentionally vague about the meaning of *concentration*, but it will become clear from the constructive proof below. A property of these solutions, u_ε, is that they tend to zero uniformly on every compact set of $\overline{\Omega}\backslash\{X_0\}$, while there exists $\delta > 0$ and some point $X_\varepsilon \in \overline{\Omega}$ such that $u_\varepsilon(X_\varepsilon) \geq \delta$ as $\varepsilon \to 0$.

We notice that the functional I_ε, if p is subcritical, satisfies the assumptions of the mountain-pass theorem (see [9]). It can be shown that the profile of the mountain-pass solution of $(\tilde{P}_\varepsilon)$ is a soliton of (1) in the half-space, and it peaks at a point of $\partial\Omega$ with maximal mean curvature. Furthermore, there are solutions with interior and multiple peaks: in particular in [35] it was shown that for any couple of integers (k, l) there exist solutions with k boundary peaks and l interior ones. The energy (i.e. the corresponding value of I_ε) of spike-layers (resp. multiple spikes) is of order ε^n, which is proportional to the volume of their *support*, heuristically identified with a ball (resp. multiple balls) of radius ε centered at the peak (resp. at each peak). Roughly speaking, concentration at the boundary occurs at critical points of the mean curvature, while concentration at the interior occurs at singular points of the distance function from the boundary. Methods of construction rely mainly on min-max arguments, finite-dimensional reductions and gluing techniques (for multiple spikes).

In some cases, see for example [26], [28], [37], [58], it is possible to construct stationary solutions of (GM) starting from spike-layer solutions of $(\tilde{P}_\varepsilon)$ and using perturbation arguments. In these references some stability conditions are also given: we notice that non-trivial solutions of $(\tilde{P}_\varepsilon)$ are always unstable, but their counterparts in (GM) may gain stability through the coupling with a second equation.

There are other motivations for the study of the above kind of equations: $(\tilde{P}_\varepsilon)$ arises also as limit of different reaction-diffusion systems (with chemotaxis for example, as shown in [55]). Another motivation comes from the Nonlinear Schrödinger equation

$$i\hbar \frac{\partial\psi}{\partial t} = -\hbar^2\Delta\psi + V(x)\psi - |\psi|^{p-1}\psi \qquad \text{in } \mathbb{R}^n, \qquad (2)$$

where ψ is a complex-valued function (the *wave function*), V is a potential and p is an exponent greater than 1. Indeed, if one looks for *standing waves*, namely solutions of the form $\psi(x, t) = e^{-\frac{i\omega t}{\hbar}}u(x)$, for some real function u, then the latter will satisfy

$$-\varepsilon^2\Delta u + V(x)u = u^p \qquad \text{in } \mathbb{R}^n, \qquad (3)$$

where we have set $\varepsilon = \hbar$ and we absorbed the constant ω into the potential V. Therefore, the problem very similar to $(\tilde{P}_\varepsilon)$, apart from the addition of a potential term. Also for this problem we have highly peaked solutions, which in general concentrate at critical points of the potential V: about this subject we refer the reader to the (still incomplete) list of papers [1], [2], [3], [7], [8], [10], [17], [24], [25], [30], [32], [38], [59] and to the bibliographies therein.

Apart from existence of solutions concentrating at points, one may ask whether there exist others which scale only in some of the variables, and not all of them, which therefore concentrate at higher dimensional sets, like curves or manifolds. Indeed under generic assumptions, see [55], if $\Omega \subseteq \mathbb{R}^n$ and $k = 1, \ldots, n-1$, it was conjectured that there exist solutions of $(\tilde{P}_\varepsilon)$ concentrating at k-dimensional sets. The phenomenon was known for particular domains with some symmetry: for these and related issues see e.g. [5], [6], [11], [12], [13], [22], [23], [51], [54], [60]. In these cases, some of the techniques for studying concentration at points can be modified, but these adaptations do not work for asymmetric situations. In [49], [50] the first general result in this direction was obtained. It was indeed shown that there exist solutions to $(\tilde{P}_\varepsilon)$ which scale in one variable only and which concentrate near the whole boundary of the domain. Their profile, as one may expect, is given by the solution of (1) in one dimension, namely

$$\begin{cases} -w_0'' + w_0 = w_0^p & \text{in } \mathbb{R}_+; \\ w_0(x) \to 0 & \text{as } x \to +\infty; \\ w_0'(0) = 0. \end{cases} \qquad (4)$$

The result in [49], [50] is the following.

Theorem 3 *Let $\Omega \subseteq \mathbb{R}^n$ be a smooth bounded domain, and let $p > 1$. Then there exists a sequence $\varepsilon_j \to 0$ and a sequence of solutions u_{ε_j} of (P_{ε_j}) with the following properties*

(i) *u_{ε_j} concentrates near $\partial\Omega$ as $j \to +\infty$, namely for every $r > 0$ one has $\int_{\Omega^r} \left(\varepsilon_j^2 |\nabla u_{\varepsilon_j}|^2 + u_{\varepsilon_j}^2 \right) \to 0$ as $j \to +\infty$, where $\Omega^r = \{x \in \Omega : dist(x, \partial\Omega) \geq r\}$;*

(ii) *if $x_0 \in \partial\Omega$ and if ν_0 denotes the interior unit normal to $\partial\Omega$ at x_0, then for every $k \in \mathbb{N}$ one has $u_{\varepsilon_j}(\varepsilon_j(x - x_0)) \to w_0(\langle x, \nu_0 \rangle)$ in $C_{loc}^k(V_0)$, where $V_0 = \{x \in \mathbb{R}^n : \langle x, \nu_0 \rangle > 0\}$, and where w_0 is the solution of (4).*

Below we will give an almost complete proof of this result: for the moment we just illustrate two main features. The first is that the exponent p can be any real number grater than 1, and no upper bound is required, differently from Theorems 1 and 2: the reason is that the limit profile of the solutions is one-dimensional, and for solving (4) no restriction on p is needed. The second is that in Theorem 3 existence is proved only along a sequence $\varepsilon_j \to 0$: this is due to an underlying resonance phenomenon, described below, and is peculiar of higher-dimensional concentration.

The latter result has been extended to the case of limit sets of dimension $k = 1, \ldots, n - 2$, first in [48] for $n = 3$ and $k = 1$, and then in [46] for the general case.

Theorem 4 *Let $\Omega \subseteq \mathbb{R}^N$, $N \geq 3$, be a smooth and bounded domain, and let $K \subseteq \partial\Omega$ be a compact embedded non-degenerate minimal submanifold of dimension $k \in \{1, \ldots, n-2\}$. Then, if $p \in \left(1, \frac{n-k+2}{n-k-2}\right)$, there exists a sequence $\varepsilon_j \to 0$ such that (P_{ε_j}) admits positive solutions u_{ε_j} concentrating along K as $j \to \infty$. Precisely there exists a positive constant C, depending on Ω, K and p such that for any $x \in \Omega$ $u_{\varepsilon_j}(x) \leq Ce^{-\frac{dist(x,K)}{C\varepsilon_j}}$; moreover for any $q \in K$, in a system of coordinates $(\overline{y}, \zeta)$, with $\overline{y}$ coordinates on K and ζ coordinates normal to K, for any integer m one has $u_{\varepsilon_j}(0, \varepsilon_j \cdot) \overset{C_{loc}^m(\mathbb{R}_+^{n-k})}{\longrightarrow} w_0(\cdot)$, where $w_0 : \mathbb{R}_+^{n-k} \to \mathbb{R}$ is the unique radial solution of*

$$
\begin{cases}
-\Delta u + u = u^p & in\ \mathbb{R}_+^{n-k}, \\
\frac{\partial u}{\partial \nu} = 0 & on\ \partial\mathbb{R}_+^{n-k}, \\
u > 0, u \in H^1(\mathbb{R}_+^{n-k}).
\end{cases}
\tag{5}
$$

By *minimal submanifold* we mean that K is stationary for the area functional, namely its mean curvature in $\partial\Omega$ vanishes: this condition is indeed quite natural, since a heuristic expansion for the energy of solutions concentrated near a submanifold $\tilde{K}$ gives $C_0 \, area(\tilde{K}) \varepsilon^{n-k} + o(\varepsilon^{n-k})$, with C_0 a given positive constant. Therefore, since the energy of solutions should be extremal, also the area of K has to be. We notice that also the bound on the exponent p is natural, since $p \in \left(1, \frac{n-k+2}{n-k-2}\right)$ is a necessary and sufficient condition for having solutions to (5) which decay to zero at infinity.

For reasons of brevity, we will not give the proof of the latter theorem, referring to [46] for complete details. Further results in this direction can be found in [29], [47], [52], [65].

Part I
Concentration at Points

In this part we investigate the case of concentration at points, and prove Theorems 1 and 2. We first present a general perturbative technique introduced by Ambrosetti and Badiale, and then apply it to our specific problem.

We are interested in finding solutions to $(\tilde{P}_\varepsilon)$ with a specific asymptotic profile, so it is convenient to make the change of variables $x \mapsto \varepsilon x$ and to study the problem in the dilated domain

$$\Omega_\varepsilon := \frac{1}{\varepsilon}\Omega.$$

After this change of variables the problem becomes

$$\begin{cases} -\Delta u + u = u^p & \text{in } \Omega_\varepsilon, \\ \frac{\partial u}{\partial \nu} = 0 & \text{on } \partial\Omega_\varepsilon, \\ u > 0 & \text{in } \Omega_\varepsilon. \end{cases} \qquad (P_\varepsilon)$$

The corresponding Euler functional (for $p \in (1, \frac{n+2}{n-2})$) is then

$$I_\varepsilon(u) = \frac{1}{2}\int_{\Omega_\varepsilon} \left(|\nabla u|^2 + u^2\right) - \frac{1}{p+1}\int_{\Omega_\varepsilon} |u|^{p+1}; \qquad u \in H^1(\Omega_\varepsilon). \quad (6)$$

As already mentioned, the limit profiles for solutions to (P_ε) when ε tends to zero is given by the function in (1). We next review the existence theory for the latter equation: the arguments are rather standard but for the reader's convenience we sketch the proof below. Solutions to (1) in $\mathbb{R}^n$ can be found looking for minima of the Sobolev quotient (up to a Lagrange multiplier which can be eliminated multiplying the minimizer by a uitable positive constant)

$$\min_{u \in H^1(\mathbb{R}^n)} \frac{\int_{\mathbb{R}^n} \left(|\nabla u|^2 + u^2\right)}{\left(\int_{\mathbb{R}^n} |u|^{p+1}\right)^{\frac{2}{p+1}}}. \qquad (7)$$

Since a spherical decreasing rearrangement decreases the above ratio, see [40], we can restrict ourselves to the class $H^1_r(\mathbb{R}^n)$, the functions in $H^1(\mathbb{R}^n)$ which have radial symmetry and are radially non-increasing[1]. In the class of radial functions we have the following result.

---

[1] If one wants to avoid the use of rearrangements, it is simply possible to find minima of (7) within the radial class, but the information that they are global minima will be lost.

Lemma 5 *Let $n \geq 3$. Then there exists $c_n > 0$, depending only on n, such that for all $u \in H_r^1(\mathbb{R}^n)$*

$$|u(r)| \leq c_n r^{(1-n)/2} \|u\|, \qquad \forall\, r \geq 1. \tag{8}$$

PROOF. By density, we can suppose that $u \in C_0^\infty(\mathbb{R}^n)$. If the prime symbol denotes the derivative with respect to r, we have $(r^{n-1}u^2)' = 2r^{n-1}uu' + (n-1)r^{n-2}u^2$, whence $(r^{n-1}u^2)' \geq 2r^{n-1}uu'$. Integrating over $[r, \infty)$ we find

$$r^{n-1}u^2(r) \leq -2\int_r^\infty r^{n-1}uu'\,dr \leq c\|u\|^2,$$

where c depends on n, only, proving (8). ∎

Corollary 6 *[61] The embedding of $H_r^1(\mathbb{R}^n)$ into $L^q(\mathbb{R}^n)$, $n \geq 3$, is compact for all $2 < q < 2^*$. For $n = 2$, the embedding is compact for all $q > 2$.*

PROOF. Suppose $n \geq 3$, and let $(u_k)_k \subseteq H_r^1(\mathbb{R}^n)$ be such that $u_k \rightharpoonup 0$ as $k \to +\infty$. From (8) it follows that

$$|u_k(r)| \leq C_1 r^{(1-n)/2}\|u_k\| \leq C_2 r^{(1-n)/2}.$$

Since $q > 2$ we deduce that, given $\varepsilon > 0$, there exists $C_3 > 0$ and $\overline{R} > 0$ such that $|u_k(r)|^q \leq C_3\varepsilon\,|u_k(r)|^2$, for all $r \geq \overline{R}$. This implies

$$\int_{|x|\geq\overline{R}} |u_k(x)|^q \leq C_3\varepsilon \int_{|x|\geq\overline{R}} |u_k(x)|^2 \leq C_3\varepsilon\,\|u_k\|^2 \leq C_4\varepsilon. \tag{9}$$

Moreover, from the standard Sobolev embedding Theorem, we have that $u_k \to 0$ strongly in $L^q(B_{\overline{R}})$, for every $2 \leq q < 2^*$. Thus there exists $k_0 > 0$ such that for all $k \geq k_0$ one has

$$\int_{|x|\leq\overline{R}} |u_k(x)|^q \leq \varepsilon.$$

This and (9) imply that $\int_{\mathbb{R}^n} |u_k(x)|^q \leq C_5\varepsilon$ for $k \geq k_0$, proving that $u_k \to 0$, strongly in $L^q(\mathbb{R}^n)$, for every $2 < q < 2^*$. For $n = 2$ the proof is identical. ∎

Proposition 7 *If $p \in \left(1, \frac{n+2}{n-2}\right)$ problem (1) admits a positive solution U.*

PROOF. We consider a radial sequence $(u_l)_l$ which is minimizing for (7): by the Sobolev embeddings the infimum is bounded from below by a positive constant. We can assume the H^1 norm of each function u_l is 1, so by the latter bound we find that $\int_{\mathbb{R}}^n |u_l|^{p+1} \geq \delta_0 > 0$ for every l. By the compactness of the embedding for radial functions, u_l converges strongly to a non-zero function U_0, and also

$$\|U_0\| \leq \liminf_l \|u_l\|.$$

Therefore

$$\frac{\int_{\mathbb{R}^n} \left(|\nabla U_0|^2 + U_0^2\right)}{\left(\int_{\mathbb{R}^n} |U_0|^{p+1}\right)^{\frac{2}{p+1}}} \leq \liminf_l \frac{\int_{\mathbb{R}^n} \left(|\nabla u_l|^2 + u_l^2\right)}{\left(\int_{\mathbb{R}^n} |u_l|^{p+1}\right)^{\frac{2}{p+1}}},$$

so U_0 realizes the minimum in (7). Finally U can be obtained from U_0 simply by scaling (to remove the Lagrange multiplier).   ∎

Remark 8 *It is possible to prove that the radial solutions are unique (see [41]) and decay to zero exponentially: more precisely satisfy the property*

$$\lim_{r \to +\infty} e^r r^{\frac{n-1}{2}} U(r) = \alpha_{n,p}; \qquad \lim_{r \to +\infty} \frac{U'(r)}{U(r)} = -\lim_{r \to +\infty} \frac{U''(r)}{U(r)} = -1, \quad (10)$$

where $\alpha_{n,p}$ is a positive constant depending only on n and p.

2 Perturbation in Critical Point Theory

In this and in the subsequent section we will discuss the existence of critical points for a class of functionals which are perturbative in nature. Given a Hilbert space $\mathcal{H}$ (which might as well depend on ε), we are interested in functionals $I_\varepsilon : \mathcal{H} \to \mathbb{R}$ of class C^2 which satisfy the following properties

i) there exists a smooth finite-dimensional manifold (compact or not) $Z_\varepsilon \subseteq \mathcal{H}$ such that $\|I_\varepsilon'(z)\| \leq C\varepsilon$ for every $z \in Z_\varepsilon$ and for some fixed constant C (independent of z and ε): moreover $\|I_\varepsilon''(z)[q]\| \leq C\varepsilon\|q\|$ for every $z \in Z_\varepsilon$ and every $q \in T_z Z_\varepsilon$;

ii) there exist C, $\alpha \in (0,1)$ and $r_0 > 0$ (independent of ε) such that $\|I_\varepsilon''\|_{C^\alpha} \leq C$ in the subset $\{u : dist(u, Z_\varepsilon) < r_0\}$;

iii) letting P_z, $z \in Z_\varepsilon$, denote the projection onto the orthogonal complement of $T_z Z_\varepsilon$, there exists $C > 0$ (independent of z and ε) such that $P_z I_\varepsilon''(z)$, restricted to $(T_z Z_\varepsilon)^\perp$, is invertible from $(T_z Z_\varepsilon)^\perp$ into itself, and the inverse operator satisfies $\|(P_z I_\varepsilon''(z))^{-1}\| \leq C$.

Example 9 *One can consider functionals of the form*

$$I_\varepsilon(u) = I_0(u) + \varepsilon\, G(u). \qquad (11)$$

where $I_0 \in C^{2,\alpha}(\mathcal{H}, \mathbb{R})$, is called the unperturbed functional *and $G \in C^{2,\alpha}(\mathcal{H}, \mathbb{R})$ is a* perturbation. *Suppose there exists a d-dimensional smooth manifold Z such that all $z \in Z$ is a critical point of I_0. The set Z will be called a* critical manifold *(of I_0). Let $T_z Z$ denote the tangent space to Z at z. Since Z is a critical manifold then for every $z \in Z$ one has $I_0'(z) = 0$. Differentiating this identity, we get*

$$(I_0''(z)[v]|\varphi) = 0, \quad \forall\, v \in T_z Z, \ \forall\, \varphi \in \mathcal{H},$$

and this shows that all $v \in T_z Z$ is a solution of the linearized equation $I_0''(z)[v] = 0$, namely that $v \in Ker[I_0''(z)]$, so $T_z Z \subseteq Ker[I_0''(z)]$. In particular, $I_0''(z)$ has a non trivial kernel (whose dimension is at least d) and hence all the $z \in Z$ are degenerate critical points of I_0. We shall require that this degeneracy is minimal: precisely we will suppose that

(ND) $\qquad T_z Z = Ker[I_0''(z)], \quad \forall z \in Z.$

In addition to (ND) we will assume that

(Fr) $\qquad$ for all $z \in Z$, $I_0''(z)$ is an index 0 Fredholm map.[2]

The last two properties imply that $I_0''(z)$ is invertible from $(T_z Z)^\perp$ onto itself. Therefore, if I_0 and G are bounded (near Z) in $C^{2,\alpha}$ norm and if we take $Z_\varepsilon = Z$, properties $i) - iii)$ above will follow.

2.1 The Finite-Dimensional Reduction

First some notation is in order. Let us set $W = (T_z Z_\varepsilon)^\perp$ and let $(q_i)_{1 \le i \le d}$ be an orthonormal set (locally smooth) such that $T_z Z_\varepsilon = span\{q_1, \ldots, q_d\}$. In the sequel we will always assume that Z_ε has a (local) C^2 parametric representation (with uniformly bounded second derivatives) $z = z_\xi$, $\xi \in \mathbb{R}^d$. Furthermore, we also suppose that $q_i = \partial_{\xi_i} z_\xi / \|\partial_{\xi_i} z_\xi\|$. This will be verified in our applications.

We look for critical points of I_ε in the form $u = z + w$ with $z \in Z_\varepsilon$ and $w \in W$. If $P : \mathcal{H} \to W$ denotes the orthogonal projection onto W, the equation $I_\varepsilon'(z + w) = 0$ is equivalent to the following system

$$\begin{cases} P I_\varepsilon'(z + w) = 0 & \text{(the \textit{auxiliary equation})}; \\ (Id - P) I_\varepsilon'(z + w) = 0, & \text{(the \textit{bifurcation equation})}. \end{cases} \qquad (12)$$

Proposition 10 Let $i)$–$iii)$ hold. Then there exists $\varepsilon_0 > 0$ with the following property: for all $|\varepsilon| < \varepsilon_0$ and for all $z \in Z_\varepsilon$, the auxiliary equation in (12) has a unique solution $w = w_\varepsilon(z)$ such that:

(j) $w_\varepsilon(z) \in W = (T_z Z_\varepsilon)^\perp$, is of class C^1 with respect to $z \in Z_\varepsilon$ and $w_\varepsilon(z) \to 0$ as $|\varepsilon| \to 0$, uniformly with respect to $z \in Z_\varepsilon$, together with its derivative with respect to z, w_ε';

(jj) quantitatively, one has that $\|w_\varepsilon(z)\| = O(\varepsilon)$ as $\varepsilon \to 0$, for all $z \in Z_c$.

PROOF. Property $iii)$ allows us to apply the contraction mapping theorem to the auxiliary equation. In fact, by the invertibility of $P I_\varepsilon''(z)$ we can rewrite it tautologically as

$$w = -(P I_\varepsilon''(z))^{-1} \left[P I_\varepsilon'(z) + (P I_\varepsilon'(z + w) - P I_\varepsilon'(z) - P I_\varepsilon''(z)[w]) \right] := G_{\varepsilon,z}(w).$$

---

[2] A linear map $T \in L(\mathcal{H}, \mathcal{H})$ is *Fredholm* if the kernel is finite-dimensional and if the image is closed and has finite codimension. The *index* of T is $\dim(Ker[T]) - \operatorname{codim}(Im[T])$.

We claim next that the latter map is a contraction on a suitable subset of $\mathcal{H}$. In fact, for the second term of G_ε we can write that

$$\|PI'_\varepsilon(z+w) - PI'_\varepsilon(z) - PI''_\varepsilon(z)[w]\| = \left\| P \int_0^1 (I''_\varepsilon(z+sw) - I''_\varepsilon(z))[w]ds \right\|$$
$$\leq C\|w\|^{1+\alpha},$$

and therefore by $i)$ and $iii)$ we have

$$\|G_{z,\varepsilon}(w)\| \leq C^2\varepsilon + C^2\|w\|^{1+\alpha}; \qquad \|w\| \leq r_0.$$

Similarly, one also finds

$$\|G_{z,\varepsilon}(w_1) - G_{z,\varepsilon}(w_2)\| \leq C^2(\|w_1\|^\alpha + \|w_2\|^\alpha)\|w_1 - w_2\|.$$

By the last two equations, if we fix $C_1 > 0$ sufficiently large and let $B_\varepsilon = \{w \in W : \|w\| \leq C_1\varepsilon\}$, we can check that $G_{z,\varepsilon}$ is a contraction in B_ε, so for every $z \in Z_\varepsilon$ we obtain a function w satisfying (jj) (for brevity, in the sequel the dependence on z will be assumed understood).

Let us now show that also the derivatives of w with respect to ξ are bounded. Indeed (here we are sketchy), for the components of w tangent to Z_ε we can argue as follows: since $(w|\partial_\xi z) = 0$ for every ξ, differentiating with respect to ξ we find that

$$(\partial_\xi w|\partial_\xi z) = -(w|\partial_\xi^2 z).$$

Since $\|w\| = O(\varepsilon)$ and since $\partial_\xi^2 z$ is bounded (we are assuming Z_ε has a C^2-controlled parameterization), the tangent components of $\partial_\xi w$ are bounded in norm by $C\varepsilon$.

About the normal components, we can differentiate the relation $(I'_\varepsilon(z+w)|\partial_\xi z) = 0$ with respect to ξ, to find that

$$(I''_\varepsilon(z+w)[\partial_\xi z + \partial_\xi w]|\partial_\xi z) + (I'_\varepsilon(z+w)|\partial_\xi^2 z) = 0,$$

which implies

$$I''_\varepsilon(z+w)[\partial_\xi w] = -(I'_\varepsilon(z+w)|\partial_\xi^2 z) - (I''_\varepsilon(z+w)[\partial_\xi z]|\partial_\xi z).$$

Since I''_ε is locally Hölder continuous and since $\|w\| = O(\varepsilon)$, by $iii)$ we have that

$$\|P\partial_\xi w\| \leq C\|I'_\varepsilon(z+w)\| + \|I''_\varepsilon(z+w)[\partial_\xi z]\| \leq C\varepsilon^\alpha.$$

For the latter inequality we used again $\|w\| = O(\varepsilon)$, together with the Lipschitzianity of I'_ε and $i)$). ∎

2.2 Existence of Critical Points

We shall now provide conditions for solving the bifurcation equation in (12). In order to do this, let us define the *reduced functional* $\Phi_\varepsilon : Z \to \mathbb{R}$ by setting

$$\Phi_\varepsilon(z) = I_\varepsilon(z + w_\varepsilon(z)). \tag{13}$$

From a geometric point of view the argument can be outlined as follows. Consider the manifold $\tilde{Z}_\varepsilon = \{z + w_\varepsilon(z) \,:\, z \in Z_\varepsilon\}$. If z_ε is a critical point of Φ_ε, it follows that $u_\varepsilon = z_\varepsilon + w(z_\varepsilon) \in \tilde{Z}_\varepsilon$ is a critical point of I_ε constrained on $\tilde{Z}_\varepsilon$ and thus u_ε satisfies $I'_\varepsilon(u_\varepsilon) \perp T_{u_\varepsilon}\tilde{Z}_\varepsilon$. Moreover the definition of w_ε, see Proposition 10, implies that $I'_\varepsilon(z + w_\varepsilon(z)) \in T_z Z_\varepsilon$. In particular, $I'_\varepsilon(u_\varepsilon) \in T_{z_\varepsilon} Z_\varepsilon$. Since, for $|\varepsilon|$ small, $T_{u_\varepsilon}\tilde{Z}_\varepsilon$ and $T_{z_\varepsilon} Z_\varepsilon$ are close, see (j) in Proposition 10, it follows that $I'_\varepsilon(u_\varepsilon) = 0$. A manifold with these properties is called a *natural constraint* for I_ε.

Theorem 11 *Suppose we are in the situation of Proposition 10, and let us assume that Φ_ε has, for $|\varepsilon|$ sufficiently small, a critical point z_ε. Then $u_\varepsilon = z_\varepsilon + w_\varepsilon(z_\varepsilon)$ is a critical point of I_ε.*

PROOF. We use the previous notation and, to be short, we write below D_i for D_{ξ_i}, etc. Let ξ_ε be such that $z_\varepsilon = z_{\xi_\varepsilon}$, and set $q_i^\varepsilon = \partial z / \partial \xi_i|_{\xi_\varepsilon}$.

From Proposition 10 we infer that there exists $\varepsilon_0 > 0$ such that the auxiliary equation in (12) has a solution $w_\varepsilon(z_\varepsilon)$, defined for $|\varepsilon| < \varepsilon_0$. In particular, from (j) and by continuity, one has that

$$\lim_{|\varepsilon| \to 0} (D_i w_\varepsilon(z_\varepsilon) \,|\, q_j^\varepsilon) = 0, \qquad i, j = 1, \ldots, d.$$

Let us consider the matrix $B^\varepsilon = (b_{ij}^\varepsilon)_{ij}$, where

$$b_{ij}^\varepsilon = \left(D_i w_\varepsilon(z_\varepsilon) \,|\, q_j^\varepsilon\right).$$

From the above arguments we can choose $0 < \varepsilon_1 < \varepsilon_0$, such that

$$|det(B^\varepsilon)| < 1, \quad \forall\, |\varepsilon| < \varepsilon_1, \tag{14}$$

Fix $\varepsilon > 0$ such that $|\varepsilon| < \min\{\varepsilon_0, \varepsilon_1\}$. Since z_ε is a critical point of Φ_ε we get

$$(I'_\varepsilon(z_\varepsilon + w_\varepsilon(z_\varepsilon)) \,|\, q_i^\varepsilon + D_i w_\varepsilon(z_\varepsilon)) = 0, \qquad i = 1, \ldots, d.$$

From $PI'_\varepsilon(z + w_\varepsilon(z_\varepsilon)) = 0$, we deduce that $I'_\varepsilon(z_\varepsilon + w_\varepsilon(z_\varepsilon)) = \sum A_{i,\varepsilon} q_i^\varepsilon$, where

$$A_{i,\varepsilon} = (I'_\varepsilon(z_\varepsilon + w_\varepsilon(z_\varepsilon)) \,|\, q_i^\varepsilon).$$

Then we find

$$\left(\sum_j A_{j,\varepsilon}\, q_j^\varepsilon \mid q_i^\varepsilon + D_i w_\varepsilon(z_\varepsilon) \right) = 0, \qquad i = 1, \dots, d,$$

namely

$$A_{i,\varepsilon} + \sum_j A_{j,\varepsilon}\left(q_j^\varepsilon \mid D_i w_\varepsilon(z_\varepsilon) \right) = A_{i,\varepsilon} + \sum_j A_{j,\varepsilon} b_{ij}^\varepsilon = 0, \quad i = 1, \dots, d. \quad (15)$$

Equation (15) is a $(d \times d)$ linear system for which the matrix of coefficients, $Id_{\mathbb{R}^d} + B^\varepsilon$, has entries of the form $\delta_{ij} + b_{ij}^\varepsilon$, where δ_{ij} is the Kronecker symbol and b_{ij}^ε are defined above and satisfy (14). Then, for $|\varepsilon| < \varepsilon_1$, $Id_{\mathbb{R}^d} + B^\varepsilon$ is invertible, thus (15) has the trivial solution only: $A_{i,\varepsilon} = 0$ for all $i = 1, \dots, d$. Since the $A_{i,\varepsilon}$ are the components of $\Phi_\varepsilon(z_\varepsilon)$, the conclusion follows. ∎

We next provide a criterion for applying Theorem 11, based on expanding I_ε on Z_ε in powers of ε. We will assume a *slow dependence* of G in the parameter ξ, since this applies to our concrete case ($\tilde{P}_\varepsilon$).

Theorem 12 *Suppose the assumptions of Proposition 10 hold, and that for ε small there is a local parameterization $\xi \in \frac{1}{\varepsilon}\mathcal{U} \subseteq \mathbb{R}^d$ of Z_ε such that, as $\varepsilon \to 0$, I_ε admits the expansion*

$$I_\varepsilon(z_\xi) = C_0 + \varepsilon G(\varepsilon \xi) + o(\varepsilon), \qquad \xi \in \frac{1}{\varepsilon}\mathcal{U}$$

for some function $G : \mathcal{U} \to \mathbb{R}$. Then we still have the expansion

$$\Phi_\varepsilon(z_\xi) = C_0 + \varepsilon G(\varepsilon \xi) + o(\varepsilon) \qquad as\ \varepsilon \to 0. \quad (16)$$

Moreover, if $\overline{\xi} \in \mathcal{U}$ is a strict local maximum or minimum of G, then for $|\varepsilon|$ small the functional I_ε has a critical point u_ε. Furthermore, if $\overline{\xi}$ is isolated, we can take $u_\varepsilon - z_{\overline{\xi}/\varepsilon} = o(1/\varepsilon)$ as $\varepsilon \to 0$.

Remark 13 *The last statement asserts that, once we scale back in ε, the solution concentrates near $\overline{\xi}$.*

PROOF. To prove (16) we simply use property $i)$ and $jj)$ in Proposition 10, together with a Taylor expansion of I_ε, to get

$$I_\varepsilon(z_\xi + w_\varepsilon) = I_\varepsilon(z_\xi) + I_\varepsilon'(z)[w_\varepsilon] + O(\|w_\varepsilon\|^2) = C_0 + \varepsilon G(\xi) + o(\varepsilon),$$

so the first assertion follows.

We will prove the second one when $\overline{\xi}$ is a minimum of G: the other case is completely similar. Let $\gamma > 0$ and let $\mathcal{U}_\delta$ be a δ-neighborhood of $\overline{\xi}$ such that

$$G(\xi) \geq G(\overline{\xi}) + \gamma, \quad \forall\, \xi \in \partial \mathcal{U}_\delta.$$

Using (16) we find, for $|\varepsilon|$ small

$$\Phi_\varepsilon(z_\xi) - \Phi_\varepsilon(z_{\overline{\xi}}) = \varepsilon \left(G(\xi) - G(\overline{\xi}) \right) + o(\varepsilon).$$

Then, there exists $\varepsilon_1 > 0$ small such that for every $\xi \in \partial \mathcal{U}_\delta$ one has

$$\begin{cases} \Phi_\varepsilon(z_\xi) - \Phi_\varepsilon(z_{\overline{\xi}}) > 0 & \text{if } 0 < \varepsilon < \varepsilon_1, \\ \Phi_\varepsilon(z_\xi) - \Phi_\varepsilon(z_{\overline{\xi}}) < 0 & \text{if } -\varepsilon_1 < \varepsilon < 0. \end{cases}$$

In the former situation $\Phi_\varepsilon(z_\xi)$ has a local minimum for ξ in $\mathcal{U}_\delta$, in the latter a local maximum. In any case, Φ_ε has a critical point z_{ξ_ε} with $\xi_\varepsilon \in \mathcal{U}_\delta$ and hence, by Theorem 11, $u_\varepsilon = z_\varepsilon + w_\varepsilon(z_\varepsilon)$ is a critical point of I_ε. If $\overline{\xi}$ is an isolated minimum or maximum of G we can take δ arbitrarily small and hence $z_\varepsilon \to z_{\overline{\xi}}$ as well as $u_\varepsilon \to z_{\overline{\xi}}$.   ∎

The last statement in Theorem 12 can be extended to the more general situation in which $\hat{\xi}$ is a critical point of G satisfying

(G') $\exists\ \mathcal{N} \subset \mathbb{R}^d$ open bounded such that the topological degree $d(G', \mathcal{N}, 0)$ is different from zero.

For the definition of topological degree see for example [4]. Let us point out that if (G') holds then G has a critical point in $\mathcal{N}$. Moreover, if G has either a strict local maximum (or minimum), or any non-degenerate critical point $\overline{\xi}$, we can take as $\mathcal{N}$ the ball $B_r(\overline{\xi})$ with $r \ll 1$, and (G') holds.

Theorem 14 *Suppose we are under the (first) assumptions of Theorem 12, that $\|I'_\varepsilon(z)\| \leq C\varepsilon^2$ for all $z \in Z_\varepsilon$ (C fixed) and that we have the expansion*

$$\frac{\partial}{\partial \xi} I_\varepsilon(z_\xi) = \varepsilon^2 G'(\varepsilon\xi) + o(\varepsilon^2), \qquad\qquad \xi \in \mathcal{U} \ as\ \varepsilon \to 0. \qquad (17)$$

Assume (G') holds: then for $|\varepsilon|$ small the functional I_ε has a critical point u_ε and there exists $\hat{\xi} \in \mathcal{N}$, $G'(\hat{\xi}) = 0$, such that $u_{\varepsilon_j} \to z_{\hat{\xi}}$ for some $\varepsilon_j \to 0$. Therefore if, in addition, $\mathcal{N}$ contains only an isolated critical point $\hat{\xi}$ of G', then $u_\varepsilon - z_{\hat{\xi}/\varepsilon} = o(1/\varepsilon)$ as $\varepsilon \to 0$.

PROOF. The arguments of Proposition 10 still apply, but yield the improved estimates $\|w\| \leq C\varepsilon^2$ and $\|w'\| \leq C\varepsilon^{2\alpha}$ (we just perform the contraction in a smaller set). From the definition of Φ_ε we infer that, for all $v \in T_z Z_\varepsilon$

$$(\Phi'_\varepsilon(z) \,|\, v) = (I'_\varepsilon(z + w_\varepsilon) \,|\, v + w'_\varepsilon). \qquad (18)$$

Therefore, we can write that

$$(\Phi'_\varepsilon(z) \,|\, v) = I'_\varepsilon(z)[v] + I''_\varepsilon(z)[w_\varepsilon, v] + o(\|w_\varepsilon\|)\|v\| + I'_\varepsilon(z)[w'_\varepsilon] + O(\|w_\varepsilon\|\,\|w'_\varepsilon\|).$$

Now, using (17), the fact that $\|w_\varepsilon\| = O(\varepsilon^2)$, $\|w_\varepsilon'\| = O(\varepsilon^{2\alpha})\|v\|$ and $\|I_\varepsilon''(z)[v]\| \leq C\varepsilon^2\|v\|$ (see Proposition 10 and i), which has to be modified with an extra ε) we have that

$$(\Phi_\varepsilon'(z)\,|v\,) = \varepsilon^2 G'(\varepsilon\xi)[v] + o(\varepsilon) \qquad \text{in } \mathcal{N}. \tag{19}$$

Then the continuity property of the topological degree and (G') yield, for $|\varepsilon|$ small,

$$d(\Phi_\varepsilon', \mathcal{N}, 0) = d(G', \mathcal{N}, 0) \neq 0.$$

This implies that, for $|\varepsilon|$ small, the equation $\Phi_\varepsilon'(z) = 0$ has a solution in $\mathcal{N}$. The convergence result follows from (19). ∎

3 Application to the Study of $\tilde{P}_\varepsilon$

In this section we apply the above methods to the Neumann problem $(\tilde{P}_\varepsilon)$. As we mentioned, the limit profile of the solutions we are interested in solves equation (1): the latter has variational structure, and solutions are critical points of the functional

$$\overline{I}(u) = \frac{1}{2} \int_{\mathbb{R}^n} \left(|\nabla u|^2 + u^2 \right) - \frac{1}{p+1} \int_{\mathbb{R}^n} |u|^{p+1}. \tag{20}$$

We call $\overline{I}_+$ the functional corresponding to (1) in $\mathbb{R}_+^n$, namely

$$\overline{I}_+(u) = \frac{1}{2} \int_{\mathbb{R}_+^n} \left(|\nabla u|^2 + u^2 \right) - \frac{1}{p+1} \int_{\mathbb{R}_+^n} |u|^{p+1}. \tag{21}$$

Recall that we are assuming here $p < \frac{n+2}{n-2}$. Our next goal is to characterize the spectrum and some eigenfunctions of $\overline{I}''(U)$ or of $\overline{I}_+''(U)$.

3.1 Study of $Ker[\overline{I}''(U)]$ and $Ker[\overline{I}_+''(U)]$

For $\xi \in \mathbb{R}^n$ we set

$$U_\xi = U(\cdot - \xi); \qquad Z = \{U_\xi \ : \ \xi \in \mathbb{R}^n\}.$$

The first goal of this section is to prove the following result.

Lemma 15 Z is non-degenerate for $\overline{I}$, namely the following properties are true:

(ND) $\qquad T_{U_\xi} Z = Ker[\overline{I}''(U_\xi)], \quad \forall\, \xi \in \mathbb{R}^n;$

(Fr) $\qquad \overline{I}''(U_\xi)$ is an index 0 Fredholm map, for all $\xi \in \mathbb{R}^n$.

PROOF. It is sufficient to prove the lemma for $\xi = 0$, hence taking $U_\xi = U$: the case of a general ξ will follow immediately. The proof will be carried out in several steps.

Step 1. In order to characterize $Ker[\overline{I}''(U)]$, let us introduce some notation. We set

$$r = |x|; \qquad\qquad \vartheta = \frac{x}{|x|} \in S^{n-1}$$

and let Δ_r, resp. $\Delta_{S^{n-1}}$ denote the Laplace operator in radial coordinates, resp. the Laplace-Beltrami operator on S^{n-1}

$$\Delta_r = \frac{\partial^2}{\partial r^2} + \frac{n-1}{r}\frac{\partial}{\partial r}; \qquad\qquad \Delta_{S^{n-1}} = \frac{1}{\sqrt{g}}\sum \frac{\partial}{\partial y_j}\left(\sqrt{g}\, g^{ij}\frac{\partial}{\partial y_i}\right).$$

In the latter formula standard notation is used: $ds^2 = g_{ij}dy^i dy^j$ denotes the standard metric on S^{n-1}, $g = det(g_{ij})$ and $[g^{ij}] = [g_{ij}]^{-1}$. Consider the spherical harmonics $Y_k(\vartheta)$ satisfying

$$-\Delta_{S^{n-1}}Y_k = \lambda_k Y_k, \tag{22}$$

and recall that this equation has a sequence of eigenvalues

$$\lambda_k = k(k+n-2), \qquad\qquad k = 0,1,2,\ldots$$

whose multiplicity is given by $N_k - N_{k-2}$ where

$$N_k = \frac{(n+k-1)!}{(n-1)!\,k!}, \quad (k \geq 0); \qquad N_k = 0,\ \forall\, k < 0,$$

see [16]. In particular, one has that

$$\lambda_0 = 0 \quad \text{has multiplicity 1} \quad \text{and} \quad \lambda_1 = n-1 \quad \text{has multiplicity } n.$$

Every $v \in \mathcal{H}$ can be written in the form

$$v(x) = \sum_{k=0}^{\infty} \psi_k(r)Y_k(\vartheta), \qquad \text{where } \psi_k(r) = \int_{S^{n-1}} u(r\vartheta)Y_k(\vartheta)d\vartheta \in H^1(\mathbb{R}),$$

and moreover

$$\Delta(\psi_k Y_k) = Y_k(\vartheta)\Delta_r\psi_k(r) + \frac{1}{r^2}\psi_k(r)\Delta_{S^{n-1}}Y_k(\vartheta). \tag{23}$$

Recall that $v \in \mathcal{H}$ belongs to $Ker[\overline{I}''(U)]$ if and only if

$$-\Delta v + v = pU^{p-1}(x)v, \quad v \in \mathcal{H}. \tag{24}$$

Substituting (23) and (22) into (24) we get the following equations for ψ_k

$$A_k(\psi_k) := -\psi_k'' - \frac{n-1}{r}\psi_k' + \psi_k + \frac{\lambda_k}{r^2}\psi_k - pU^{p-1}\psi_k = 0, \quad k = 0, 1, 2, \ldots.$$

Step 2. Let us first consider the case $k = 0$. Since $\lambda_0 = 0$ we infer that ψ_0 satisfies

$$A_0(\psi_0) = -\psi_0'' - \frac{n-1}{r}\psi_0' + \psi_0 - pU^{p-1}\psi_0 = 0.$$

It has been shown in [41] that all the non trivial solutions to $A_0(u) = 0$ are unbounded. Since we are looking for solutions $\psi_0 \in H^1(\mathbb{R})$, it follows that $\psi_0 = 0$.

Step 3. For $k = 1$ one has that $\lambda_1 = n - 1$ and we find

$$A_1(\psi_1) = -\psi_1'' - \frac{n-1}{r}\psi_1' + \psi_1 + \frac{n-1}{r^2}\psi_1 - pU^{p-1}\psi_1 = 0.$$

Let $\widehat{U}(r)$ denote the function such that $U(x) = \widehat{U}(|x|)$. Since $U(x)$ satisfies $-\Delta U + U = U^p$, then $\widehat{U}$ solves

$$-\widehat{U}'' - \frac{n-1}{r}\widehat{U}' + \widehat{U} = \widehat{U}^p.$$

Differentiating in r, we get

$$-(\widehat{U}')'' - \frac{n-1}{r}(\widehat{U}')' + \frac{n-1}{r^2}\widehat{U}' + \widehat{U}' = p\widehat{U}^{p-1}\widehat{U}'. \tag{25}$$

In other words, $\widehat{U}'(r)$ satisfies $A_1(\widehat{U}') = 0$, and $\widehat{U}' \in H^1(\mathbb{R})$. Let us look for a second solution of $A_1(\psi_1) = 0$ in the form $\psi_1(r) = c(r)\widehat{U}'(r)$. By a straight calculation, we find that $c(r)$ solves

$$-c''\widehat{U}' - 2c'(\widehat{U}')' - \frac{n-1}{r}c'\widehat{U}' = 0.$$

If $c(r)$ is not constant, it follows that

$$-\frac{c''}{c'} = 2\frac{\widehat{U}''}{\widehat{U}'} + \frac{n-1}{r}.$$

This yields

$$c(r) \sim \frac{1}{r^{n-1}\widehat{U}'^2}, \qquad (r \to +\infty),$$

which implies $c(r) \to +\infty$ as $r \to +\infty$. Therefore, the family of solutions of $A_1(\psi_1) = 0$, with $\psi_1 \in H^1(\mathbb{R})$, is given by $\psi_1(r) = \bar{c}\widehat{U}'(r)$, for some $\bar{c} \in \mathbb{R}$.

Step 4. Let us show that the equation $A_k(\psi_k) = 0$ has only the trivial solution in $H^1(\mathbb{R})$, provided that $k \geq 2$. Actually, the equation $A_1(u) = 0$ has the solution $\widehat{U}'$ which does not change sign in $(0, \infty)$. Thus, by standard arguments, A_1 is a non-negative operator. From

$$\lambda_k = (n + k - 2)k = \lambda_1 + \delta_k; \qquad \delta_k = k(n + k - 2) - (n - 1),$$

we infer that

$$A_k = A_1 + \frac{\delta_k}{r^2}.$$

Since $\delta_k > 0$ whenever $k \geq 2$, it follows that A_k is a *positive* operator for any $k \geq 2$. Thus $A_k(\psi_k) = 0$ implies that $\psi_k = 0$.

Conclusion. Collecting all the previous information, we deduce that any $v \in Ker[\overline{I}''(U)]$ has to be a constant multiple of $\widehat{U}'(r)Y_1(\vartheta)$. Here Y_1 satisfies

$$-\Delta_{S^{n-1}}Y_1 = \lambda_1 Y_1 = (n - 1)Y_1,$$

namely it belongs to the kernel of the operator $-\Delta_{S^{n-1}} - \lambda_1 Id$. Recalling that such a kernel is n dimensional, and letting $Y_{1,1}, \ldots, Y_{1,n}$ denote a basis of it, we finally find that

$$v \in span\{\widehat{U}'Y_{1,i} : 1 \leq i \leq n\} = span\{U_{x_i} : 1 \leq i \leq n\} = T_U Z.$$

This proves that (ND) holds. It is also easy to check that the operator $\overline{I}''(U)$ is a compact perturbation of the Identity, showing that (Fr) holds too. This completes the proof of Lemma 15. ∎

Remark 16 *U is a mountain pass solution of (1), so the spectrum of $\overline{I}''(U)$ has exactly one negative simple eigenvalue, $p - 1$, with eigenspace spanned by U itself, and denoted by $\langle U \rangle$. Moreover, we have shown in the preceding lemma that $\lambda = 0$ is an eigenvalue with multiplicity n and eigenspace spanned by $D_i U$, $i = 1, 2, \ldots, n$. Furthermore, there exists $\kappa > 0$ such that*

$$(\overline{I}''(U)v|v) \geq \kappa \|v\|^2, \qquad \forall v \perp \langle U \rangle \oplus T_U Z, \tag{26}$$

and hence the rest of the spectrum is positive.

Corollary 17 *Let U be as above and consider the functional $\overline{I}_+$ given in (21). Then for every $\xi \in \mathbb{R}^{n-1}$, $U(\cdot - (\xi, 0))$ is a critical point of $\overline{I}_+$. Moreover, the kernel of $\overline{I}''_+(U)$ is generated by $\frac{\partial U}{\partial x_1}, \ldots, \frac{\partial U}{\partial x_{n-1}}$. The operator has only one negative eigenvalue, and therefore there exists $\delta > 0$ such that*

$$\overline{I}''_+(U)[v, v] \geq \delta \|v\|^2 \qquad for\ all\ v \in H^1(\mathbb{R}^n_+), v \perp_+ U, \frac{\partial U}{\partial x_1}, \ldots, \frac{\partial U}{\partial x_{n-1}},$$

where we have used the symbol $\perp_+$ to denote orthogonality with respect to the H^1 scalar product in $\mathbb{R}_{\cdot+}$.

PROOF. Given any $v \in H^1(\mathbb{R}^n_+)$, we define the function $\overline{v} \in H^1(\mathbb{R}^n)$ by an even extension across $\partial\mathbb{R}^n_+$, namely we set

$$\overline{v}(x', x_n) = \begin{cases} v(x', x_n), & \text{for } x_n > 0; \\ v(x', -x_n) & \text{for } x_n < 0. \end{cases}$$

We prove first the following claim.

Claim. Suppose $v \in H^1(\mathbb{R}^n_+)$ is an eigenfunction of $\overline{I}''_+(U)$ with eigenvalue λ. Then the function $\overline{v}$ is an eigenfunction of $\overline{I}''(U)$ with eigenvalue λ.

In order to prove the claim, we notice that the function v satisfies the equation

$$\begin{cases} -\Delta v + v - pU^{p-1}v = \lambda(-\Delta v + v), & \text{in } \mathbb{R}^n_+; \\ \frac{\partial v}{\partial \nu} = 0, & \text{on } \partial\mathbb{R}^n_+. \end{cases}$$

Similarly, by symmetry, there holds

$$\begin{cases} -\Delta \overline{v} + \overline{v} - pU^{p-1}\overline{v} = \lambda(-\Delta \overline{v} + \overline{v}), & \text{in } \mathbb{R}^n_-; \\ \frac{\partial \overline{v}}{\partial \nu} = 0, & \text{on } \partial\mathbb{R}^n_-, \end{cases}$$

where we have set $\mathbb{R}^n_- = \{(x', x_n) \; : \; x' \in \mathbb{R}^{n-1}, x_n < 0\}$. Then, considering any function $w \in H^1(\mathbb{R}^n)$, integrating by parts and using the Neumann boundary condition one finds

$$\begin{aligned} \overline{I}''(U)[\overline{v}, w] &= \int_{\mathbb{R}^n} (\langle \nabla\overline{v}, \nabla w \rangle + \overline{v}w) - \frac{1}{p+1}\int_{\mathbb{R}^n} U^{p-1}\overline{v}w \\ &= \lambda \int_{\mathbb{R}^n_+} (-\Delta v + v)w + \lambda \int_{\mathbb{R}^n_-} (-\Delta \overline{v} + \overline{v}) \\ &= \lambda \int_{\mathbb{R}^n_+} (\langle \nabla v, \nabla w \rangle + vw) + \lambda \int_{\mathbb{R}^n_-} (\langle \nabla\overline{v}, \nabla w \rangle + \overline{v}w) \\ &= \lambda(\overline{v}|w)_{H^1(\mathbb{R}^n)}. \end{aligned}$$

This proves the above claim.

We know that the functions $\partial_{\xi_1}U, \dots, \partial_{\xi_{n-1}}U$ belong to the kernel of $\overline{I}''_+(U)$. Suppose by contradiction that there exists a non zero element v in the kernel of $\overline{I}''_+(U)$, orthogonal to $\partial_{\xi_1}U, \dots, \partial_{\xi_{n-1}}U$. Then, by the above claim, its even extension $\overline{v}$ would belong to the kernel of $\overline{I}''(U)$. On the other hand we know that the only element in the kernel of $\overline{I}''(U)$ which is orthogonal to $\partial_{\xi_1}U, \dots, \partial_{\xi_{n-1}}U$ is $\partial_{\xi_n}U$. Since $\partial_{\xi_n}U$ is odd with respect to x_n, while $\overline{v}$ is even with respect to x_n, we get a contradiction. This concludes the proof. ∎

3.2 Proof of Theorem 1

Let us describe $\partial\Omega_\varepsilon$ near a generic point $X \in \partial\Omega_\varepsilon$. Without loss of generality, we can assume that $X = 0 \in \mathbb{R}^n$, that $\{x_n = 0\}$ is the tangent plane of $\partial\Omega_\varepsilon$ (or $\partial\Omega$) at X, and that $\nu(X) = (0,\ldots,0,-1)$. In a neighborhood of X, let $x_n = \psi(x')$ be a local parametrization of $\partial\Omega$. Then one has

$$x_n = \psi(x') := \tfrac{1}{2}\langle A_X x', x'\rangle + C_X(x') + O(|x'|^4); \qquad |x'| < \mu_0, \qquad (27)$$

where A_X is the hessian of ψ at 0 and C_X is a cubic polynomial, which is given precisely by

$$C_X(x') = \tfrac{1}{6} \sum_{i,j,k} \partial^3_{ijk}\psi|_0 x'_i x'_j x'_k. \qquad (28)$$

We have clearly $H(X) = \frac{1}{n-1}tr A_X$. On the other hand, $\partial\Omega_\varepsilon$ is parameterized by $y_n = \psi_\varepsilon(x') := \frac{1}{\varepsilon}\psi(\varepsilon x')$, for which the following expansion holds

$$\psi_\varepsilon(x') = \frac{\varepsilon}{2}\langle A_X x', x'\rangle + \varepsilon^2 C_X(x') + \varepsilon^3 O(|x'|^4);$$

$$\partial_i \psi_\varepsilon(x') = \varepsilon(A_X x')_i + \varepsilon^2 Q^i_X(x') + \varepsilon^3 O(|x'|^3), \qquad (29)$$

where Q^i_X are quadratic forms in x' given by (see (28))

$$Q^i_X(x') = \frac{1}{2}\sum_{j,k} \partial^3_{ijk}\psi|_0 x'_j x'_k.$$

In particular, from the Schwartz's Lemma, it follows that

$$(Q^i_X)_{jk} = (Q^i_X)_{kj} = (Q^j_X)_{ik} \qquad \text{for every } i,j,k. \qquad (30)$$

Concerning the outer normal ν, we have also

$$\nu = \frac{\left(\frac{\partial\psi_\varepsilon}{\partial x_1},\ldots,\frac{\partial\psi_\varepsilon}{\partial x_{n-1}},-1\right)}{\sqrt{1+|\nabla\psi_\varepsilon|^2}} = \left(\varepsilon(A_X x') + \varepsilon^2 Q_X(x'), -1 + \frac{1}{2}\varepsilon^2|Ax'|^2\right)$$

$$+ \varepsilon^3 O(|x'|^3). \qquad (31)$$

Since $\partial\Omega_\varepsilon$ is almost flat for ε small and since the function U is radial, for $X \in \partial\Omega_\varepsilon$ we have $\frac{\partial}{\partial\nu}U(\cdot - X) \sim 0$. Thus $U(\cdot - X)$ is an approximate solution to (P_ε). Hence, a natural choice of the manifold Z_ε could be the following

$$Z_\varepsilon = \{U(\cdot - X) := U_X \ : \ X \in \partial\Omega_\varepsilon\}.$$

We show next that the abstract setting of the previous section can be applied with this choice of Z_ε and for I_ε given by (6).

Proposition 18 *With the above choices of I_ε and Z_ε, the assumptions $i)$ – $iii)$ in Section 2 hold true.*

In order to prove this proposition, we need the following preliminary result.

Lemma 19 *There exists $\overline{\delta} > 0$ such that for ε small one has*

$$I_\varepsilon''(U_X)[v,v] \geq \overline{\delta}\|v\|^2 \qquad \text{for every } v \perp U_X, \frac{\partial U_X}{\partial X}.$$

PROOF. Let $R \gg 1$ and consider a radial smooth function $\chi_R : \mathbb{R}^n \to \mathbb{R}$ such that

$$\begin{cases} \chi_R(x) = 1, & \text{in } B_R(0); \\ \chi_R(x) = 0 & \text{in } \mathbb{R}^n \backslash B_{2R}(0); \\ |\nabla \chi_R| \leq \frac{2}{R} & \text{in } B_{2R}(0) \backslash B_R(0), \end{cases} \qquad (32)$$

and we set

$$v_1(x) = \chi_R(x - X)v(x); \qquad v_2 = (1 - \chi_R)(x - X)v(x).$$

A straight computation yields

$$\|v\|^2 = \|v_1\|^2 + \|v_2\|^2 + 2\int_{\Omega_\varepsilon} [\langle \nabla v_1, \nabla v_2 \rangle + v_1 v_2].$$

We write $\int_{\Omega_\varepsilon} [\langle \nabla v_1, \nabla v_2 \rangle + v_1 v_2] = \tau_1 + \tau_2$, where

$$\tau_1 = \int_{\Omega_\varepsilon} \chi_R(1 - \chi_R)(v^2 + |\nabla v|^2);$$

$$\tau_2 = \int_{\Omega_\varepsilon} v_2 \langle \nabla v, \nabla \chi_R \rangle - v_1 \langle \nabla v, \nabla \chi_R \rangle - v^2 |\nabla \chi_R|^2.$$

Since the integrand in τ_2 is supported in $\{R \leq |x| \leq 2R\}$, using (32) and the Hölder's inequality we deduce that $|\tau_2| = o_R(1)\|v\|^2$. As a consequence we have

$$\|v\|^2 = \|v_1\|^2 + \|v_2\|^2 + 2\tau_1 + o_R(1)\|v\|^2. \qquad (33)$$

After these preliminaries, let us evaluate $I_\varepsilon''(U_X)[v,v] = \sigma_1 + \sigma_2 + \sigma_3$, where

$$\sigma_1 = I_\varepsilon''(U_X)[v_1, v_1]; \qquad \sigma_2 = I_\varepsilon''(U_X)[v_2, v_2]; \qquad \sigma_3 = 2I_\varepsilon''(U_X)[v_1, v_2].$$

Since U_X decays exponentially away from X, we get immediately

$$\sigma_2 \geq C^{-1}\|v_2\|^2 + o_R(1)\|v\|^2; \qquad \sigma_3 \geq C^{-1}\tau_1 + o_R(1)\|v\|^2, \qquad (34)$$

hence it is sufficient to estimate the term σ_1. By the exponential decay of U_X and the fact that $(v|U_X) = (v|\frac{\partial U_X}{\partial X}) = 0$ one easily finds

$$(v_1|U_X) = -(v_2|U_X) = o_R(1)\|v\|;$$
$$\left(v_1\Big|\frac{\partial U_X}{\partial X}\right) = -\left(v_2\Big|\frac{\partial U_X}{\partial X}\right) = o_R(1)\|v\|\left\|\frac{\partial U_X}{\partial X}\right\|. \tag{35}$$

Therefore, using Corollary 17 we obtain

$$\overline{I}''_+(U)[v_1,v_1] \geq \delta\|v_1\|_+^2 + o_{\varepsilon,R}(1)\|v\|^2$$

(to be rigorous, the function v_1 needs to be modified for being defined in $\mathbb{R}^n_+$. An easy way to do it is to stretch the boundary of the domain near X into a hyperplane, but since $\partial\Omega_\varepsilon$ is almost flat this requires a small modification of the H^1 norm of v_1). Using this reasoning then one finds

$$\sigma_1 \geq \overline{I}''_+(U)[v_1,v_1] + o_\varepsilon(1)\|v_1\|^2 \geq \delta\|v_1\|_+^2 + o_{\varepsilon,R}(1)\|v\|^2$$
$$\geq \delta\|v_1\|^2 + o_{\varepsilon,R}(1)\|v\|^2. \tag{36}$$

In conclusion, from (34) and (36) we deduce

$$I''_\varepsilon(U_X)[v,v] \geq \delta\|v\|_1^2 + \|v_2\|^2 + I_v + o_{\varepsilon,R}(1)\|v\|^2 \geq \frac{\delta}{2}\|v\|^2,$$

provided R is taken large and ε is sufficiently small. This concludes the proof. ∎

PROOF OF PROPOSITION 18. By the previous lemma, we need to prove only i) and ii). To show i), we consider an arbitrary function $v \in H^1(\Omega_\varepsilon)$. Integrating by parts and using (1) we find that

$$I'_\varepsilon(U_X)[v] = \int_{\partial\Omega_\varepsilon} \frac{\partial U_X}{\partial \nu} v d\sigma + \int_{\Omega_\varepsilon} \left(-\Delta_g U_X + U_X - U_X^p\right) v dy = \int_{\partial\Omega_\varepsilon} \frac{\partial U_X}{\partial \nu} v d\sigma.$$

For μ_0 small and positive, we divide next $\partial\Omega_\varepsilon$ into its intersection with $B_{\frac{\mu_0}{\varepsilon}}(X)$ and its complement. In the latter set we use the exponential decay of $\check{U}_X$ and its derivatives, plus the trace embedding for v to obtain

$$\left|\int_{\partial\Omega_\varepsilon \setminus B_{\frac{\mu_0}{\varepsilon}}(X)} \frac{\partial U_X}{\partial \nu} v d\sigma\right| \leq C e^{-\frac{\mu_0}{\varepsilon}} \|v\|. \tag{37}$$

In $B_{\frac{\mu_0}{\varepsilon}}(X)$ we use instead the coordinates x' defined above, and the function ψ_ε to parameterize $\partial\Omega_\varepsilon$. From (29), (31) and the fact that $\nabla U(x) = U'\frac{x}{|x|}$ one finds that

$$\frac{\partial U_X}{\partial \nu} = U'(x) \left\langle \frac{(x', \varepsilon/2x' A_X x' + O(\varepsilon^2)|x'|^2)}{(|x'|^2 + O(\varepsilon^2)|x'|^4)^{\frac{1}{2}}}, \left[(\varepsilon A_X x', -1) + O(\varepsilon^2)|x'|^2\right] \right\rangle$$

$$= \left(\frac{1}{2}\varepsilon \frac{\langle x', A_X x' \rangle}{|x'|} U'(|x'|) + O(\varepsilon^2 e^{-|x'|}) \right) = O(\varepsilon|x|e^{-|x'|}). \tag{38}$$

This estimate, jointly with the trace embedding, yields

$$\left| \int_{\partial\Omega_\varepsilon \cap B_{\frac{\mu_0}{\varepsilon}}(X)} \frac{\partial U_X}{\partial \nu} v \, d\sigma \right| \leq C e^{-\frac{\mu_0}{\varepsilon}} \|v\|.$$

This equation and (37) imply that $\|I_\varepsilon'(U_X)\| \leq C\varepsilon$. Similar estimates imply $\|I_\varepsilon''(z)[q]\| \leq C\varepsilon\|q\|$ for every $z \in Z_\varepsilon$ and every $q \in T_z Z_\varepsilon$.

Property $ii)$ is rather standard. In fact, given two functions u_1, u_2 we have that

$$I_\varepsilon''(u_1)[v, w] - I_\varepsilon''(u_2)[v, w] = -p \int_{\Omega_\varepsilon} \left(|u_1|^{p-1} - |u_2|^{p-1} \right) vw.$$

Using the Hölder inequality and the Sobolev embeddings we get

$$|I_\varepsilon''(u_1)[v, w] - I_\varepsilon''(u_2)[v, w]| \leq C \left(\int_{O_\varepsilon} \left| |u_1|^{p-1} - |u_2|^{p-1} \right|^{\frac{p+1}{p-1}} \right)^{\frac{p-1}{p+1}} \|v\| \, \|w\|.$$

From the elementary inequality

$$\left| |u_1|^{p-1} - |u_2|^{p-1} \right|^{\frac{p+1}{p-1}} \leq C \left(|u_1 - u_2| + |u_1 - u_2|^{p-1} \right)$$
$$\leq C|u_1 - u_2|^{\frac{p+1}{p-1}} + C|u_1 - u_2|^{p+1}$$

and again the Sobolev embeddings one finds that

$$\|I_\varepsilon''(u_1) - I_\varepsilon''(u_2)\| \leq C \left(\|u_1 - u_2\| + \|u_1 - u_2\|^{p-1} \right),$$

therefore when the $u_i's$ are bounded (e.g. in an r_0-strip around Z_ε) we obtain local Hölderianity of I_ε''. ∎

We now show that also Theorem 12 applies to this case.

Lemma 20 *For ε small the following expansion holds*

$$I_\varepsilon(U_X) = C_0 - C_1 \varepsilon H(\varepsilon X) + O(\varepsilon^2),$$

where

$$C_0 = \left(\frac{1}{2} - \frac{1}{p+1} \right) \int_{\mathbb{R}^n_+} U^{p+1}; \qquad C_1 = C_1(n, p) > 0.$$

PROOF. Integrating by parts and using (1) one finds

$$I_\varepsilon(U_X) = \frac{1}{2} \int_{\partial\Omega_\varepsilon} U_X \frac{\partial U_X}{\partial \nu} + \left(\frac{1}{2} - \frac{1}{p+1}\right) \int_{\partial\Omega_\varepsilon} |U_X|^{p+1}.$$

Employing (38) we also deduce

$$\frac{1}{2} \int_{\partial\Omega_\varepsilon} U_X \frac{\partial U_X}{\partial \nu} = \frac{1}{4}\varepsilon\sigma_{n-2}H(\varepsilon X) \int_0^\infty U(r)U'(r)r^{n-1} dr + o(\varepsilon),$$

where σ_{n-2} is the volume of S^{n-2}.

On the other hand, from (29) we find that

$$\int_{\Omega_\varepsilon} |U_X|^{p+1} = \int_{\mathbb{R}^n_+} |U|^{p+1} - \frac{1}{2}\varepsilon\sigma_{n-2} \int_{\mathbb{R}^{n-1}} U^{p+1}\langle x', A_X x'\rangle dx' + o(\varepsilon)$$

$$= \int_{\mathbb{R}^n_+} |U|^{p+1} - \frac{1}{2}\varepsilon\sigma_{n-2}H(\varepsilon X) \int_0^\infty U^{p+1}(r)r^n dr + o(\varepsilon).$$

In conclusion we obtain the expansion

$$I_\varepsilon(U_X) =$$

$$C_0 + \varepsilon\sigma_{n-2}H(\varepsilon X)\left[\frac{1}{4}\int_0^\infty U(r)U'(r)r^{n-1} dr - \frac{1}{2}\left(\frac{1}{2} - \frac{1}{p+1}\right)\int_0^\infty U^{p+1}(r)r^n dr\right].$$

Since $U'(r) < 0$, the coefficient of $\varepsilon H(\varepsilon X)$ is negative, and we obtain the conclusion. ∎

PROOF OF THEOREM 1 It is sufficient to apply Theorem 12, Proposition 18 and Lemma 20. ∎

3.3 Proof of Theorem 2

Our goal is to apply Theorem 14, and therefore we need to find a more accurate approximate solution. We first prove the following technical lemma.

Lemma 21 Let $T = (a_{ij})$ be an $(n-1) \times (n-1)$ symmetric matrix, and consider the following problem

$$\begin{cases} Lw = -2\langle Tx', \nabla_{x'}\partial_{x_n} U\rangle - trT \, \partial_{x_n} U & \text{in } \mathbb{R}^n_+; \\ \frac{\partial}{\partial x_n} w = \langle Tx', \nabla_{x'} U\rangle & \text{on } \partial\mathbb{R}^n_+, \end{cases} \tag{39}$$

where L is the operator

$$Lu = -\Delta u + u - pU^{p-1}u.$$

Then (39) *admits a solution* $\overline{w}_T$, *which is even in the variables* x' *and satisfies the following decay estimates*

$$|\overline{w}_T(x)| + |\nabla \overline{w}_T(x)| + |\nabla^2 \overline{w}_T(x)| \leq \overline{C}|T|_\infty (1 + |x|^{\overline{C}})e^{-|x|}, \qquad (40)$$

where $\overline{C}$ *is a constant depending only on* n *and* p, *and* $|T|_\infty := \max_{ij} |a_{ij}|$.

PROOF. Problem (39) can be reformulated as

$$\overline{I}''_+(U)[w] = v_T, \qquad (41)$$

where v_T is an element of $H^1(\mathbb{R}^n_+)$ defined by duality as

$$(v_T|v)_{H^1(\mathbb{R}^n_+)} = \int_{\mathbb{R}^n_+} \left(-\langle 2Tx', \nabla_{x'}\partial_{x_n}U - trT\partial_{x_n}U \rangle \right) v - \int_{\partial\mathbb{R}^n_+} \langle Tx', \nabla_{x'}U \rangle v.$$

By Proposition 17, equation (41) is solvable if and only if v_T is orthogonal to $\frac{\partial U}{\partial x_1}, \ldots, \frac{\partial U}{\partial x_{n-1}}$, which is the case since

$$\left(v_T, \frac{\partial U}{\partial x_i} \right)_{H^1(\mathbb{R}^n_+)} = -\int_{\mathbb{R}^n_+} (2\langle Tx', \nabla_{x'}\partial_{x_n}U \rangle + trT\partial_{x_n}U) \frac{\partial U}{\partial x_i}$$

$$+ \int_{\partial\mathbb{R}^n_+} \langle Tx', \nabla_{x'}U \rangle \frac{\partial U}{\partial x_i}, \qquad i = i, \ldots, n-1.$$

Indeed, all the integrals in the last formula vanish because $\frac{\partial U}{\partial x_i}$ is odd in x' and the other functions are even, by the symmetry of T. The decay in (40) follows from (10) and standard elliptic estimates. ∎

Given μ_0 as in (27), we introduce a new set of coordinates on $B_{\frac{\mu_0}{\varepsilon}}(X) \cap \Omega_\varepsilon$. Let

$$y' = x'; \qquad y_n = x_n - \psi_\varepsilon(x'). \qquad (42)$$

The advantage of these coordinates is that $\partial\Omega_\varepsilon$ identifies with $\{y_n = 0\}$, but the corresponding metric coefficients g_{ij} will not be constant anymore: indeed we have

$$(g_{ij}) = \left(\langle \frac{\partial x}{\partial y_i}, \frac{\partial x}{\partial y_i} \rangle \right) = \begin{pmatrix} \delta_{ij} + \frac{\partial\psi_\varepsilon}{\partial y_i}\frac{\partial\psi_\varepsilon}{\partial y_j} & \begin{matrix} \frac{\partial\psi_\varepsilon}{\partial y_1} \\ \vdots \\ \frac{\partial\psi_\varepsilon}{\partial y_{n-1}} \end{matrix} \\ \begin{matrix} \frac{\partial\psi_\varepsilon}{\partial y_1} & \cdots & \frac{\partial\psi_\varepsilon}{\partial y_{n-1}} \end{matrix} & 1 \end{pmatrix}.$$

From the estimates in (29) it follows that

$$g_{ij} = Id + \varepsilon A + \varepsilon^2 B + O(\varepsilon^3 |y'|^3), \tag{43}$$

and

$$\partial_{y_k}(g_{ij}) = \varepsilon \partial_{y_k} A + \varepsilon^2 \partial_{y_k} B + O(\varepsilon^3 |y'|^2),$$

where

$$A = \begin{pmatrix} 0 & A_X y' \\ (A_X y')^t & 0 \end{pmatrix}; \qquad B = \begin{pmatrix} A_X y' \otimes A_X y' & Q_X(y') \\ (Q_X(y'))^t & 0 \end{pmatrix}_3$$

It is also easy to check that the inverse matrix (g^{ij}) is of the form $g^{ij} = Id - \varepsilon A + \varepsilon^2 C + O(\varepsilon^3 |y'|^3)$, where

$$C = \begin{pmatrix} 0 & -Q_X(y') \\ -(Q_X(y'))^t & |A_X y'|^2 \end{pmatrix},$$

and

$$\partial_{y_k}(g^{ij}) = -\varepsilon \partial_{y_k} A + \varepsilon^2 \partial_{y_k} C + O(\varepsilon^3 |y'|^2).$$

Furthermore, since the transformation (42) preserves the volume, one has

$$\det g \equiv 1.$$

We also recall that the Laplace operator in a general system of coordinates is given by the expression

$$\Delta_g u = \frac{1}{\sqrt{\det g}} \partial_j \left(g^{ij} \sqrt{\det g} \right) \partial_i u + g^{ij} \partial_{ij}^2 u,$$

so in our situation we get

$$\Delta_g u = g^{ij} u_{ij} + \partial_i (g^{ij}) \partial_j u.$$

In particular, by (43), for any smooth function u we find

$$\begin{aligned}
\Delta_g u = \Delta u &- \varepsilon \left(2\langle A_X y', \nabla_{y'} \partial_{y_n} u \rangle + tr A_X \partial_{y_n} u \right) \\
&+ \varepsilon^2 \left(-2\langle Q_X, \nabla_{y'} \partial_{y_n} u \rangle + |A_X y'|^2 \partial_{y_n y_n}^2 u - div Q_X \partial_{y_n} u \right) \\
&+ O(\varepsilon^3 |y'|^2)|\nabla u| + O(\varepsilon^3 |y'|^3)|\nabla^2 u|.
\end{aligned} \tag{44}$$

Here A_X is the Hessian of ψ at $x' = 0$, see the above notation. Now we choose a cut-off function ψ_{μ_0} with the properties

$$\begin{cases} \psi_{\mu_0}(x) = 1 & \text{in } B_{\frac{\mu_0}{4}}; \\ \psi_{\mu_0}(x) = 0 & \text{in } \mathbb{R}_+^n \setminus B_{\frac{\mu_0}{2}}; \\ |\nabla \psi_{\mu_0}| + |\nabla^2 \psi_{\mu_0}| \le C & \text{in } B_{\frac{\mu_0}{2}} \setminus B_{\frac{\mu_0}{4}}, \end{cases}$$

---

[3] If the vector v has components $(v_i)_i$, the notation $v \otimes v$ denotes the square matrix with entries $(v_i v_j)_{ij}$.

and for any $X \in \partial\Omega$ we define the following function, in the coordinates (y', y_n)

$$z_{\varepsilon,X}(y) = \psi_{\mu_0}(\varepsilon y)(U(y) + \varepsilon \overline{w}_{A_X}(y)). \qquad (45)$$

where $\overline{w}_{A_X}$ is given by Lemma 21 with $T = A_X$. We also give the expression of the unit outer normal to $\partial\Omega_\varepsilon$, $\tilde{\nu}$, in the new coordinates y. Letting ν_i (resp. $\tilde{\nu}_i$) be the components of ν (resp. $\tilde{\nu}$), from $\nu = \sum_{i=1}^n \nu^i \frac{\partial}{\partial x^i} = \sum_{i=1}^n \tilde{\nu}^i \frac{\partial}{\partial y^i}$, we have $\tilde{\nu}_k = \sum_{i=1}^n \nu^i \frac{\partial y^k}{\partial x^i}$. This implies

$$\tilde{\nu}^k = \nu^k, \quad k = 1, \dots, n-1; \qquad \tilde{\nu}^n = \sum_{i=1}^{n-1} \nu^i \frac{\partial \psi_\varepsilon}{\partial y^i} + \nu^n.$$

From (29) and the subsequent formulas we find

$$\tilde{\nu} = \left(\varepsilon A_X(y') + \varepsilon^2 Q_X(y'), -1 + \tfrac{3}{2}\varepsilon^2 |A_X(y')|^2\right) + \varepsilon^3 O(|y'|^3). \qquad (46)$$

Finally the area-element of $\partial\Omega_\varepsilon$ can be expanded as

$$d\sigma = (1 + O(\varepsilon^2 |y'|^2))dy'. \qquad (47)$$

Next, we estimate the gradient of I_ε at $z_{\varepsilon,X}$ showing that the $z_{\varepsilon,X}$'s constitute, as X varies on $\partial\Omega_\varepsilon$, a manifold of pseudo-critical points of I_ε with a better accuracy than U_X.

Lemma 22 *There exists $C > 0$ such that for ε small*

$$\|I_\varepsilon'(z_{\varepsilon,X})\| \leq C\varepsilon^2; \qquad \text{for all } X \in \partial\Omega_\varepsilon.$$

PROOF. Let $v \in H^1(\Omega_\varepsilon)$. Since the function $z_{\varepsilon,X}$ is supported in $B_{\frac{\mu_0}{2\varepsilon}}(X)$, see (45), we can use the coordinates y in this set, and we obtain

$$I_\varepsilon'(z_{\varepsilon,X})[v] = \int_{\partial\Omega_\varepsilon} \frac{\partial z_{\varepsilon,X}}{\partial \tilde{\nu}} v d\sigma + \int_{\Omega_\varepsilon} \left(-\Delta_g z_{\varepsilon,X} + z_{\varepsilon,X} - z_{\varepsilon,X}^p\right) v dy. \qquad (48)$$

Let us now evaluate $\frac{\partial z_{\varepsilon,X}}{\partial \tilde{\nu}}$: one has

$$\frac{\partial z_{\varepsilon,X}}{\partial \tilde{\nu}} = (U + \varepsilon \overline{w}_{A_X})\langle \nabla \psi_{\mu_0}(\varepsilon y), \tilde{\nu}\rangle + \psi_{\mu_0}(\varepsilon y)\langle \nabla(U + \varepsilon \overline{w}_{A_X}), \tilde{\nu}\rangle.$$

Since $\nabla\psi_{\mu_0}(\varepsilon \cdot)$ is supported in $\mathbb{R}^n \setminus B_{\frac{\mu_0}{4\varepsilon}}$, and both $U, \overline{w}_{A_X}$ decay exponentially to zero, we have

$$|(U + \varepsilon \overline{w}_{A_X})\langle \nabla \psi_{\mu_0}(\varepsilon y), \tilde{\nu}\rangle| \leq C(1 + |y|^C)e^{-\frac{1}{C\varepsilon}}e^{-|y|}.$$

On the other hand, from the boundary condition in (39) and from (46), the terms of order ε in the scalar product $\psi_{\mu_0}(\varepsilon y)\langle \nabla(U + \varepsilon \overline{w}_{A_X}), \tilde{\nu}\rangle$ cancel and we obtain

$$\frac{\partial z_{\varepsilon,X}}{\partial \tilde{\nu}} = O(\varepsilon^2 |y'||\nabla w|) + O(\varepsilon^2 |y'|^2 |\nabla U|); \qquad |y| \le \frac{\mu_0}{4\varepsilon};$$

$$\left| \frac{\partial z_{\varepsilon,X}}{\partial \tilde{\nu}} \right| \le C e^{-|y|} + \overline{C} \varepsilon (1 + |y|^C) e^{-|y|} \le C \varepsilon^{-C} e^{-\frac{1}{C\varepsilon}}; \qquad \frac{\mu_0}{4\varepsilon} \le |y| \le \frac{\mu_0}{2\varepsilon}.$$

The last two estimates, (47), and the trace Sobolev inequalities readily imply

$$\left| \int_{\partial \Omega_\varepsilon} \frac{\partial z_{\varepsilon,X}}{\partial \tilde{\nu}} v d\sigma \right| \le C \varepsilon^2 \|v\|. \tag{49}$$

On the other hand, using (40), (44) and the decay of U, the volume integrand can be estimated as

$$\left| -\Delta_g z_{\varepsilon,X} + z_{\varepsilon,X} - z_{\varepsilon,X}^p \right| \le C \varepsilon^2 \left(|y'||\nabla U| + |y'|^2 |\nabla^2 U| + |\nabla w| + |y'||\nabla^2 w| \right)$$
$$+ \left| |U + \varepsilon w|^{p-1} (U + \varepsilon w) - U^p - p \varepsilon U^{p-1} w \right|,$$

for $|y| \le \left(\frac{1}{4\varepsilon \overline{C} \sup_X \|A_X\|} \right)^{\frac{1}{C}}$, and

$$\left| -\Delta_g z_{\varepsilon,X} + z_{\varepsilon,X} - z_{\varepsilon,X}^p \right| \le C(1 + |y'|^{\overline{C}}) e^{-|y'|}$$
$$\le C \varepsilon^{-C} e^{-\frac{1}{C\varepsilon}},$$

for $\left(\frac{1}{4\varepsilon \overline{C} \sup_X \|A_X\|} \right)^{\frac{1}{C}} \le |y| \le \frac{\mu_0}{2\varepsilon}$. We notice that the following inequality holds true

$$\left| (a + b)^p - a^p - p a^{p-1} b \right| \le C b^2; \qquad a > 0, |b| \le \frac{a}{2}.$$

In particular, by (40) we have

$$\varepsilon |\overline{w}(y)| \le \frac{U(y)}{2}; \qquad \text{for } |y| \le \left(\frac{1}{4\varepsilon \overline{C} \sup_X \|A_X\|} \right)^{\frac{1}{C}},$$

hence it follows that

$$\left| -\Delta_g z_{\varepsilon,X} + z_{\varepsilon,X} - z_{\varepsilon,X}^p \right| \le C \varepsilon^2 (1 + |y|^C) e^{-|y|};$$

$$|y| \le \left(\frac{1}{4\varepsilon \overline{C} \sup_X \|A_X\|} \right)^{\frac{1}{C}}.$$

Then, using the Hölder inequality we easily find

$$\left| \int_{\Omega_\varepsilon} \left(-\Delta_g z_{\varepsilon,X} + z_{\varepsilon,X} - z_{\varepsilon,X}^p \right) v dy \right| \le C \varepsilon^2 \|v\|. \tag{50}$$

From (49) and (50) we obtain the conclusion. ∎

We also need to compute $\frac{\partial z_{\varepsilon,X}}{\partial X}$ in the coordinates y introduced in (42). We notice that in the definition of $z_{\varepsilon,X}$, see (45), not only the analytic expression of this function depends on X, but also the choice of the coordinates y. Therefore, when we differentiate in X, we have to take also this dependence into account. First we derive the variation in X of the coordinates x (introduced before (27)) of a given point in Ω. Using the dot to denote the differentiation with respect to X, one can prove that

$$\dot{x}' = \frac{\partial}{\partial X} x'_X = -\dot{X}; \qquad \dot{x}_n = \frac{\partial}{\partial X}(x_n)_X = -\langle x', \mathbf{H}_\varepsilon \dot{X}\rangle, \qquad (51)$$

where $\mathbf{H}_\varepsilon = \varepsilon A_X$ is the second fundamental form of Ω_ε. The second equation in (51) is obtained by computing the variation of the distance of a fixed point in $\mathbb{R}^n$ from a moving tangent plane to Ω_ε. Similarly, we get a dependence on X of the coordinates y. To emphasize the dependence of $z_{\varepsilon,X}$ on X we write

$$z_{\varepsilon,X} = U(y_X) + \varepsilon \overline{w}_{A_X}(y_X); \qquad y_X = (x'_X, (x_n)_X - \psi_\varepsilon(x'_X)). \qquad (52)$$

Since the set Ω_ε is a dilation of Ω, the derivatives of A_X and ψ_ε with respect to X are of order ε (if $\dot{X}$ is of order 1). More precisely, if we set $\tilde{X} = \varepsilon X$, then we have

$$\frac{\partial A_X}{\partial X} = \varepsilon \frac{\partial A_{\tilde{X}}}{\partial \tilde{X}}; \qquad \frac{\partial \psi_\varepsilon}{\partial X} = \varepsilon \frac{\partial \psi}{\partial \tilde{X}},$$

where ψ is given in (27). Differentiating (52) with respect to X and using (51) it follows that, in the coordinates y

$$\dot{z}_{\varepsilon,X} = -\langle \dot{X}, \nabla_{y'} U\rangle + O(\varepsilon) \quad \text{in } H^1(\mathbb{R}^n_+). \qquad (53)$$

In this spirit, we also compute the variation of the matrix A_X, see (27), with respect to X. Differentiating the equation $x_n = \psi_\varepsilon(x')$ with respect to X and using (51) we find

$$-\langle x', \mathbf{H}_\varepsilon \dot{X}\rangle = \frac{1}{2}\varepsilon^2 \langle \frac{\partial A_{\tilde{X}}}{\partial \tilde{X}} x', x'\rangle - \varepsilon \langle A_X x', \dot{X}\rangle - \varepsilon^2 \sum_{i=1}^{n-1} Q_i \dot{X}_i.$$

If $e_1, \ldots, e_{n-1}$ is an orthonormal system of tangent vectors to $\partial \Omega$ with $e_i = \frac{\partial \tilde{X}}{\partial x_i}$, the last equation implies

$$\langle \frac{\partial A_{\tilde{X}}}{\partial e_i} x', x'\rangle = 2Q_X^i(x'), \qquad \text{namely} \qquad (Q_X^i)_{jk} = \left(\frac{\partial A_{\tilde{X}}}{\partial e_i}\right)_{jk}. \qquad (54)$$

By the symmetries in (30), we have in particular

$$\left(\frac{\partial A_{\tilde{X}}}{\partial e_j}\right)_{ij} = \left(\frac{\partial A_{\tilde{X}}}{\partial e_i}\right)_{jj} \qquad \text{for every } i, j. \qquad (55)$$

The proof of Proposition 18 can be easily modified if one takes $Z_\varepsilon = \{z_{\varepsilon,X} : X \in \partial\Omega_\varepsilon\}$: our next goal is to expand $I_\varepsilon(z_{\varepsilon,X})$ up to order ε^2 so that, taking Lemma 22 into account, we can apply Theorem 14.

Lemma 23 *For ε small the following expansion holds*

$$I_\varepsilon(z_{\varepsilon,X}) = C_0 - \tilde{C}_1 \varepsilon H(\varepsilon X) + O(\varepsilon^2),$$

where

$$C_0 = \left(\frac{1}{2} - \frac{1}{p+1}\right) \int_{\mathbb{R}^n_+} U^{p+1}, \quad \tilde{C}_1 = \left(\int_0^\infty r^n U_r^2 dr\right) \int_{S^n_+} y_n |y'|^2 d\sigma.$$

PROOF. To be short, we will often write z instead of $z_{\varepsilon,X}$ and w instead of $w(\varepsilon, X)$. Since z is supported in $B_{\frac{\mu_0}{2\varepsilon}}(X)$, we can use the coordinates y and integrate by parts to get

$$I_\varepsilon(z) = \frac{1}{2}\int_{\partial\mathbb{R}^n_+} z\frac{\partial z}{\partial \tilde\nu} + \frac{1}{2}\int_{\mathbb{R}^n_+} z\left(-\Delta_g z + z\right) - \frac{1}{p+1}\int_{\mathbb{R}^n_+} |z|^{p+1}.$$

Using the definition of z given in (45), as well as the expression of the Δ_g in (44) we find

$$\frac{1}{2}\int_{\mathbb{R}^n_+} z\left(-\Delta_g z + z\right) - \frac{1}{p+1}\int_{\mathbb{R}^n_+} |z|^{p+1} = \left(\frac{1}{2} - \frac{1}{p+1}\right)\int_{\mathbb{R}^n_+} U^{p+1}$$

$$+\varepsilon \int_{\mathbb{R}^n_+} U\langle A_X y', \nabla_{y'} \partial_{y_n} U\rangle + \frac{\varepsilon}{2} tr A_X \int_{\mathbb{R}^n_+} U\partial_{y_n} U + O(\varepsilon^2).$$

Moreover, from (46) we get

$$\frac{1}{2}\int_{\partial\mathbb{R}^n_+} z\frac{\partial z}{\partial \tilde\nu} = \frac{\varepsilon}{2}\int_{\partial\mathbb{R}^n_+} U\langle A_X y', \nabla_{y'} U\rangle + O(\varepsilon^2).$$

Putting together the preceding formulas we have

$$I_\varepsilon(z) = \left(\frac{1}{2} - \frac{1}{p+1}\right)\int_{\mathbb{R}^n_+} U^{p+1} + \frac{\varepsilon}{2}\int_{\partial\mathbb{R}^n_+} U\langle A_X y', \nabla_{y'} U\rangle$$

$$+\varepsilon \int_{\mathbb{R}^n_+} U\langle A_X y', \nabla_{y'}\partial_{y_n} U\rangle + \frac{\varepsilon}{2} tr A_X \int_{\mathbb{R}^n_+} U\partial_{y_n} U + O(\varepsilon^2).$$

Integrating by parts (more than once if needed), we find that the three terms of order ε are given by

$$\frac{1}{4}\int_{\partial\mathbb{R}^n_+}\langle A_X y', \nabla_{y'} U^2\rangle + \int_{\mathbb{R}^n_+} U\langle A_X y', \nabla_{y'}\partial_{y_n} U\rangle + \frac{1}{4}tr A_X\int_{\mathbb{R}^n_+}\partial_{y_n} U^2$$

$$= -\frac{1}{2}tr A_X\int_{\partial\mathbb{R}^n_+} U^2 - \int_{\partial\mathbb{R}^n_+} U\langle A_X y', \nabla_{y'} U\rangle - \int_{\mathbb{R}^n_+}\partial_{y_n} U\langle A_X y', \nabla_{y'} U\rangle$$

$$= -\int_{\mathbb{R}^n_+}\partial_{y_n} U\langle A_X y', \nabla_{y'} U\rangle.$$

Now we notice that, since U is radial, one has

$$\partial_{y_n} U = \frac{y_n}{|y|} U_r; \qquad \nabla_{y'} U = \frac{y'}{|y|} U_r,$$

and hence

$$\int_{\mathbb{R}^n_+}\partial_{y_n} U\langle A_X(y'), \nabla_{y'} U\rangle = -\int_{\mathbb{R}^n_+}\frac{y_n\langle A_X(y'), y'\rangle}{|y|^2} U_r^2\,dy.$$

Now it is sufficient to express the last integral in radial coordinates. This concludes the proof. ∎

Lemma 24 *For ε small the following expansion holds*

$$\frac{\partial}{\partial X} I_\varepsilon(z_{\varepsilon,X}) = -\tilde{C}_1\varepsilon^2 H'(\varepsilon X) + o(\varepsilon^2),$$

where $\tilde{C}_1$ is the constant given in the previous Lemma.

PROOF. We have

$$I_\varepsilon'(z)[\partial_X z] = \int_{\mathbb{R}^n_+}\left(-\Delta_g z + z - |z|^p\right)\partial_X z + \int_{\partial\mathbb{R}^n_+}\partial_X z\frac{\partial}{\partial\nu} z\,d\sigma.$$

Notice that, by our construction, the terms $-\Delta_g z + z - |z|^p$ and $\frac{\partial}{\partial\nu} z$ are of order ε^2, hence it is sufficient to take the product only with the 0-th order term of $\partial_X z$, see (53). In this way we obtain that $I_\varepsilon'(z)[\partial_X z] = (\alpha_1 + \alpha_2)\varepsilon^2 + o(\varepsilon^2)$, where

$$\alpha_1 = \int_{\mathbb{R}^n_+}\Big[2\langle Q, \nabla_{y'}\partial_{y_n} U\rangle - |A_p y'|^2\partial^2_{y_n y_n} U + divQ\partial_{y_n} U + 2\langle A_X y', \nabla_{y'}\partial_{y_n}\overline{w}\rangle$$

$$+ tr A_X\partial_{y_n}\overline{w} - \frac{1}{2}p(p-1)U^{p-2}\overline{w}^2\Big]\partial_X U;$$

$$\alpha_2 = \int_{\partial\mathbb{R}^n_+}\langle Q, \nabla_{y'} U\rangle\partial_X U + \int_{\partial\mathbb{R}^n_+}\langle A_X y', \nabla_{y'}\overline{w}\rangle\partial_X U.$$

Since the function $\overline{w}$ is even in y', all the terms containing it vanish identically and so does the term $|A_p y'|^2\partial^2_{y_n y_n} U\partial_X U$, hence we get

$$\alpha_1 = \int_{\mathbb{R}^n_+}\left[2\langle Q, \nabla_{y'}\partial_{y_n} U\rangle + divQ\partial_{y_n} U\right]\partial_X U.$$

On the other hand, the boundary integral α_2 is given by

$$\alpha_2 = \int_{\partial \mathbb{R}^n_+} \langle Q, \nabla_{y'} U \rangle \partial_X U,$$

again by the oddness of $\overline{w}$. In conclusion we have

$$\alpha_1 + \alpha_2 = \int_{\mathbb{R}^n_+} [2\langle Q, \nabla_{y'} \partial_{y_n} U \rangle + div Q \partial_{y_n} U] \partial_X U + \int_{\partial \mathbb{R}^n_+} \langle Q, \nabla_{y'} U \rangle \partial_X U,$$

which we rewrite as

$$2 \sum_j \int_{\mathbb{R}^n_+} Q_j(x') \partial_j \partial_{y_n} U \partial_i U + \sum_j \int_{\mathbb{R}^n_+} \partial_j Q_j(x') \partial_{y_n} U \partial_i U + \sum_j \int_{\partial \mathbb{R}^n_+} Q_j(x') \partial_j U \partial_i U.$$

If we integrate by parts in the variable y_j we find

$$\alpha_1 + \alpha_2 = \sum_j \int_{\mathbb{R}^n_+} Q_j(x') \partial_j \partial_{y_n} U \partial_i U - \sum_j \int_{\mathbb{R}^n_+} Q_j(x') \partial_{y_n} U \partial_j \partial_i U$$
$$+ \sum_j \int_{\partial \mathbb{R}^n_+} Q_j(x') \partial_j U \partial_i U.$$

Then, if we integrate by parts in the variable y_n and in y_i we obtain

$$\sum_j \int_{\mathbb{R}^n_+} \partial_{y_i} \left(Q^j_X(y') \right) \partial_{y_j} U \partial_{y_n} U = \sum_j \left(\frac{\partial A_X}{\partial e_j} \right)_{ij} \int_{\mathbb{R}^n_+} y_j \partial_{y_j} U \partial_{y_n} U.$$

By the symmetry in (55) and using radial variables we finally get

$$\alpha_1 + \alpha_2 = \sum_j \left(\frac{\partial A_X}{\partial e_i} \right)_{jj} \int_{\mathbb{R}^n_+} y_j \partial_{y_j} U \partial_{y_n} U = \frac{\partial H}{\partial e_i} C_1,$$

which concludes the proof (recall that $\partial_i U = -\partial_{X_i} z$). ∎

PROOF OF THEOREM 2. We define the new manifold of approximate solutions

$$Z_\varepsilon = \{ z_{\varepsilon, X} : X \in \partial \Omega_\varepsilon \}.$$

As we already noticed, since the difference between $z_{\varepsilon, X}$ and U_X is of order $O(\varepsilon)$, one has the counterpart of Proposition 18 for this new choice of Z_ε.

Then, to prove the theorem is sufficient to apply Lemmas 22, 23, 24 and Theorem 14. In fact, using a Taylor expansion for H, one can find a small positive number δ_0 such that

$$H' \neq 0 \text{ on } \partial B_{\delta_0}(X_0) \qquad \text{and} \qquad \deg(H', B_{\delta_0}(X_0), 0) = (-1)^{\text{sgn } \det H''(X_0)}. \tag{56}$$

This implies the conclusion. ∎

Part II
Higher-Dimensional Concentration

The solutions in Theorem 3 are boundary-layers and scale qualitatively in the following way, as ε tends to zero

$$u_\varepsilon(x', x_n) \sim w_0(x_n/\varepsilon); \qquad x' \in \mathbb{R}^{n-1}, x_n \in \mathbb{R}_+,$$

where now w_0 is the solution of the problem (4).

In the case of spike-layers, both the energy and the Morse index stay bounded as ε goes to zero. Viceversa, by the results in [20], if the Morse index of a family of solutions stays bounded as $\varepsilon \to 0$, these solutions must concentrate at a finite number of points (at least in low dimensions).

The proof of Theorem 3 relies on a local inversion argument, via a contraction mapping. The main difficulty is that, since the Morse index of the solutions is changing with ε, the linearized operator $I_\varepsilon''(u_\varepsilon)$ will not be invertible for all the values of ε in any interval of the form $(0, \varepsilon_0)$. We get indeed invertibility at least along a sequence ε_j, with the norm of the inverse operator blowing-up at the rate ε_j^{n-1}, see Proposition 36. This produces a resonance phenomenon which can be described as follows.

Let $\gamma \in (0, 1)$, and let ν denote the inner unit normal to $\partial\Omega_\varepsilon$. Consider the neighborhood Σ_ε of $\partial\Omega_\varepsilon$ defined as

$$\Sigma_\varepsilon = \left\{ x' + x_n\nu \ : \ x' \in \partial\Omega_\varepsilon, x_n \in (0, \varepsilon^{-\gamma}) \right\}.$$

Let $(\varphi_l)_l$ denote the eigenfunctions of $-\Delta_{\hat{g}}$, where $\Delta_{\hat{g}}$ is the Laplace-Beltrami operator on $\partial\Omega$, and let $(\lambda_l)_l$ be the corresponding (non-negative) eigenvalues.

Let also z_σ be an eigenfunction of $I_\varepsilon''(u_\varepsilon)$ with eigenvalue σ. Then we can decompose z_σ in Fourier series (in the variable x') in the following way

$$z_\sigma(x', x_n) = \sum_{l=0}^\infty \varphi_l(\varepsilon x') z_{\sigma,l}(x_n); \qquad x' \in \partial\Omega_\varepsilon, x_n \in (0, \varepsilon^{-\gamma}).$$

If the eigenvalue σ is close to zero, it turns out that all the modes $\varphi_l z_{\sigma,l}$ are negligible, except those for which $\lambda_l \sim \varepsilon^{-2}$, see also formula (57). This is stated rigorously in Proposition 33. It follows that, qualitatively, the resonant eigenvalues of $I_\varepsilon''(u_\varepsilon)$ have eigenfunctions with more and more oscillations along $\partial\Omega_\varepsilon$ as ε tends to zero.

We describe below the general procedure employed here to tackle the problem. As we mentioned above, we are going to use the contraction mapping theorem, once two preliminary steps are accomplished.

Step 1: finding an approximate solution. Given an arbitrary positive number θ, we are able to find an approximate solution $u_{k,\varepsilon}$ of (P_ε) for which $\|I_\varepsilon'(u_{k,\varepsilon})\| = O(\varepsilon^\theta)$. The function $u_{k,\varepsilon}$ is constructed essentially by power series in ε with k iterations, where the number k depends on n, p and θ. Thus $u_{k,\varepsilon}$ has roughly the following form

$$u_{k,\varepsilon}(x', x_n) = w_0(x_n) + \varepsilon\tilde{w}_1(\varepsilon x', x_n) + \ldots \varepsilon^k\tilde{w}_k(\varepsilon x', x_n),$$

$$x' \in \partial\Omega_\varepsilon, \qquad x_n \in J_\varepsilon = [0, \varepsilon^{-\gamma}].$$

Here $\tilde{w}_1, \ldots, \tilde{w}_k$ are smooth functions on $\partial\Omega \times J_\varepsilon$, which are defined inductively in their index. Basically each function $\tilde{w}_i$, which depends on $\tilde{w}_0, \ldots,$ $\tilde{w}_{i-1}$ and the geometry of Ω, is obtained by an inversion argument. Despite the resonance phenomena of the operator I''_ε, here we can perform this inversion because we are assuming a smooth dependence on the variable $\varepsilon x'$. In some sense, since resonance occurs only at high frequencies, see the comments above, smooth functions on $\partial\Omega$ (scaled to Ω_ε) are not affected by this phenomenon since their Fourier modes are mainly low-frequency ones. The rigorous derivation of $u_{k,\varepsilon}$ is performed in Subsection 5: in this step the smoothness of Ω is essential to construct $u_{k,\varepsilon}$.

In [49], for $n = 2$, we were able to satisfy the above requirement only for $\theta = \frac{3}{2}$. To deal with a general n we need a better approximation since both the energy of the solutions and the norm of the inverse operator grow faster when the dimension n is larger.

Step 2: inverting the linearized operator $I''_\varepsilon(u_{k,\varepsilon})$. This is a rather delicate issue since, as we remarked before, the linearized operator is not invertible for all the values of ε. Qualitatively, using the Weyl's asymptotic formula, one finds that the l-th eigenvalue σ_l of $I''_\varepsilon(u_{k,\varepsilon})$ is given by

$$\sigma_l \sim -1 + \varepsilon^2 l^{\frac{2}{n-1}}, \tag{57}$$

see Subsection 6 for details. It follows that $\sigma_l \sim 0$ for $l \sim \varepsilon^{1-n}$, and that the average distance between two consecutive eigenvalues close to zero is of order ε^{n-1}. We show indeed that along a sequence ε_j the spectrum of the linearized operator stays away from zero of order ε_j^{n-1}, and hence we get an inverse operator with norm proportional to $\varepsilon_j^{-(n-1)}$. The way to prove this fact relies on a first rough comparison of the eigenvalues with those (essentially known, see Proposition 29) of a model problem, obtaining an estimate of the Morse index of the solutions. Then, Kato's Theorem allows us to choose the values of ε appropriately and to invert the linear operator along a sequence ε_j. Notice that Kato's Theorem requires some information not only on the eigenvalues, but also on the eigenfunctions, see Subsection 4.

Final step: the contraction argument. If the operator $I''_\varepsilon(u_{k,\varepsilon})$ is invertible, a function u_ε of the form $u_\varepsilon = u_{k,\varepsilon} + w$ is a solution of (P_ε) if and only if w is a fixed point of the operator F_ε, where F_ε is defined as

$$F_\varepsilon(w) = -I''_\varepsilon(u_{k,\varepsilon})^{-1}\left[I'_\varepsilon(u_{k,\varepsilon}) + N_\varepsilon(w)\right].$$

Here $N_\varepsilon(w)$ is a superlinear term satisfying $\|N_\varepsilon(w)\| \leq O\left(\|w\|\right)^{\min\{2,p\}}$. Since along the sequence ε_j the norm of the operator $I_\varepsilon''(u_{k,\varepsilon})^{-1}$ is of order ε^{-n}, we need to choose θ to be sufficiently large (see Step 1), depending also on p, in order to get a contraction.

A further difficulty in dealing with the case $n \geq 3$ is that the exponent p could also be supercritical. This case is tackled with a truncation argument, proving a priori estimates in L^∞. These are based on a combination of norm estimates, obtained using Step 2, and elliptic regularity theory.

The outline of the second part of these notes is the following. In Section 4 we collect some preliminary facts and we study a family of auxiliary one-dimensional problems, proving some continuity and monotonicity properties. Section 5 is devoted to the construction of the approximate solution $u_{k,\varepsilon}$. In Section 6 we give a characterization of the eigenfunctions and the eigenvalues of the operator T_{Σ_ε}, which basically coincide with those of $I_\varepsilon''(u_{k,\varepsilon})$, see Proposition 37. In Section 7 we finally prove Theorem 3.

4 Some Preliminary Facts

In this section we treat the linear theory of some auxiliary one-dimensional problems. Below, C denotes a large positive constant. For convenience, we allow C to vary among formulas (also within the same line) and to assume larger and larger values. The number γ will be fixed in Subsection 5.

Consider the problem

$$\begin{cases} -u'' + u = u^p & \text{in } \mathbb{R}_+; \\ u > 0 & \text{in } \mathbb{R}_+; \\ u'(0) = 0, \end{cases} \tag{58}$$

with $p > 1$. This is a particular case of (1) for $n = 1$. Therefore, calling the solution w_0, this satisfies the properties

$$\begin{cases} w_0'(r) < 0, & \text{for all } r > 0, \\ \lim_{r \to \infty} e^r w_0(r) = \alpha_p > 0, & \lim_{r \to \infty} \frac{w_0'(r)}{w_0(r)} = -1, \end{cases} \tag{59}$$

where α_p is a positive constant depending only on p. Using some ODE analysis, one can see that all the solutions of (58) in $H^1(\mathbb{R}_+)$ coincide with w_0.

The Euler functional of (58) is

$$\overline{I}_+(u) = \frac{1}{2} \int_{\mathbb{R}_+} \left((u')^2 + u^2\right) - \frac{1}{p+1} \int_{\mathbb{R}_+} |u|^{p+1}, \qquad u \in H^1(\mathbb{R}_+). \tag{60}$$

Since in one dimension we have no translation invariance once we impose Neumann conditions, Corollary 17 becomes the following.

Proposition 25 *The function w_0 is a non-degenerate critical point of $\overline{I}_+$. Precisely, there exists a positive constant C such that*

$$\overline{I}_+''(w_0)[w_0, w_0] \leq -C^{-1}\|w_0\|_0^2; \qquad \overline{I}_+''(w_0)[v, v] \geq C^{-1}\|v\|_0^2,$$

$$\text{for all } v \in H^1(\mathbb{R}_+), \quad v \perp w_0,$$

where $\| \cdot \|_0$ denotes the standard norm of $H^1(\mathbb{R}_+)$. As a consequence, we have $\mu < 0$ and $\tau > 0$, where μ and τ are respectively the first and the second eigenvalues of $\overline{I}_+''(w_0)$; furthermore μ is simple.

Our next goal is to characterize the eigenvalues σ (in particular the first two) of the following problem

$$\begin{cases} (1 - \sigma)(-u'' + (1 + \alpha)u) = pw_0^{p-1}u, & \text{in } \mathbb{R}_+; \\ u'(0) = 0, \end{cases} \tag{61}$$

where α is a positive parameter. Equation (61) arises in the study of a model for the linearization of (P_ε) near approximate solutions after performing a Fourier decomposition, see equation (81) below.

In order to study the eigenvalues of (61), it is convenient to introduce the following norm $\| \cdot \|_\alpha$ on $H^1(\mathbb{R}_+)$

$$\|u\|_\alpha^2 = \int_{\mathbb{R}_+} \left((u')^2 + (1 + \alpha)u^2\right); \qquad u \in H^1(\mathbb{R}_+). \tag{62}$$

Let H_α be the Hilbert space consisting of the functions in $H^1(\mathbb{R}_+)$, endowed with this norm: we denote by $(,)_\alpha$ the corresponding scalar product. We also define by duality the operator $T_\alpha : H_\alpha \to H_\alpha$ in the following way

$$(T_\alpha u, v)_\alpha = \int_{\mathbb{R}_+} \left(u'v' + (1 + \alpha)uv - pw_0^{p-1}uv\right); \qquad u, v \in H_\alpha.$$

Then equation (61) can be written in the abstract form

$$T_\alpha u = \sigma u; \qquad u \in H_\alpha : \tag{63}$$

note that when $\alpha = 0$, T_α coincides with $\overline{I}_+''(w_0)$. We have the following result.

Proposition 26 *Let μ_α and τ_α denote respectively the first and the second eigenvalues of T_α. Then μ_α is simple and the following properties hold true*

(i) $\alpha \mapsto \mu_\alpha$ is smooth and strictly increasing;
(ii) $\alpha \mapsto \tau_\alpha$ is non-decreasing.

The eigenfunction v_α of T_α corresponding to μ_α, normalized with $\|v_\alpha\|_\alpha = 1$, can be chosen to be positive and strictly decreasing on $\mathbb{R}_+$. The map $\alpha \mapsto v_\alpha$ is smooth from $\mathbb{R}$ into $H^1(\mathbb{R}_+)$. Furthermore, $\mu_\alpha > 0$ for α sufficiently large.

PROOF. The simplicity on μ_α and the (weak) monotonicity of $\alpha \mapsto \mu_\alpha$, $\alpha \mapsto \tau_\alpha$ can be proved as follows. By the Courant-Fischer formula we have that

$$\mu_\alpha = \inf_{u \in H_\alpha} \frac{\int_{\mathbb{R}_+} \left((u')^2 + (1+\alpha)u^2 - pw_0^{p-1}u^2 \right)}{\int_{\mathbb{R}_+} ((u')^2 + (1+\alpha)u^2)}. \tag{64}$$

If u is a minimizer, also the decreasing rearrangement of $|u|$ is, and therefore eigenfunctions corresponding to the first eigenvalue have constant sign. By orthogonality they are then unique. The (weak) monotonicity in α is a simple consequence of the monotonicity of the quantity

$$\frac{\int_{\mathbb{R}_+} \left((u')^2 + (1+\alpha)u^2 - pw_0^{p-1}u^2 \right)}{\int_{\mathbb{R}_+} ((u')^2 + (1+\alpha)u^2)}.$$

The smoothness of $\alpha \mapsto \mu_\alpha$ and $\alpha \mapsto v_\alpha$ can be deduced as follows: consider the map $\Psi_\alpha : H^1(\mathbb{R}_+) \to H_\alpha$ defined as

$$(\Psi_\alpha u)(x) = u(x); \qquad u \in H^1(\mathbb{R}_+).$$

Ψ_α is nothing but the identity as a map between functions, and is an isomorphism between Sobolev spaces. We also consider the operator $\mathcal{T}_\alpha : H^1(\mathbb{R}_+) \to H^1(\mathbb{R}_+)$

$$\mathcal{T}_\alpha = \Psi_\alpha^{-1} \circ T_\alpha \circ \Psi_\alpha.$$

$\mathcal{T}_\alpha$ depends smoothly on α and, by conjugacy to T_α, the first eigenvalue of $\mathcal{T}_\alpha$ coincides with μ_α and is simple. Hence, to get the smoothness of μ_α and v_α, it is sufficient to apply the implicit function theorem.

We now compute the derivative of μ_α with respect to α. The function v_α satisfies the equation

$$\begin{cases} (1 - \mu_\alpha)\left(-v_\alpha'' + (1+\alpha)v_\alpha\right) = pw_0^{p-1}v_\alpha & \text{in } \mathbb{R}_+; \\ v_\alpha'(0) = 0. \end{cases} \tag{65}$$

Differentiating the equation $\|v_\alpha\|_\alpha^2 = 1$, with respect to α, we find

$$\frac{d}{d\alpha}\|v_\alpha\|_\alpha^2 = 0, \qquad \text{which implies} \qquad \left(\frac{dv_\alpha}{d\alpha}, v_\alpha\right)_\alpha = -\int_{\mathbb{R}_+} v_\alpha^2. \tag{66}$$

On the other hand, differentiating (65) with respect to α we obtain

$$
\begin{cases}
-\frac{d\mu_\alpha}{d\alpha}\left(-v_\alpha'' + (1+\alpha)v_\alpha\right) + (1-\mu_\alpha)\left(-\left(\frac{dv_\alpha}{d\alpha}\right)'' + (1+\alpha)\frac{dv_\alpha}{d\alpha} + v_\alpha\right) = pw_0^{p-1}\frac{dv_\alpha}{d\alpha} & \text{in } \mathbb{R}_+; \\
\left(\frac{dv_\alpha}{d\alpha}\right)'(0) = 0.
\end{cases}
$$

(67)

Multiplying (67) by v_α, integrating by parts and using (66), one gets

$$
\frac{d\mu_\alpha}{d\alpha} = (1-\mu_\alpha)\int_{\mathbb{R}_+} v_\alpha^2 > 0.
\tag{68}
$$

Note that, since $T_\alpha \leq Id$, every eigenvalue of T_α is strictly less than 1, and in particular $(1-\mu_\alpha) > 0$. This proves the strict monotonicity of $\alpha \mapsto \mu_\alpha$. It remains to prove that $\mu_\alpha > 0$ for α large: in order to do this we show that $\mu_\alpha \to 1$ as $\alpha \to +\infty$. Fixing any $\delta > 0$, it is sufficient to notice that

$$
(u')^2 + \left((1+\alpha) - pw_0^{p-1}\right)u^2 \geq (1-\delta)\left[(u')^2 + (1+\alpha)u^2\right] \qquad \text{for all } u,
$$

provided α is sufficiently large, and to use (64). This concludes the proof. ∎

Given $\alpha > 0$, we also define the function $\tilde{F}(\alpha)$ as

$$
\tilde{F}(\alpha) = 2\alpha(1-\mu_\alpha)\int_{\mathbb{R}_+} v_\alpha^2 > 0.
\tag{69}
$$

Note that, by the smoothness of $\alpha \mapsto v_\alpha$, also the function $\alpha \mapsto \tilde{F}(\alpha)$ is smooth.

We also need to consider a variant of the eigenvalue problem (61). For $\gamma \in (0,1)$, set

$$
J_\varepsilon = \left[0, \varepsilon^{-\gamma}\right]; \qquad H_\varepsilon^1 = \left\{u \in H^1(J_\varepsilon) \,:\, u\left(\varepsilon^{-\gamma}\right) = 0\right\}.
$$

We let $H_{\alpha,\varepsilon}$ denote the space H_ε^1 endowed with the norm

$$
\|u\|_{\alpha,\varepsilon}^2 = \int_{J_\varepsilon} \left((u')^2 + (1+\alpha)u^2\right); \qquad u \in H_\varepsilon^1,
$$

and $(\,,\,)_{\alpha,\varepsilon}$ the corresponding scalar product. Similarly, we define $T_{\alpha,\varepsilon}$ by

$$
(T_{\alpha,\varepsilon}u, v)_{\alpha,\varepsilon} = \int_{J_\varepsilon} \left(u'v' + (1+\alpha)uv - pw_0^{p-1}uv\right); \qquad u, v \in H_{\alpha,\varepsilon}.
$$

The operator $T_{\alpha,\varepsilon}$ satisfies properties analogous to T_α. We list them in the next proposition, which also gives a comparison between the first eigenvalues and eigenfunctions of T_α and $T_{\alpha,\varepsilon}$.

Proposition 27 *Let $\mu_{\alpha,\varepsilon}$ and $\tau_{\alpha,\varepsilon}$ denote respectively the first and the second eigenvalues of $T_{\alpha,\varepsilon}$. Then $\mu_{\alpha,\varepsilon}$ is simple and the following properties hold true*

(i) $\alpha \mapsto \mu_{\alpha,\varepsilon}$ is smooth and strictly increasing;
(ii) $\alpha \mapsto \tau_{\alpha,\varepsilon}$ is non-decreasing.

The eigenfunction $v_{\alpha,\varepsilon}$ of $T_{\alpha,\varepsilon}$ corresponding to $\mu_{\alpha,\varepsilon}$, normalized with $\|v_{\alpha,\varepsilon}\|_{\alpha,\varepsilon} = 1$, can be chosen to be positive and strictly decreasing on $\mathbb{R}_+$. Moreover there exist a large constant C and a small constant δ depending only on p such that

$$|\mu_\alpha - \mu_{\alpha,\varepsilon}| \le Ce^{-\delta\varepsilon^{-\gamma}}; \quad \|v_\alpha - v_{\alpha,\varepsilon}\|_{H^1(\mathbb{R}_+)} \le C\varepsilon^{-\frac{1}{2}\gamma}e^{-\delta\varepsilon^{-\gamma}}, \quad \text{for } \mu_\alpha \in \left[-\frac{1}{4}\tau, \frac{1}{4}\tau\right]$$

and for ε small. The function $v_{\alpha,\varepsilon}$ in this formula has been set identically 0 outside $[0, \varepsilon^{-\gamma}]$.

PROOF. Properties **(i)**–**(ii)** and of the monotonicity of $v_{\alpha,\varepsilon}$ can be deduced as for Proposition 26.
    Also, we have

$$\mu_\alpha = \inf_{u \in H_\alpha} \frac{(T_\alpha u, u)_\alpha}{\|u\|_\alpha^2}; \qquad\qquad \mu_{\alpha,\varepsilon} = \inf_{u \in H_{\alpha,\varepsilon}} \frac{(T_{\alpha,\varepsilon}u, u)_{\alpha,\varepsilon}}{\|u\|_{\alpha,\varepsilon}^2}. \qquad (70)$$

Since the norms $\|\cdot\|_\alpha$ and $\|\cdot\|_{\alpha,\varepsilon}$ (and the operators T_α, $T_{\alpha,\varepsilon}$) coincide on $H_{\alpha,\varepsilon}$, we have $\mu_\alpha \le \mu_{\alpha,\varepsilon}$. Conversely, since $\mu_\alpha \to 1$ as $\alpha \to +\infty$ (see the proof of Proposition 26), $\mu_\alpha \in \left[-\frac{1}{4}\tau, \frac{1}{4}\tau\right]$ implies that α lies in a bounded set of $\mathbb{R}_+$. Note that, since $\bar{I}''(w_0) \le Id$ on $H^1(\mathbb{R}_+)$, τ is strictly less than 1, and so is $\frac{\tau}{4}$.
    Using equation (61), we deduce the following decay for v_α

$$|v_\alpha(t)| + |v_\alpha'(t)| \le Ce^{-\delta t}, \qquad t \ge 0. \qquad (71)$$

Let us define a non-increasing cut-off function φ_ε satisfying the following properties

$$\begin{cases} \varphi_\varepsilon(t) = 1 & \text{for } t \in \left[0, \frac{1}{2}\varepsilon^{-\gamma}\right]; \\ \varphi_\varepsilon(t) = 0 & \text{for } t \in \left[\frac{3}{4}\varepsilon^{-\gamma}, \varepsilon^{-\gamma}\right]; \\ |\varphi_\varepsilon'(t)| \le C\varepsilon^\gamma; \quad |\varphi_\varepsilon''(t)| \le C\varepsilon^{2\gamma}. \end{cases} \qquad (72)$$

By (71) and some elementary computations, one finds

$$\left|\|\varphi_\varepsilon v_\alpha\|_{\alpha,\varepsilon}^2 - 1\right| \le Ce^{-\delta\varepsilon^{-\gamma}}; \qquad\qquad |(T_{\alpha,\varepsilon}(\varphi_\varepsilon v_\alpha), \varphi_\varepsilon v_\alpha) - \mu_\alpha| \le Ce^{-\delta\varepsilon^{-\gamma}}.$$

Since $\varphi_\varepsilon v_\alpha$ belongs to $H_{\alpha,\varepsilon}$, (70) and the last formulas yield $\mu_{\alpha,\varepsilon} \le \mu_\alpha + Ce^{-\delta\varepsilon^{-\gamma}}$. This proves the first inequality in the statement.

Formula (71) implies

$$\int_{\mathbb{R}_+ \setminus J_\varepsilon} \left((v_\alpha')^2 + v_\alpha^2 \right) \leq C \varepsilon^{-\delta \varepsilon^{-\gamma}}.$$

To obtain the second inequality in the statement we set

$$u(t) = v_{\alpha,\varepsilon}(t) - v_\alpha(t) + \varepsilon^{2\gamma} v_\alpha(\varepsilon^{-\gamma}) t^2; \qquad t \in J_\varepsilon.$$

Subtracting the differential equations satisfied by $v_{\alpha,\varepsilon}$ and v_α, we obtain

$$\begin{cases} (1 - \mu_{\alpha,\varepsilon})(-u'' + (1+\alpha)u) - p w_0^{p-1} u = f_1 + f_2 & \text{in } J_\varepsilon; \\ u'(0) = 0 & u \in H_{\alpha,\varepsilon}, \end{cases} \tag{73}$$

where f_1 and f_2 are defined by duality as

$$(f_1, v) = (\mu_{\alpha,\varepsilon} - \mu_\alpha) \int_{J_\varepsilon} (-\Delta v_\alpha + (1+\alpha)v_\alpha) v \, dt, \qquad \forall v \in H_{\alpha,\varepsilon};$$

$$(f_2, v) = \varepsilon^{2\gamma} v_\alpha(\varepsilon^{-\gamma}) \int_{J_\varepsilon} \left[(1 - \mu_{\alpha,\varepsilon}) \left((1+\alpha)t^2 - 2 \right) - p w_0^{p-1} t^2 \right] v \, dt, \qquad \forall v \in H_{\alpha,\varepsilon}.$$

One can check that

$$\|f_1\|_{H_{\alpha,\varepsilon}} \leq C \varepsilon^{-\delta \varepsilon^{-\gamma}}; \qquad \|f_2\|_{H_{\alpha,\varepsilon}} \leq C \varepsilon^{-\frac{1}{2}\gamma} e^{-\delta \varepsilon^{-\gamma}}. \tag{74}$$

We write

$$u = \beta v_{\alpha,\varepsilon} + \overline{u}, \qquad \text{where} \qquad \beta \in \mathbb{R} \quad \text{and} \quad (\overline{u}, v_{\alpha,\varepsilon})_{H_{\alpha,\varepsilon}} = 0. \tag{75}$$

Since $T_{\alpha,\varepsilon}$ is invertible on $(v_{\alpha,\varepsilon})^\perp$, (74) implies

$$\|\overline{u}\|_{H_{\alpha,\varepsilon}} \leq C \varepsilon^{-\frac{1}{2}\gamma} e^{-\delta \varepsilon^{-\gamma}}. \tag{76}$$

Equation (75) can also be written as

$$(1 - \beta) v_{\alpha,\varepsilon} = \overline{u} + \left(v_\alpha - \varepsilon^{2\gamma} v_\alpha(\varepsilon^{-\gamma}) t^2 \right).$$

Taking the norm of both sides, using (76) and some elementary estimates we get

$$|1 - \beta| = 1 + O \left(\varepsilon^{-\frac{1}{2}\gamma} e^{-\delta \varepsilon^{-\gamma}} \right).$$

This equation implies $\beta \sim 0$ or $\beta \sim 2$, from which we deduce (see (75)) that either $v_{\alpha,\varepsilon} \sim v_\alpha$, or $v_{\alpha,\varepsilon} \sim -v_\alpha$. Since we are assuming that $v_{\alpha,\varepsilon}$ is non-increasing, it must be $\beta \sim 0$, and more precisely $|\beta| \leq C \varepsilon^{-\frac{1}{2}\gamma} e^{-\delta \varepsilon^{-\gamma}}$. Then (75) and (76) yield the conclusion. ∎

Remark 28 *Using the Courant-Fischer method, one can also prove that* $\tau_{\alpha,\varepsilon} \geq \tau_\alpha$ *for every* α *and* ε.

In order to study equation (P_ε) we employ Fourier analysis. For $\gamma > 0$, let us define the set S_ε and the metric g_0 on S_ε as

$$S_\varepsilon = \partial\Omega_\varepsilon \times J_\varepsilon = \partial\Omega_\varepsilon \times \left[0, \varepsilon^{-\gamma}\right], \qquad g_0 = \overline{g}_\varepsilon \otimes (dt)^2, \qquad (77)$$

where $\overline{g}_\varepsilon$ is the metric on $\partial\Omega_\varepsilon$ induced by $\mathbb{R}^n$. We also define the functional space

$$H_{S_\varepsilon} = \left\{ u \in H^1(S_\varepsilon) \ : \ u(x', \varepsilon^{-\gamma}) = 0 \text{ for all } x' \in \partial\Omega_\varepsilon \right\},$$

endowed with the norm

$$\|u\|_{H_{S_\varepsilon}}^2 = \int_{S_\varepsilon} \left(|\nabla_{g_0} u|^2 + u^2\right); \qquad u \in H_{S_\varepsilon}.$$

Let $\hat{g}$ be the metric on $\partial\Omega$ induced by that of $\mathbb{R}^n$, and let $\Delta_{\hat{g}}$ be the Laplace-Beltrami operator. Let $(\lambda_l)_l$ denote the eigenvalues of $-\Delta_{\hat{g}}$, counted in non-decreasing order and with their multiplicity. Let also $(\varphi_l)_l$ denote the corresponding eigenfunctions.

Given $u \in H_{S_\varepsilon}$, we can decompose it in Fourier components (in the variable x') as follows

$$u(x', x_n) = \sum_{l=0}^{\infty} \varphi_l(\varepsilon x') u_l(x_n), \qquad x' \in \partial\Omega_\varepsilon, x_n \in J_\varepsilon. \qquad (78)$$

Using this decomposition, one finds

$$\|u\|_{H_{S_\varepsilon}}^2 = \frac{1}{\varepsilon^{n-1}} \sum_l \int_{J_\varepsilon} \left((u_l')^2 + (1 + \varepsilon^2\lambda_l)u_l^2\right) = \frac{1}{\varepsilon^{n-1}} \sum_l \|u_l\|_{\varepsilon^2\mu_l}^2.$$

Writing for brevity $\|\cdot\|_{l,\varepsilon}$ instead of $\|\cdot\|_{\varepsilon^2\mu_l}$, the last equation becomes

$$\|u\|_{H_{S_\varepsilon}}^2 = \frac{1}{\varepsilon^{n-1}} \sum_l \|u_l\|_{l,\varepsilon}^2. \qquad (79)$$

This is how the norms $\|\cdot\|_\alpha$ introduced above enter in our study; in particular we will choose α belonging to the discrete set $(\varepsilon^2\lambda_l)_l$.

We are also interested in an eigenvalue problem of the form

$$T_{S_\varepsilon} u = \sigma u; \qquad u \in H_{S_\varepsilon}, \qquad (80)$$

where T_{S_ε} is defined by

$$(T_{S_\varepsilon} u, v) = \int_{S_\varepsilon} \left(\langle \nabla_{g_0} u, \nabla_{g_0} v \rangle + uv - p w_0^{p-1} uv\right) dV_{g_0}, \qquad u, v \in H_{S_\varepsilon}.$$

Writing u as in (78), equation (80) is equivalent to

$$\begin{cases} (1-\sigma)\left(-u_l'' + (1+\varepsilon^2\lambda_l)u_l\right) = pw_0^{p-1}u_l & \text{in } J_\varepsilon; \\ u_l'(0) = 0, & \end{cases} \qquad \text{for all } l. \qquad (81)$$

Since T_{S_ε} represents a model for the study of $I_\varepsilon''(u_{k,\varepsilon})$, it was essential for us to perform a spectral analysis of the operators T_α. The spectrum of T_{S_ε} is characterized in the next proposition. We recall the definition of the (positive) number τ in Proposition 25.

Proposition 29 *Let $\sigma < \frac{\tau}{4}$ be an eigenvalue of T_{S_ε}. Then $\sigma = \mu_{l,\varepsilon}$ for some index l. The corresponding eigenfunctions u are of the form*

$$u(x', x_n) = \sum_{\{l \,:\, \mu_{l,\varepsilon}=\sigma\}} \alpha_l \varphi_l(\varepsilon x') v_{l,\varepsilon}(x_n), \qquad x' \in \partial\Omega_\varepsilon, x_n \in J_\varepsilon, \quad (82)$$

where $(\alpha_l)_l$ are arbitrary constants. Viceversa, every function of the form (82) is an eigenfunction of T_{S_ε} with eigenvalue σ. In particular the eigenvalues of T_{S_ε} which are smaller than τ coincide with the numbers $(\mu_{l,\varepsilon})_l$ which are smaller than τ.

PROOF.   Let u be an eigenfunction of T_{S_ε} with eigenvalue σ. Then we can write

$$u(x', x_n) = \sum_l \left(\alpha_l v_{l,\varepsilon}(x_n) + \hat{v}_l(x_n)\right)\varphi_l(\varepsilon x'),$$

where α_l are real numbers, and where $(\hat{v}_l, v_{l,\varepsilon})_{l,\varepsilon} = 0$ for all l. We have

$$\sum_l \left(\alpha_l\mu_{l,\varepsilon} v_{l,\varepsilon}(x_n) + T_{l,\varepsilon}\hat{v}_l(x_n)\right)\varphi_l(\varepsilon x') = T_{S_\varepsilon}u = \sigma u$$

$$= \sigma\sum_l \left(\alpha_l v_{l,\varepsilon}(x_n) + \hat{v}_l(x_n)\right)\varphi_l(\varepsilon x').$$

By the uniqueness of the Fourier decomposition, the last equation implies

$$\mu_{l,\varepsilon} = \sigma \quad \text{if} \quad \alpha_l \neq 0; \quad \text{and} \quad T_{l,\varepsilon}\hat{v}_l = \sigma\hat{v}_l \quad \text{for all } l. \qquad (83)$$

Proposition 27 $ii)$ and Remark 28 yield

$$\tau_{\alpha,\varepsilon} \geq \tau_\alpha \geq \tau, \qquad (84)$$

which means that $T_{\alpha,\varepsilon} \geq \tau Id$ on the subspace of $H_{\alpha,\varepsilon}$ orthogonal to $v_{\alpha,\varepsilon}$. Therefore, the second equation in (83) and the fact that $\sigma < \frac{\tau}{4}$ imply $\hat{v}_l = 0$ for all l. Moreover, since $\lambda_l \to \mu_{l,\varepsilon}$ is a monotone function, the first equation in (84) shows that all the indices l for which $\alpha_l \neq 0$ correspond to the same value λ_l. This concludes the proof. ∎

We finally recall the following theorem due to T. Kato, ([39], page 444) which is fundamental to us in order to obtain invertibility, see Proposition 36.

Theorem 30 *Let $T(\chi)$ denote a differentiable family of operators from an Hilbert space X into itself, where χ belongs to an interval containing 0. Let $T(0)$ be a self-adjoint operator of the form* Identity *- compact and let $\sigma(0) = \sigma_0 \neq 1$ be an eigenvalue of $T(0)$. Then the eigenvalue $\sigma(\chi)$ is differentiable at 0 with respect to χ. The derivative of σ is given by*

$$\frac{\partial \sigma}{\partial \chi} = \left\{ \text{eigenvalues of } P_{\sigma_0} \circ \frac{\partial T}{\partial \chi}(0) \circ P_{\sigma_0} \right\},$$

where $P_{\sigma_0} : X \to X_{\sigma_0}$ denotes the projection onto the σ_0 eigenspace X_{σ_0} of $T(0)$.

Remark 31 *We note that, when perturbing the operator $T(0)$, the eigenvalue σ_0 can split in several ones, so in general σ_0 possesses a multivalued derivative. Anyway, since the operator $T(0)$ is of the form* Identity *- compact, and since σ_0 is different from 1, σ_0 is an isolated eigenvalue and the projection P_{σ_0} has a finite dimensional range, therefore the splitting of σ_0 is always finite.*

5 An Approximate Solution

In this subsection we construct approximate solutions $u_{k,\varepsilon}$ of (P_ε) up to an arbitrary order in ε. Basically, we expand (P_ε) in powers of ε, solve it term by term, and then use suitable truncations.

Proposition 32 *Consider the Euler functional I_ε defined in (6) and associated to problem (P_ε). Then for any $k \in \mathbb{N}$ there exists a function $u_{k,\varepsilon} : S_\varepsilon \to \mathbb{R}$ with the following properties*

$$\|I_\varepsilon'(u_{k,\varepsilon})\| \leq C_k \varepsilon^{k+1-\frac{n-1}{2}}; \qquad u_{k,\varepsilon} \geq 0 \quad in \ \Omega_\varepsilon \qquad \frac{\partial u_{k,\varepsilon}}{\partial \nu} = 0 \quad on \ \partial \Omega_\varepsilon, \tag{85}$$

where C_k depends only on Ω, p and k. Moreover one has

$$\begin{cases} \|(\nabla')^{(m)} u_{k,\varepsilon}(x', x_n)\|' \leq C_m \varepsilon^m e^{-x_n}; \\ \|(\nabla')^{(m)} \partial_{x_n} u_{k,\varepsilon}(x', x_n)\|' \leq C_m \varepsilon^m e^{-x_n}; \qquad x' \in \partial \Omega_\varepsilon, x_n \in J_\varepsilon, , m = 0, 1, \ldots, \\ \|(\nabla')^{(m)} \partial_{x_n}^2 u_{k,\varepsilon}(x', x_n)\|' \leq C_m \varepsilon^m e^{-x_n}; \end{cases}$$
$$\tag{86}$$

where ∇' and $\|\cdot\|'$ denote the derivative and the norm taken with respect to the variable x' (freezing x_n), and C_m is a constant depending only on Ω, p and m.

We parameterize the set S_ε, see equation (77), using coordinates x' on $\partial\Omega_\varepsilon$ and x_n in J_ε. Let ν denote the unit inner normal to $\partial\Omega$, and define the map $\Gamma_\varepsilon : \partial\Omega_\varepsilon \times J_\varepsilon \to \mathbb{R}^n$ by

$$\Gamma_\varepsilon(x', x_n) = x' + x_n \nu(\varepsilon x').$$

We let the upper-case indices $A, B, C, \ldots$ run from 1 to n, and the lower-case indices $i, j, k, \ldots$ run from 1 to $n-1$. Using some local coordinates $(x_i)_{i=1,\ldots,n-1}$ on $\partial\Omega_\varepsilon$, and letting φ_ε be the corresponding immersion into $\mathbb{R}^n$, we have

$$
\begin{cases}
\dfrac{\partial\Gamma_\varepsilon}{\partial x_i}(x', x_n) = \dfrac{\partial\varphi_\varepsilon}{\partial x_i}(x') + \varepsilon x_n \dfrac{\partial\nu}{\partial x_i}(\varepsilon x') \\[2mm]
\qquad = \dfrac{\partial\varphi_\varepsilon}{\partial x_i}(x') + \varepsilon x_n H_i^j(\varepsilon x') \dfrac{\partial\varphi_\varepsilon}{\partial x_j}(x') \qquad \text{for } i = 1, \ldots, n-1; \\[2mm]
\dfrac{\partial\Gamma_\varepsilon}{\partial x_n}(x', x_n) = \nu(\varepsilon x').
\end{cases}
$$

where (H_i^j) are the coefficients of the mean-curvature operator on $\partial\Omega$. Let also $\overline{g}_{ij}$ be the coefficients of the metric on $\partial\Omega_\varepsilon$ in the above coordinates (x', x_n). Then, letting $g = g_\varepsilon$ denote the metric on Ω_ε induced by $\mathbb{R}^n$, we have

$$g_{AB} = \left\langle \frac{\partial\Gamma_\varepsilon}{\partial x_A}, \frac{\partial\Gamma_\varepsilon}{\partial x_B} \right\rangle = \begin{pmatrix} (g_{ij}) & 0 \\ 0 & 1 \end{pmatrix}, \tag{87}$$

where

$$
\begin{aligned}
g_{ij} &= \left\langle \frac{\partial\varphi_\varepsilon}{\partial x_i}(x') + \varepsilon x_n H_i^k(\varepsilon x') \frac{\partial\varphi_\varepsilon}{\partial x_k}(x'); \frac{\partial\varphi_\varepsilon}{\partial x_j}(x') + \varepsilon x_n H_j^l(\varepsilon x') \frac{\partial\varphi_\varepsilon}{\partial x_l}(x') \right\rangle \\
&= \overline{g}_{ij} + \varepsilon x_n \left(H_i^k \overline{g}_{kj} + H_j^l \overline{g}_{il} \right) + \varepsilon^2 x_n^2 H_i^k H_j^l \overline{g}_{kl}.
\end{aligned}
$$

Note that also the inverse matrix (g^{AB}) decomposes as

$$g^{AB} = \begin{pmatrix} (g^{ij}) & 0 \\ 0 & 1 \end{pmatrix}.$$

We begin by finding a first-order approximation for u_ε, to show the ideas of the general procedure. We define the following map $u \mapsto \tilde{u}$ from functions on $\partial\Omega_\varepsilon$ (resp. $\partial\Omega_\varepsilon \times \mathbb{R}_+$) into functions on $\partial\Omega$ (resp. $\partial\Omega \times \mathbb{R}_+$)

$$
\begin{aligned}
\tilde{u}(\varepsilon x') &= u(x') \qquad (\text{resp. } \tilde{u}(\varepsilon x', x_n) = u(x', x_n)), \\
x' &\in \partial\Omega_\varepsilon \qquad (\text{resp. } (x', x_n) \in \partial\Omega_\varepsilon \times \mathbb{R}_+).
\end{aligned}
\tag{88}
$$

Using the above parametrization, we look for an approximate solution $u_{1,\varepsilon}$ of the form

$$u_{1,\varepsilon}(x', x_n) = w_0 + \varepsilon \tilde{w}_1(\varepsilon x', x_n),$$

where the function $\tilde{w}_1$ is defined on $\partial\Omega \times \mathbb{R}_+$, consistently with (88). We note that, from the above decomposition of g_{AB} (and g^{AB}), one finds

$$-\Delta_g u = -g^{AB} u_{AB} - \frac{1}{\sqrt{\det g}} \partial_A \left(g^{AB} \sqrt{\det g} \right) u_B$$

$$= -u_{nn} - g^{ij} u_{ij} - \frac{1}{\sqrt{\det g}} \partial_n \left(\sqrt{\det g} \right) u_n - \frac{1}{\sqrt{\det g}} \partial_i \left(g^{ij} \sqrt{\det g} \right) u_j.$$

We have, formally

$$\det g = det(\overline{g}^{-1} g) \det \overline{g} = (\det \overline{g}) \left(1 + \varepsilon x_n \operatorname{tr} \left(\overline{g}^{-1} \alpha \right) \right) + o(\varepsilon), \qquad (89)$$

where

$$\alpha_{ij} = H_i^k \overline{g}_{kj} + H_j^l \overline{g}_{il}.$$

We have

$$(\overline{g}^{-1} \alpha)_{is} = \overline{g}^{sj} \alpha_{ij} = \overline{g}^{sj} \left(H_i^k \overline{g}_{kj} + H_j^l \overline{g}_{il} \right),$$

and hence

$$\operatorname{tr} \left(\overline{g}^{-1} \alpha \right) = \overline{g}^{ij} \left(H_i^k \overline{g}_{kj} + H_j^l \overline{g}_{il} \right) = 2 \overline{g}^{ij} H_i^k \overline{g}_{kj} = 2 H_i^i. \qquad (90)$$

We recall that the quantity H_i^i represents the opposite of the mean curvature of $\partial\Omega$ (by our choice of ν), and in particular it is independent of the choice of coordinates.

Using (89), (90), writing formally $(w_0 + \varepsilon w_1)^p$ as $w_0^p + p \varepsilon w_0^{p-1} w_1 + o(\varepsilon)$, and expanding $-\Delta_{g_\varepsilon} u_{1,\varepsilon} + u_{1,\varepsilon} = u_{1,\varepsilon}^p$ up to first order in ε, we obtain the following equation for w_1

$$\begin{cases} -w_1'' + w_1 - p w_0^{p-1} w_1 = H_1^1 w_0' & \text{in } \mathbb{R}_+; \\ w_1'(0) = 0. \end{cases} \qquad (91)$$

By Proposition 25 and the subsequent comment there, equation (91) can be solved for any right-hand side in $L^2(\mathbb{R}_+^2)$, as in this case (see equation (59)). The resulting expression for w_1 will be independent of the choice of coordinates on $\partial\Omega_\varepsilon$.

The construction of a better approximate solution is performed by expanding formally the equation $-\Delta_{g_\varepsilon} u + u = u^p$ in powers of ε up to order k. A rigorous proof is in [50].

Given smooth functions $\tilde{w}_1, \ldots \tilde{w}_k : \partial\Omega \times \mathbb{R}_+$, let

$$\hat{u}_{k,\varepsilon}(x', x_n) = w_0(x_n) + \varepsilon \tilde{w}_1(\varepsilon x', x_n) + \ldots \varepsilon^k \tilde{w}_k(\varepsilon x', x_n); \quad x' \in \partial\Omega_\varepsilon, x_n \in \mathbb{R}_+.$$

Expanding formally the equation $-\Delta_g u + u = u^p$ (imposing Neumann boundary conditions) in powers of ε, we find that $\tilde{w}_i$ satisfies the following recurrence formula, for i running from 1 to k

$$
\begin{cases}
-\tilde{w}_i'' + \tilde{w}_i - p w_0^{p-1} \tilde{w}_i = \displaystyle\sum_{j=1}^{i} x_n^{j-1} \tilde{G}_i(x') \tilde{w}_{i-j}'(\varepsilon x', x_n) \\
\qquad\qquad - \displaystyle\sum_{j=0}^{i-2} (L_{i-j-2}\tilde{w}_j)(\varepsilon x', x_n) \qquad\qquad\qquad \text{in } \mathbb{R}_+; \\
\qquad\qquad + \displaystyle\sum_{\substack{j_1,\dots,j_{i-1} \\ \sum l j_l = i}} C_{i,j_1,\dots,j_i} w_0^{p-\sum j_l} \tilde{w}_1^{j_1} \cdots \tilde{w}_{i-1}^{j_{i-1}}(\varepsilon x', x_n) \\
\tilde{w}_k'(0) = 0,
\end{cases}
$$

$$(92)$$

where the derivatives are taken with respect to the variable x_n, the coefficients $C_{i,j_1,\dots,j_i}$ are constants, $\tilde{G}_i : \partial\Omega \to \mathbb{R}$ are smooth functions and L_{i-j-2} are linear second-order differential operators.

Proceeding by induction and using standard estimates on solutions of ODE's, we find that there exist polynomials $P_i(t)$ such that

$$
\begin{cases}
\left\| (\nabla')^{(m)} \tilde{w}_i(x', x_n) \right\|' \leq C_{i,m} P_i(x_n) e^{-x_n}; \\
\left\| (\nabla')^{(m)} \tilde{w}_i'(x', x_n) \right\|' \leq C_{i,m} P_i(x_n) e^{-x_n}; \qquad x' \in \partial\Omega_\varepsilon, x_n \in \mathbb{R}_+, m = 0, 1, \dots, \\
\left\| (\nabla')^{(m)} \tilde{w}_i''(x', x_n) \right\|' \leq C_{i,m} P_i(x_n) e^{-x_n};
\end{cases}
$$

$$(93)$$

where ∇' denotes the derivative with respect to x', and $C_{i,m}$ is a constant depending only on Ω, p, i and m. In particular these decay estimates ensure that the right-hand side in (92) belongs to $L^2(\mathbb{R}_+)$, and (92) is solvable for all $i = 1, \dots, k$.

In conclusion, $u_{k,\varepsilon}$ is obtained by multiplying $\hat{u}_{k,\varepsilon}$ by a cutoff function supported in $\{x_n \in [0, \varepsilon^{-\gamma}]\}$: to guarantee convergence of the Taylor expansions up to the desired order, the constant γ has to be taken sufficiently small, depending on k, n and p (see Subsection 3.2 in [50]).

6 Eigenvalues of the Linearized Equation

Let Σ_ε be defined as

$$
\Sigma_\varepsilon = \Gamma_\varepsilon(S_\varepsilon) := \Gamma_\varepsilon(\partial\Omega_\varepsilon \times J_\varepsilon) = \left\{ x \in \mathbb{R}^n \ : \ \text{dist}(x, \partial\Omega_\varepsilon) < \varepsilon^{-\gamma} \right\}.
$$

We endow Σ_ε with the metric g_ε induced by the inclusion in $\mathbb{R}^n$, and introduce the functional space

$$
H_{\Sigma_\varepsilon} = \left\{ u \in H^1(\Sigma_\varepsilon) \ : \ u\left(\Gamma_\varepsilon(x', \varepsilon^{-\gamma})\right) = 0 \text{ for all } x' \in \partial\Omega_\varepsilon \right\},
$$

with its natural norm

$$\|u\|_{H_{\Sigma_\varepsilon}}^2 = \int_{\Sigma_\varepsilon} \left(|\nabla_{g_\varepsilon} u|^2 + u^2 \right); \qquad u \in H_{\Sigma_\varepsilon},$$

and the corresponding scalar product $(\ , \ )_{H_{\Sigma_\varepsilon}}$. Using the expression of the metric g_ε, see equation (87) and the subsequent formulas, one finds

$$\left(1 - C\varepsilon^{1-\gamma} \right) \|u\|_{H_{S_\varepsilon}} \leq \|u\|_{H_{\Sigma_\varepsilon}} \leq \left(1 + C\varepsilon^{1-\gamma} \right) \|u\|_{H_{S_\varepsilon}}; \qquad \text{for all } u \in H_{S_\varepsilon}, \tag{94}$$

where C is a constant depending only on Ω.

Let $u_{k,\varepsilon}$ be the function constructed in Subsection 5. Then we define $T_{\Sigma_\varepsilon} : H_{\Sigma_\varepsilon} \to H_{\Sigma_\varepsilon}$ in the following way

$$(T_{\Sigma_\varepsilon} u, v)_{H_{\Sigma_\varepsilon}} = \int_{\Sigma_\varepsilon} \left(\langle \nabla_{g_\varepsilon} u, \nabla_{g_\varepsilon} v \rangle + uv - p u_{k,\varepsilon}^{p-1} uv \right) dV_{g_\varepsilon}; \qquad u, v \in H_{\Sigma_\varepsilon}. \tag{95}$$

The goal of this subsection is to study the eigenvalues σ of the problem

$$T_{\Sigma_\varepsilon} u = \sigma u, \qquad u \in H_{\Sigma_\varepsilon}. \tag{96}$$

In the next proposition we characterize the eigenfunctions of T_{Σ_ε} when the eigenvalue σ is close to zero. We recall the definition of the functions φ_l and $v_{l,\varepsilon}$ from Subsection 4. Proposition 33 is the counterpart of Proposition 29 for a non-flat metric.

Proposition 33 *Let $\gamma > 0$ be sufficiently small, and let $\mathcal{L} \in (0, \tau/4)$. Suppose $\sigma \in [-\tau/4, \tau/4]$, and let u be an eigenfunction of T_{Σ_ε} with eigenvalue σ. Then one has*

$$\left\| u - \sum_{\{l : \mu_{l,\varepsilon} \in [\sigma - \mathcal{L}, \sigma + \mathcal{L}]\}} \alpha_l \varphi_l v_{l,\varepsilon} \right\|_{H_{\Sigma_\varepsilon}} \leq C \frac{\varepsilon^{1-\gamma}}{\mathcal{L}} \|u\|_{H_{\Sigma_\varepsilon}}, \tag{97}$$

for some coefficients $(\alpha_l)_l$ and for some constant C depending only on Ω and p.

Let $\overline{u}_{k,\varepsilon} : \Omega \to \mathbb{R}$ be the function defined by

$$\overline{u}_{k,\varepsilon}(x) = u_{k,\varepsilon}(\varepsilon x), \qquad x \in \Omega.$$

We have next the following result.

Proposition 34 *The eigenvalues of the operator T_{Σ_ε} are differentiable with respect to ε, the eigenvalues being considered a possibly multivalued function of ε. If $\sigma(\varepsilon)$ is such an eigenvalue, then*

$$\frac{\partial \sigma}{\partial \varepsilon} = \{ eigenvalues \ of \ Q_\sigma \}, \tag{98}$$

where $Q_\sigma : H_\sigma \times H_\sigma \to \mathbb{R}$ is the quadratic form given by

$$Q_\sigma(u,v) = (1-\sigma)\frac{2}{\varepsilon} \int_{\Sigma_\varepsilon} \langle \nabla u, \nabla v \rangle - p(p-1) \int_{\Sigma_\varepsilon} uvu_{k,\varepsilon}^{p-2} \frac{\partial \overline{u}_{k,\varepsilon}}{\partial \varepsilon}(\varepsilon \cdot). \quad (99)$$

Here $H_\sigma \subseteq H_{\Sigma_\varepsilon}$ denotes the eigenspace of T_{Σ_ε} corresponding to σ and the function $u_{k,\varepsilon}$ is defined in Subsection 5.

This formula can be derived heuristically differentiating in ε (96) (scaled to Ω). A rigorous proof, using Kato's theorem, can be found in [49]. We apply Proposition 34 to the eigenvalues σ which are close to 0.

Proposition 35 *Let σ be an eigenvalue of T_{Σ_ε} belonging to the interval $[-\tau/4, \tau/4]$. Then one has*

$$\left| \frac{\partial \sigma}{\partial \varepsilon} - \frac{1}{\varepsilon} \tilde{F}(\sigma) \right| \leq C\varepsilon^{-\frac{1+\gamma}{2}},$$

where γ is sufficiently small, $\tilde{F}(\cdot)$ is given in (69), and C is a constant depending only on Ω and p.

We are now in position to prove the following proposition, which characterizes the spectrum of T_{Σ_ε}.

Proposition 36 *Let $u_{k,\varepsilon}$ and T_{Σ_ε} be as above. Then there exists a positive constant C, depending on p, Ω and k with the following property. For a suitable sequence $\varepsilon_j \to 0$, the operator $T_{\Sigma_{\varepsilon_j}} : H_{\Sigma_{\varepsilon_j}} \to H_{\Sigma_{\varepsilon_j}}$ is invertible and the inverse operator satisfies*

$$\left\| T_{\Sigma_{\varepsilon_j}}^{-1} \right\| \leq \frac{C}{\varepsilon_j^{n-1}}, \qquad \text{for all } j \in \mathbb{N}.$$

PROOF. Let $(\sigma_j)_j$ (resp. $(s_j)_j$) denote the eigenvalues of T_{Σ_ε} (resp. T_{S_ε}), counted in increasing order with their multiplicity. Let also

$$N(\Sigma_\varepsilon) = \sharp\{j \,:\, \sigma_j \leq 0\}; \qquad\qquad (\text{resp. } N(S_\varepsilon) = \sharp\{j \,:\, s_j \leq 0\}).$$

denote the number of non-positive eigenvalues of T_{Σ_ε} (resp. T_{S_ε}). Recall that, from Proposition 29, the eigenvalues of T_{S_ε} coincide with the numbers $(\mu_{j,\varepsilon})_j := \mu_{\varepsilon^2 \lambda_j, \varepsilon}$, where $(\lambda_j)_j$ are the eigenvalues of the Laplace-Beltrami operator $-\Delta_{\hat{g}}$, see also equations (78), (81). We also recall the Weyl's asymptotic formula, see e.g. [18] page 169

$$\lambda_j \sim \frac{(2\pi)^2}{\omega_{n-1}^{\frac{2}{n-1}}} \left(\frac{j}{Vol(\partial\Omega)} \right)^{\frac{2}{n-1}} = C_{n,\Omega} j^{\frac{2}{n-1}}, \qquad \text{as } j \to +\infty. \quad (100)$$

From the strict monotonicity of $\alpha \mapsto \mu_\alpha := \mu(\alpha)$ and from the last formula we deduce

$$\sharp \left\{ j \, : \, \varepsilon^2 \lambda_j \leq \mu^{-1}(0) \right\} \sim \tilde{C}_{n,\Omega,p} \varepsilon^{1-n}. \tag{101}$$

From Proposition 27, we have

$$\left| \mu_{\varepsilon^2 \lambda_j, \varepsilon} - \mu_{\varepsilon^2 \lambda_j} \right| \leq C e^{-\delta \varepsilon^{-\gamma}}; \qquad \text{provided} \quad \mu_{\varepsilon^2 \lambda_j} \in \left[-\frac{\tau}{4}, \frac{\tau}{4} \right], \tag{102}$$

which implies

$$\sharp \left\{ j : \varepsilon^2 \lambda_j \leq \mu^{-1}(0) - C e^{-\delta \varepsilon^{-\gamma}} \right\} \leq N(S_\varepsilon) \leq \sharp \left\{ j : \varepsilon^2 \lambda_j \leq \mu^{-1}(0) + C e^{-\delta \varepsilon^{-\gamma}} \right\}$$

and hence

$$N(S_\varepsilon) \sim \tilde{C}_{n,\Omega,p} \varepsilon^{1-n}. \tag{103}$$

From the Courant-Fischer method, we get

$$\sigma_j = \inf \left\{ \sup_{u \in M} \frac{(T_{\Sigma_\varepsilon} u, u)_{H_{\Sigma_\varepsilon}}}{\|u\|_{H_{\Sigma_\varepsilon}}^2} \, : \, M \text{ subspace of } H_{\Sigma_\varepsilon}, \dim M = j \right\},$$

and a similar characterization for s_j. By (94) one finds

$$\left| (T_{\Sigma_\varepsilon} u, u)_{H_{\Sigma_\varepsilon}} - (T_{S_\varepsilon} u, u)_{H_{S_\varepsilon}} \right| \leq C \varepsilon^{1-\gamma} \|u\|_{H_{S_\varepsilon}}^2;$$

$$\left| \|u\|_{H_{\Sigma_\varepsilon}}^2 - \|u\|_{H_{S_\varepsilon}}^2 \right| \leq C \varepsilon^{1-\gamma} \|u\|_{H_{S_\varepsilon}}^2,$$

for all $u \in H_{\Sigma_\varepsilon}$ and for some constant C depending only on Ω and p. Note that, still by (94), in the last inequality we can also substitute $\| \cdot \|_{H_{S_\varepsilon}}$ with $\| \cdot \|_{H_{\Sigma_\varepsilon}}$. The last two formulas yield

$$|s_j - \sigma_j| \leq C \varepsilon^{1-\gamma}, \qquad \text{for all } j, \tag{104}$$

where C depends only on Ω and p. From (102), (104) there exists a positive constant C, depending only on Ω and p, such that

$$\sharp \left\{ j : \varepsilon^2 \lambda_j \leq \mu^{-1}(0) - C \varepsilon^{1-\gamma} \right\} \leq N(\Sigma_\varepsilon) \leq \sharp \left\{ j : \varepsilon^2 \lambda_j \leq \mu^{-1}(0) + C \varepsilon^{1-\gamma} \right\},$$

which implies

$$N(\Sigma_\varepsilon) \sim \sharp \left\{ j \, : \, \lambda_j \leq \frac{\mu^{-1}(0)}{\varepsilon^2} \right\} \sim \left(\frac{\mu^{-1}(0)}{\varepsilon^2 C_{n,\Omega}} \right)^{\frac{n-1}{2}} = \tilde{C}_{n,\Omega,p} \varepsilon^{1-n}. \tag{105}$$

For $l \in \mathbb{N}$, let $\varepsilon_l = 2^{-l}$. Then from (105) we have

$$N(\Sigma_{\varepsilon_{l+1}}) - N(\Sigma_{\varepsilon_l}) \sim \tilde{C}_{n,\Omega,p} \left(2^{(l+1)(n-1)} - 2^{l(n-1)} \right) = \tilde{C}_{n,\Omega,p} \left(2^{n-1} - 1 \right) \varepsilon_l^{1-n}. \tag{106}$$

Note that, by Proposition 35, the eigenvalues of T_{Σ_ε} close to 0 decrease when ε decreases to zero. In other words, by the last equation, the number of eigenvalues which cross 0, when ε decreases from ε_l to ε_{l+1}, is of order ε_l^{1-n}. Now we define

$$A_l = \{\varepsilon \in (\varepsilon_{l+1}, \varepsilon_l) \ : \ \ker T_{\Sigma_\varepsilon} \neq \emptyset\}; \qquad\qquad B_l = (\varepsilon_{l+1}, \varepsilon_l) \backslash A_l.$$

By Proposition 35 and (106), it follows that $\mathrm{card}(A_l) < C\varepsilon_l^{1-n}$, and hence there exists an interval (a_l, b_l) such that

$$(a_l, b_l) \subseteq B_l; \qquad\qquad |b_l - a_l| \geq C^{-1}\frac{\mathrm{meas}(B_l)}{\mathrm{card}(A_l)} \geq C^{-1}\varepsilon_l^n. \qquad (107)$$

From Proposition 35 we deduce that

$$T_{\Sigma_{\frac{a_l+b_l}{2}}} \qquad \text{is invertible and} \qquad \left\| T_{\Sigma_{\frac{a_l+b_l}{2}}}^{-1} \right\| \leq \frac{C}{\varepsilon_l^{n-1}}.$$

Now it is sufficient to set $\varepsilon_j = \frac{a_j+b_j}{2}$. This concludes the proof. ∎

7 Proof of Theorem 3

In this subsection we prove Theorem 3. We just treat the case $p \leq \frac{n+2}{n-2}$: for $p > \frac{n+2}{n-2}$, see [50] (some comments have been given at the beginning of this part). The following result can be proved as is [49], Proposition 5.6.

Proposition 37 *Let $k \in \mathbb{N}$ be fixed, and let $u_{k,\varepsilon}$ be as in Subsection 5. Let $(\sigma_j)_j$ denote the eigenvalues of T_{Σ_ε}, counted in increasing order and with their multiplicity, and let $(\hat{\sigma}_j)_j$ denote the eigenvalues of $I_\varepsilon''(u_{k,\varepsilon})$. Then there exists a positive constant C, depending only on Ω and p, such that*

$$|\sigma_j - \hat{\sigma}_j| \leq Ce^{-\delta\varepsilon^{-\frac{\gamma}{2}}}; \qquad \text{whenever } \sigma_j \in \left[-\frac{\tau}{4}, \frac{\tau}{4}\right].$$

The proof of this proposition consists in showing that the eigenfunctions of $I_\varepsilon''(u_{k,\varepsilon})$ decay exponentially away from $\partial\Omega_\varepsilon$, and and in modifying them through suitable cut-off functions. The eigenvalues σ_j and $\hat{\sigma}_j$ are then compared using the Courant-Fischer method, as in the proof of Proposition 27.

Let ε_j be as in Proposition 36: then from Proposition 37 we deduce

$$I_{\varepsilon_j}''(u_{k,\varepsilon_j}) \quad \text{is invertible and} \quad \|I_{\varepsilon_j}''(u_{k,\varepsilon_j})^{-1}\| \leq \frac{C}{\varepsilon_j^{n-1}}. \qquad (108)$$

For brevity, in the rest of the proof, we simply write ε instead of ε_j. We apply the contraction mapping theorem, looking for a solution u_ε of the form

$$u_\varepsilon = u_{k,\varepsilon} + w, \qquad w \in H^1(\Omega_\varepsilon).$$

If $I_\varepsilon''(u_{k,\varepsilon})$ is invertible (which is true along the sequence ε_j), we can write

$$I_\varepsilon''(u_{k,\varepsilon} + w) = 0 \qquad \Leftrightarrow \qquad w = -\left(I_\varepsilon''(u_{k,\varepsilon})\right)^{-1}\left[I_\varepsilon'(u_{k,\varepsilon}) + N_\varepsilon(w)\right], \quad (109)$$

where

$$N_\varepsilon(w) = I_\varepsilon'(u_{k,\varepsilon} + w) - I_\varepsilon'(u_{k,\varepsilon}) - I_\varepsilon''(u_{k,\varepsilon})[w].$$

Let us define the operator $F_\varepsilon : H^1(\Omega_\varepsilon) \to H^1(\Omega_\varepsilon)$

$$F_\varepsilon(w) = -\left(I_\varepsilon''(u_{k,\varepsilon})\right)^{-1}\left[I_\varepsilon'(u_{k,\varepsilon}) + N_\varepsilon(w)\right], \qquad w \in H^1(\Omega_\varepsilon).$$

We are going to prove that F_ε is a contraction in a suitable closed set of $H^1(\Omega_\varepsilon)$: from Proposition 32 and (108) we get

$$\|F_{\varepsilon_j}(w)\| \leq \begin{cases} C\varepsilon_j^{-(n-1)}\left(\varepsilon_j^{k+1-\frac{n-1}{2}} + \|w\|^p\right) & \text{for } p \leq 2, \\ C\varepsilon_j^{-(n-1)}\left(\varepsilon_j^{k+1-\frac{n-1}{2}} + \|w\|^2\right) & \text{for } p > 2; \end{cases} \qquad \|w\| \leq 1;$$

$$(110)$$

$$\|F_{\varepsilon_j}(w_1) - F_{\varepsilon_j}(w_2)\| \leq \begin{cases} C\varepsilon_j^{-(n-1)}\left(\|w_1\|^{p-1} + \|w_2\|^{p-1}\right)\|w_1 - w_2\| & p \leq 2, \\ C\varepsilon_j^{-(n-1)}\left(\|w_1\| + \|w_2\|\right)\|w_1 - w_2\| & p > 2; \\ \|w_1\|, \|w_2\| \leq 1. \end{cases}$$

$$(111)$$

Now we choose integers d and k such that

$$d > \begin{cases} \frac{n-1}{p-1} & \text{for } p \leq 2, \\ n-1 & \text{for } p > 2; \end{cases} \qquad k+1 > d + \frac{3}{2}(n-1), \qquad (112)$$

and we set

$$\mathcal{B}_j = \left\{ w \in H^1(\Omega_\varepsilon) \ : \ \|w\| \leq \varepsilon_j^d \right\}.$$

From (110), (111) it follows that F_{ε_j} is a contraction in $\mathcal{B}_j$ for ε_j small. Since u_ε is close in norm to $u_{k,\varepsilon} \geq 0$, the positivity of u_ε follows from standard arguments, based on the Sobolev inequalities. The points **i)** and **ii)** in the statement follow from the construction of u_ε. This concludes the proof.

Remark 38 *By the above construction and some standard estimates one has*

$$I_{\varepsilon_j}''(u_{\varepsilon_j}) = I_{\varepsilon_j}''(u_{k,\varepsilon_j}) + O\left(\varepsilon^{d\min\{1,p-1\}}\right) = I_{\varepsilon_j}''(u_{k,\varepsilon_j}) + o\left(\varepsilon_j^n\right).$$

It follows from Propositions 36, 37 that the Morse index of u_{ε_j} coincides with $N(\Sigma_{\varepsilon_j}) \sim \varepsilon_j^{1-n}$.

References

1. Ambrosetti, A., Badiale, M., Cingolani, S., Semiclassical states of nonlinear Schrödinger equations, Arch. Rational Mech. Anal. 140 (1997), 285–300.
2. Ambrosetti, A., Felli, V., Malchiodi, A., Ground states of nonlinear Schrödinger equations with potentials vanishing at infinity, J. Eur. Math. Soc. 7 (2005), 117–144.
3. Ambrosetti, A., Malchiodi, A., Perturbation Methods and Semilinear Elliptic Problems on $\mathbb{R}^n$, Birkhäuser, Progr. in Math. 240, (2005).
4. Ambrosetti, A., Malchiodi, A., Nonlinear Analysis and Semilinear Elliptic Problems, Cambridge Univ. Press, Cambridge Studies in Advanced Mathematics, No.104 (2007).
5. Ambrosetti, A., Malchiodi, A., Ni, W.M., Singularly Perturbed Elliptic Equations with Symmetry: Existence of Solutions Concentrating on Spheres, Part I, Comm. Math. Phys., 235 (2003), 427–466.
6. Ambrosetti, A., Malchiodi, A., Ni, W.M., Singularly Perturbed Elliptic Equations with Symmetry: Existence of Solutions Concentrating on Spheres, Part II, Indiana Univ. Math. J. 53 (2004), no. 2, 297–329.
7. Ambrosetti, A., Malchiodi, A., Ruiz, D., Bound states of Nonlinear Schrödinger Equations with Potentials Vanishing at Infinity, J. d'Analyse Math., 98 (2006), 317–348.
8. Ambrosetti, A., Malchiodi, A., Secchi, S., Multiplicity results for some nonlinear singularly perturbed elliptic problems on R^n, Arch. Rat. Mech. Anal. 159 (2001) 3, 253–271.
9. Ambrosetti, A., Rabinowitz, P.H., Dual variational methods in critical point theory and applications, J. Funct. Anal. 14 (1973), 349–381.
10. Arioli, G., Szulkin, A., A semilinear Schrödinger equation in the presence of a magnetic field. Arch. Ration. Mech. Anal. 170 (2003), 277–295.
11. Badiale, M., D'Aprile, T., Concentration around a sphere for a singularly perturbed Schrödinger equation. Nonlinear Anal. 49 (2002), no. 7, Ser. A: Theory Methods, 947–985.
12. Bartsch, T., Peng, S., Semiclassical symmetric Schrdinger equations: existence of solutions concentrating simultaneously on several spheres, Z Angew Math Phys, 58 (5) (2007), 778–804.
13. Benci, V., D'Aprile, T., The semiclassical limit of the nonlinear Schrödinger equation in a radial potential. J. Differential Equations 184 (2002), no. 1, 109–138.
14. Berestycki, H., Lions, P.L. Nonlinear scalar field equations. I. Existence of a ground state. Arch. Rational Mech. Anal. 82 (1983), no. 4, 313–345.
15. Berestycki, H., Lions, P.L. Nonlinear scalar field equations. II. Existence of infinitely many solutions. Arch. Rational Mech. Anal. 82 (1983), no. 4, 347–375.
16. Berezin, F.A., Shubin, M.A., The Schrdinger equation, Kluwer Academic Publishers Group, Dordrecht, 1991.
17. Byeon, J., Wang, Z.Q., Standing waves with a critical frequency for nonlinear Schrodinger equations, Arch. Rat. Mech. Anal. 165, 295-316 (2002).
18. Casten, R.G., Holland, C.J., Instability results for reaction diffusion equations with Neumann boundary conditions, J. Diff. Eq. 27 (1978), no. 2, 266–273.
19. Cingolani, S., Pistoia, A., Nonexistence of single blow-up solutions for a nonlinear Schrödinger equation involving critical Sobolev exponent, Z. Angew. Math. Phys. 55 (2004), no. 2, 201–215.
20. Dancer, E.N., Stable and finite Morse index solutions on R^n or on bounded domains with small diffusion, Trans. Amer. Math. Soc., 357 (2005), no. 3, 1225–1243.
21. Dancer, E.N., Yan, S., Multipeak solutions for a singularly perturbed Neumann problem. Pacific J. Math. 189 (1999), no. 2, 241–262.
22. Dancer, E.N., Yan, S., A new type of concentration solutions for a singularly perturbed elliptic problem, Trans Amer Math Soc, 359(4) (2007), 1765–1790.

23. D'Aprile, T., On a class of solutions with non-vanishing angular momentum for nonlinear Schrödinger equations. Diff. Int. Equ. 16 (2003), no. 3, 349–384.

24. Del Pino, M., Felmer, P., Local mountain passes for semilinear elliptic problems in unbounded domains, Calc. Var. 4 (1996), 121–137.

25. Del Pino, M., Felmer, P., Semi-classcal states for nonlinear schrödinger equations, J. Funct. Anal. 149 (1997), 245–265.

26. Del Pino, M., Felmer, P., Kowalczyk, M., Boundary spikes in the Gierer-Meinhardt system. Commun. Pure Appl. Anal. 1 (2002), no. 4, 437–456.

27. Del Pino, M., Felmer, P., Wei, J., On the role of the mean curvature in some singularly perturbed Neumann problems, S.I.A.M. J. Math. Anal. 31 (1999), 63–79.

28. Del Pino, M., Kowalczyk, M., Wei, J., Multi-bump ground states of the Gierer-Meinhardt system in $\mathbb{R}^2$. Ann. Inst. H. Poincaré Anal. Non Linéaire 20 (2003), no. 1, 53–85.

29. Del Pino, M., Kowalczyk, M., Wei, J., Concentration at curves for Nonlinear Schrödinger Equations, Comm. Pure Appl. Math., 60 (2007), no. 1, 113–146.

30. Floer, A., Weinstein, A., Nonspreading wave packets for the cubic Schrödinger equation with a bounded potential, J. Funct. Anal. 69, (1986), 397–408.

31. Gierer, A., Meinhardt, H., A theory of biological pattern formation, Kybernetik (Berlin), 12 (1972), 30–39.

32. Grossi, M. Some results on a class of nonlinear Schrödinger equations, Math. Zeit. 235 (2000), 687–705.

33. Grossi, M., Pistoia, A., Wei, J., Existence of multipeak solutions for a semilinear Neumann problem via nonsmooth critical point theory. Calc. Var. Partial Differential Equations 11 (2000), no. 2, 143–175.

34. Gui, C., Multipeak solutions for a semilinear Neumann problem, Duke Math. J., 84 (1996), 739–769.

35. Gui, C., Wei, J., On multiple mixed interior and boundary peak solutions for some singularly perturbed Neumann problems, Canad. J. Math. 52 (2000), no. 3, 522–538.

36. Gui, C., Wei, J., Winter, M., Multiple boundary peak solutions for some singularly perturbed Neumann problems, Ann. Inst. H. Poincaré Anal. Non Linéaire 17 (2000), no. 1, 47–82.

37. Iron, D., Ward, M., Wei, J., The stability of spike solutions to the one-dimensional Gierer-Meinhardt model. Phys. D 150 (2001), no. 1-2, 25–62.

38. Jeanjean, L., Tanaka, K., A positive solution for a nonlinear Schrödinger equation on $\mathbb{R}^N$. Indiana Univ. Math. J. 54 (2005), no. 2, 443–464.

39. Kato, T., Perturbation theory for linear operators. Second edition. Grundlehren der Mathematischen Wissenschaften, Band 132. Springer-Verlag, Berlin-New York, 1976.

40. Kawohl, B., Rearrangements and convexity of level sets in PDE. Lecture Notes in Mathematics, 1150. Springer-Verlag, Berlin, 1985.

41. Kwong, M.K., Uniqueness of positive solutions of $-\Delta u + u + u^p = 0$ in $\mathbb{R}^n$, Arch. Rational Mech. Anal. 105, (1989), 243–266.

42. Li, Y.Y., On a singularly perturbed equation with Neumann boundary conditions, Comm. Partial Differential Equations 23 (1998), 487–545.

43. Li, Y.Y., Nirenberg, L., The Dirichlet problem for singularly perturbed elliptic equations, Comm. Pure Appl. Math. 51 (1998), 1445–1490.

44. Lin, C.S., Ni, W.M., Takagi, I., Large amplitude stationary solutions to a chemotaxis systems, J. Differential Equations, 72 (1988), 1–27.

45. Lin, F.H., Ni, W.M., Wei, J., On the number of interior peak solutions for a singularly perturbed Neumann problem, 60 (2007), no. 2, 252–281.

46. Mahmoudi, F., Malchiodi, A., Concentration on minimal submanifolds for a singularly perturbed Neumann problem, Adv. in Math., 209-2 (2007), 460–525.

47. Mahmoudi, F., Malchiodi, A., Wei, J., Transition Layer for the Heterogeneous Allen-Cahn Equation, Annales IHP, Analyse non Lineaire, to appear.

48. Malchiodi, A., Concentration at curves for a singularly perturbed Neumann problem in three-dimensional domains, G.A.F.A., 15-6 (2005), 1162–1222.

49. Malchiodi, A., Montenegro, Boundary concentration phenomena for a singularly perturbed elliptic problem, Comm. Pure Appl. Math, 15 (2002), 1507–1568.
50. Malchiodi, A., Montenegro, Multidimensional Boundary-layers for a singularly perturbed Neumann problem, Duke Math. J. 124 (2004), no. 1, 105–143.
51. Malchiodi, A., Ni, W.M., Wei, J., Multiple clustered layer solutions for semilinear Neumann problems on a ball, Ann. Inst. H. Poincaré Anal. Non Linéaire, 22 (2005), 143–163.
52. Malchiodi, A., Wei, J., Boundary interface for the Allen-Cahn equation, J. Fixed Point Theory and Appl., 1 (2007), no. 2, 305–336.
53. Matano, H., Asymptotic behavior and stability of solutions of semilinear diffusion equations, Publ. Res. Inst. Math. Sci. 15 (1979), no. 2, 401–454.
54. Molle, R., Passaseo, D., Concentration phenomena for solutions of superlinear elliptic problems, Ann. Inst. H. Poincaré Anal. Non Linéaire 23 (2006), no. 1, 63–84.
55. Ni, W.M. Diffusion, cross-diffusion, and their spike-layer steady states. Notices Amer. Math. Soc. 45 (1998), no. 1, 9–18.
56. Ni, W.M., Takagi, I., On the shape of least-energy solution to a semilinear Neumann problem, Comm. Pure Appl. Math., 41 (1991), 819–851.
57. Ni, W.M., Takagi, I., Locating the peaks of least-energy solutions to a semilinear Neumann problem, Duke Math. J. 70, (1993), 247–281.
58. Ni, W.M., Takagi, I., Yanagida, E., Stability of least energy patterns of the shadow system for an activator-inhibitor model. Recent topics in mathematics moving toward science and engineering. Japan J. Indust. Appl. Math. 18 (2001), no. 2, 259–272.
59. Oh, Y.G. On positive Multi-lump bound states of nonlinear Schrödinger equations under multiple well potentials, Comm. Math. Phys. 131, (1990), 223–253.
60. Shi, J., Semilinear Neumann boundary value problems on a rectangle. Trans. Amer. Math. Soc. 354 (2002), no. 8, 3117–3154.
61. Strauss, W.A., Existence of solitary waves in higher dimensions. Comm. Math. Phys. 55 (1977), no. 2, 149–162.
62. Turing, A.M., The chemical basis of morphogenesis, Phil. Trans. Royal Soc. London, Series B, Biological Sciences, 237 (1952), 37–72.
63. Wang, Z.Q., On the existence of multiple, single-peaked solutions for a semilinear Neumann problem, Arch. Rational Mech. Anal., 120 (1992), 375–399.
64. Wei, J., On the boundary spike layer solutions of a singularly perturbed semilinear Neumann problem, J. Differential Equations, 134 (1997), 104–133.
65. Wei J., Yang J., Concentration on Lines for a Singularly Perturbed Neumann Problem in Two-Dimensional Domains, Indiana Univ. Math. J., to appear.

On Some Elliptic Problems in the Study of Selfdual Chern-Simons Vortices

Gabriella Tarantello

1 Introduction

In these lectures we use an approach introduced by Taubes (cf. [JT]) in the study of selfdual vortices for the abelian-Higgs model, in order to describe vortex configurations for the Chern-Simons (CS in short) theory discussed in [D1].

Notice that the abelian-Higgs model corresponds (in a non-relativistic context) to the bi-dimensional Ginzburg-Landau (GL in short) model (cf. [GL], [DGP]), for which much has been accomplished in recent years also away from the selfdual regime. In this respect, beside the seminal work of Bethuel-Brezis-Helein (cf. [BBH]), we mention for example: [BeR], [JS1],[JS2], [Lin1], [Lin2], [LR1], [LR2], [PiR], [PR] and the recent monograph by Sandier-Safarty [SS]. However, the methods and techniques introduced for the GL-model do not seem to apply as successfully for the CS-model (see the attemps of Kurzke-Sprin [KS1], [KS2] and Han-Kim in [HaK]). Thus, so far a rigorous mathematical analysis of CS-vortices has been possible only at the selfdual regime where Taubes approach applies equally well and allows one to reduce the vortex problem to the study of elliptic problems involving exponential nonlinearities. In this way it has been possible to treat many relevant selfdual theories of interest in theoretical physics by means of nonlinear analysis, see [Y1].

Here, we only wish to give a general idea on the type of analytical problems that one encounters in the study of selfdual vortex configurations, and for this purpose we shall focus our attention only on the planar Chern-Simons-Higgs model proposed in [JW] and [HKP], and refer the interested reader to the monographs [Y1] and [T8] for further developments.

G. Tarantello
Universita' di Roma Tor Vergata- Dipartimento di Matematica, via della ricerca scientifica, 00133 Rome, Italy
e-mail: tarantel@mat.uniroma2.it

S.-Y.A. Chang et al. (eds.), *Geometric Analysis and PDEs*,
Lecture Notes in Mathematics 1977, DOI: 10.1007/978-3-642-01674-5_4,
© Springer-Verlag Berlin Heidelberg 2009

The physical motivation to study CS-vortices rests upon the important property that (contrary to GL-vortices) they carry both electric and magnetic charge, a property observed for example in superconductivity (cf. [Ab], [Park] and [Sch]).

On this basis, Chern-Simons theory has been used more generally to resolve important issues in condensed matter physics, as discussed in [D1], [D2], [D3], [Fro], [FM1] and [FM2].

The mathematical reasons to treat the CS-model rests upon the fact that it introduces a new 6th order gauge theory, whose space of moduli is rather rich and, as we shall see, far from being completely characterized.

We are going to compare the abelian-Higgs and Chern-Simons model and emphasize their analogies and differences. To this purpose, we first formulate both models within the framework of abelian gauge field theory. For a more detailed physical background about gauge field theories we refer to [AH], [ChNe], [Fel], [Po] and [Q]; while an introduction on the mathematical formalism of gauge theory can be found in [Tra] and [GS], (see also [JT] and [Jo]).

Here, we shall be concerned with a gauge theory whose gauge group G (typically a compact Lie group) is specified by the group of rotation in $\mathbb{R}^2$.

Namely, recalling that the group of rotation in $\mathbb{R}^n$ is given by the orthogonal group:

$$O(n) = \{A,\ n \times n \text{ real matrix} : A^t A = \text{Id}\}$$

and

$$SO(n) = \{A \in O(n) : \det A = 1\} \text{ (special orthogonal group)}$$

we fix $G = SO(2)$.

Recalling that the complex counterpart of the groups above, are given respectively by,

$$U(n) = \{A,\ n \times n \text{ complex matrix} : A^t \overline{A} = \text{Id}\} \text{ (unitary group)}$$

and

$$SU(n) = \{A \in U(n) : \det A = 1\} \text{ (special unitary group)}$$

we see that

$$G = SO(2) = U(1) = \{z \in \mathbb{C} : |z| = 1\}.$$

Hence G is topologically equivalent to S^1 and defines an abelian group, whose group multiplication coincides with the multiplication of complex numbers. In particular, the group $U(1)$ acts on $\mathbb{C}$ just as a multiplication by a unitary complex number, and in this way we obtain a unitary rapresentation for it. Since any element of our gauge group $U(1)$ takes the form $e^{i\omega}$, we also see that the corresponding Lie algebra $u(1) = -i\mathbb{R}$.

From the physical point of view, (cf. [AH], [ChNe], [Fel], [Po] and [Q]), a $U(1)$-gauge field theory describes electromagnetic particle interaction, while a (non abelian) $SU(2)$-gauge field theory concerns with weak particle interactions and an $SU(3)$-gauge field theory pertains with strong particle interactions.

The celebrated Electroweak theory of Salam-Glashow-Weinberg (cf. [La]) unifies electromagnetic and weak particle interactions via a $SU(2) \times U(1)$-gauge field theory, and there is hope that a general unified theory of all particle interactions should involve the gauge group $SU(5)$.

Notice also that the Yang-Mills (cf. [YM]) and more generally the Yang-Mills-Higgs theory are formulated as (non-abelian) $SU(2)$-gauge field theories. It is well known (cf. [JT], [Ra]) that both theories carry selfdual solitons-type solutions as given by the Yang-Mills instantons in 4-space dimension, and the Yang-Mills-Higgs monopoles in 3-space dimension (see e.g. [ADHM], [AHS1], [AHS2], [BPST], [DK], [Wit], [AtH] and [PS]). In many respects abelian-Higgs vortices correspond to the bi-dimensional version of such static configurations. For example, it is known that they are in one to one correspondence with the class of $O(3)$-symmetric four dimensional instantons (cf. [JT]).

Other interesting features about vortices for the abelian-Higgs model can be found in [Bra1], [Bra2], [Hi], [Ga1], [Ga2], [Ga3] and [NO].

2 The Abelian-Higgs Model and The Chern-Simons-Higgs Model

We denote by $\mathbb{R}^{1,3}$ the 3-dimensional Minkowski space with metric tensor $g = tr(1, -1, -1, -1)$, and let (x^0, x^1, x^2, x^3) be an element of $\mathbb{R}^{1,3}$, so that x^0 denotes the time-variable and x^j, $j = 1, 2, 3$, the space variables. Generally, we use greek indices to denote indifferently time and space variables. As usual we will use the metric to rise (or lower) indices in the usual way (i.e. $x^\alpha = \sum\limits_{\alpha,\beta=0}^{3} g^{\alpha\beta} x_\beta$, with $g^{\alpha\beta}$ the coefficients of the inverse of g. Also, we use the standard convention that repeated lower/upper indices are summed.

The $U(1)$-gauge field theory over $\mathbb{R}^{1,3}$ is formulated in terms of the following variables:

(electromagnetic) gauge potential: $\mathcal{A} = -iA_\alpha dx^\alpha$, $A_\alpha : \mathbb{R}^{1,3} \to \mathbb{R}$, $\alpha = 0, 1, 2, 3$, given by a globally defined 1-form with values in the Lie algebra $u(1) = -i\mathbb{R}$, and viewed as a connection on the (trivial) principle bundle $\mathbb{P} = \mathbb{R}^{1,3} \times U(1)$;

(matter) Higgs field: $\varphi : \mathbb{R}^{1,3} \to \mathbb{C}$, which can be seen as a global section of the associated bundle $\mathcal{M} = \mathbb{R}^{1,3} \times \mathbb{C}$ (relative to the above mentioned representation of $U(1)$ over $\mathbb{C}$).

The gauge potential $\mathcal{A}$ and the Higgs field φ are weakly coupled by means of the covariant derivative D_A associated to A:

$$D_A = D_\alpha dx^\alpha, \text{ with } D_\alpha = \partial_\alpha - iA_\alpha, \ \alpha = 0, 1, 2, 3.$$

So that,

$$D_\alpha \varphi = \partial_\alpha \varphi - iA_\alpha \varphi, \ \alpha = 0, 1, 2, 3.$$

Also we obtain the electromagnetic gauge field F_A as the curvature corresponding to $\mathcal{A}$, namely:

$$F_A = -\frac{i}{2} F_{\alpha\beta} dx^\alpha \wedge dx^\beta$$

with

$$F_{\alpha\beta} = \partial_\beta A_\beta - \partial_\beta A_\alpha.$$

The abelian Maxwell-Higgs theory is formulated in terms of the following lagrangian density:

$$\mathcal{L}(A, \varphi) = -\frac{q^2}{4} F_{\alpha\beta} F^{\alpha\beta} + \frac{1}{2} D_\alpha \varphi \overline{D^\alpha \varphi} - V \tag{1}$$

where V defines the scalar potential, and q is a given parameter which relates to the electric charge.

We obtain the Ginzburg-Landau model when (in normalized units) we specify the scalar potential:

$$V(|\varphi|) = \frac{\lambda}{8}(1 - |\varphi|^2)^2, \tag{2}$$

(the well known double-well potential), with $\lambda > 0$ a given parameter.

In the bi-dimensional case, this model give rise to the abelian-Higgs model. More in general, we shall always take $V = V(|\varphi|)$, so that (1) is invariant under the gauge transformations:

$$\mathcal{A} \to \mathcal{A} + d\omega; \quad \varphi \to e^{-i\omega} \varphi \tag{3}$$

for any function $\omega : \mathbb{R}^{1,3} \to \mathbb{R}$ sufficiently smooth.

Hence, $\mathcal{A}$ and φ are defined only up to gauge transformations, and so they are not observable quantities. On the contrary, gauge invariant quantities (i.e. observables) are given by the gauge field F_A and the magnitude of the Higgs field $|\varphi|$.

The Euler-Lagrange equations corresponding to (1) are given as follows,

$$\begin{cases} D_\alpha D^\alpha \varphi = -2\frac{\partial V}{\partial \overline{\varphi}} & (a) \\ q^2 \partial_\nu F^{\mu\nu} = \frac{i}{2}(\overline{\varphi}D^\mu\varphi - \varphi\overline{D^\mu\varphi}) & (b) \end{cases} \tag{4}$$

and obviously they inherit the invariance under the gauge transformations (3). Furthermore, setting:

$$J^\mu = \frac{i}{2}(\overline{\varphi}D^\mu\varphi - \varphi\overline{D^\mu\varphi}), \quad \mu = 0, 1, 2, 3$$

we can intepret the vector field $J = (J^\mu)_{\mu=0,1,2,3}$ as the (conserved) <u>current</u> generated by the internal symmetries, with $\rho = J^0$ the <u>charge density</u>, and $\overrightarrow{j} = (J^k)_{k=1,2,3}$ the <u>current density</u>.

From (4) we can recover Maxwell's equations. Indeed, from F_A we obtain the <u>electric field</u>: $\overrightarrow{E} = (F^{0j})_{j=1,2,3}$, and the <u>magnetic field</u>: $\overrightarrow{B} = (B^j)_{j=1,2,3}$, with $B^j = \frac{i}{2}\varepsilon^{jkl}F_{kl}$; where ε^{jkl} is the totally anti-symmetric tensor fixed by $\varepsilon^{123} = 1$.

Therefore the four equations in (4)-(b) express the following familiar set of Maxwell's equations:

$$q^2\nabla \cdot \overrightarrow{E} = \rho \quad \& \quad q^2\partial_0\overrightarrow{E} + \overrightarrow{j} = q^2\text{curl}\overrightarrow{B}.$$

We can complete Maxwell's equations, simply by virtue of Bianchi identity which is satisfied by the curvature F_A as follows:

$$\partial^\mu F^{\alpha\beta} + \partial^\alpha F^{\beta\mu} + \partial^\beta F^{\mu\alpha} = 0. \quad (3\text{-equations})$$

The Gauss-law of the system is obtained by taking $\mu = 0$ in (4)-(b), and thus given by,

$$q^2\partial_j F^{0j} = \rho. \tag{5}$$

So, we can obtain the total <u>electric charge</u> $= \int \rho$, by integration of (5), and by neglecting boundary terms, we find $\int \rho = 0$.

In other words, the abelian Maxwell-Higgs model can only support electrically neutral configurations.

Of particular interest to us are the so called vortex-configurations, namely the <u>stationary</u> solutions of (4) with <u>finite</u> energy. Here stationary means that it is possible to specify a gauge map so that, after the gauge transformation (3), we obtain φ and the component A_α, $\alpha = 0, 1, 2, 3$ indipendent of the time-variable x^0.

In case such gauge map can be choosen to correspond to the <u>temporal gauge</u> $A_0 = 0$ (i.e. $\omega = \int^{x^0} A_0(t, x^1, \ldots, x^3)dt$), then the solution is called <u>static</u>.

To compute the energy density, we write the lagrangean density more explicitly as follows,

$$\mathcal{L} = \frac{q^2}{2}F_{0,j}^2 - \frac{q^2}{4}F_{k,j}^2 + \frac{1}{2}(|D_0\varphi|^2 - |D_j\varphi|^2) - V(|\varphi|).$$

Consequently,

$$\mathcal{E} = \frac{\partial\mathcal{L}}{\partial(\partial_0 A_\mu)}\partial_0 A_\mu + \frac{\partial\mathcal{L}}{\partial(\partial_0\varphi)}\partial_0\varphi + \frac{\partial\mathcal{L}}{\partial(\partial_0\overline{\varphi})}\partial_0\overline{\varphi} + \mathcal{L}$$

$$= \frac{q^2}{2}F_{0,j}^2 + \frac{q^2}{4}F_{k,j}^2 + \frac{1}{2}(|D_0\varphi|^2 + |D_j\varphi|^2) + V + [q^2\partial_j F^{0j} - \rho]A_0$$

$$+ \text{ (total spatial divergence terms)}. \tag{6}$$

Hence, by neglecting the total spatial divergence terms (which disappear upon integration under natural boundary conditions) and recalling (5), we obtain the following expression for the <u>Ginzburg-Landau energy</u>:

$$\mathcal{E}_{GL} = \frac{q^2}{2}F_{0,j}^2 + \frac{q^2}{4}F_{k,j}^2 + \frac{1}{2}(|D_0\varphi|^2 + |D_j\varphi|^2) + \frac{\lambda}{8}(1 - |\varphi|^2)^2. \tag{7}$$

We refer to [BBH], [PR], [SS] and reference therein, for the description of static minimizers of (7) in various situations.

However, to obtain charged vortex configurations, as they naturally occur for example in low critical temperature superconductivity, it is necessary to modify the conventional Maxwell's electrodynamics of the Ginzburg-Landau model.

The Chern-Simons theory modifies Maxwell theory in the bi-dimensional case, by introducing the Chern-Simons lagrangian as a lower order perturbation of Maxwell lagrangean.

When we deal with a bi-dimensional theory, we can consider a gauge trasformation which specifies $A_3 = 0$, so that the remaining variables A_α, $\alpha = 0, 1, 2$ and φ are indipendent of the x^3-variable ($D_3 A_\alpha = 0 = D_3\varphi$, $\alpha = 0, 1, 2$). Thus, for $A_\alpha = A_\alpha(x^0, x^1, x^2)$, $\alpha = 0, 1, 2$, the Chern-Simons lagrangean density takes the form:

$$\mathcal{L}_{CS} = -\frac{1}{4}\varepsilon^{\alpha\beta\gamma}A_\alpha F_{\beta\gamma}, \tag{8}$$

with $\varepsilon^{\alpha\beta\gamma}$ the totally anti-symmetric tensor, fixed so that $\varepsilon^{012} = 1$.

Although $\mathcal{L}_{CS}$ is <u>not</u> invariant under the gauge transformation (3), the corresponding Euler-Lagrange equations: $F_{\alpha\beta} = 0$, are gauge invariant.

The bi-dimensional Maxwell-Chern-Simons model is described by the Lagrangean density: $\mathcal{L}_{MCS} = \mathcal{L}_{MH} + \kappa\mathcal{L}_{CS}$, and defines a sensible gauge field theory, with $\kappa > 0$ the Chern-Simons coupling parameter. More precisely we have,

$$\mathcal{L}_{MCS} = \frac{q^2}{2}F_{0,j}^2 - \frac{q^2}{4}F_{k,j}^2 - \frac{k}{4}\varepsilon^{\alpha\beta\gamma}A_\alpha F_{\beta\gamma} + \frac{1}{2}(|D_0\varphi|^2 - |D_j\varphi|^2) - V(|\varphi|) \tag{9}$$

with corresponding Euler-Lagrange equation:

$$\begin{cases} D_\alpha D^\alpha \varphi = -2\frac{\partial V}{\partial \overline{\varphi}} & (a) \\ q^2 \partial_\nu F^{\mu\nu} + \frac{\kappa}{2}\varepsilon^{\mu\alpha\beta}F_{\alpha\beta} = J^\mu & (b) \end{cases} \tag{10}$$

where again $J^\mu = \frac{i}{2}(\overline{\varphi}D^\mu\varphi - \varphi\overline{D^\mu}\varphi)$, $\mu = 0, 1, 2$ defines the current density generated by the internal symmetries.

As above, letting:

$$\rho = J^0 = \frac{i}{2}(\overline{\varphi}D^0\varphi - \varphi\overline{D^0}\varphi) \tag{11}$$

the charge density, then (from (10)-(b) with $\mu = 0$) we obtain the modified Gauss law relative to the Maxwell-Chern-Simons model as given by,

$$q^2\partial_j F^{0j} + \kappa F_{12} = \rho. \tag{12}$$

From (12) we see that the Maxwell-Chern-Simons theory allows for (electrically and magnetically) charged vortices. In fact, upon integration of (12), and by neglecting as usual total spatial divergence terms, we deduce a relation between the magnetic flux $\Phi = \int F_{12}$ and the electric charge $Q = \int \rho$ as follows,

$$Q = \kappa\Phi.$$

Due to the presence of the Chern-Simons term, the study of the field equations (10) (with $\kappa \neq 0$) appears much more difficult than the abelian-Higgs field equations (4), (namely (10) with $\kappa = 0$), for which much is known especially in the bi-dimensional case (cf. [JT], [BBH], [PR] and [SS]).

So far, it has been possible to analyze with mathematical rigor only Chern-Simons vortex-configurations in the selfdual regime, by means of Taubes' approach for studying (4) with V in (2) and $\lambda = \frac{1}{q^2}$, (cf. [JT]).

3 Selfduality

As observed by Bogomolnyi [Bo], the bi-dimensional abelian-Higgs field equations admit a first order reduction. This property is ensured since the corresponding energy density,

$$\mathcal{E}_{aH} = \frac{q^2}{2}(F_{0j}^2 + F_{12}^2) + \frac{1}{2}(|D_0\varphi|^2 + |D_1\varphi|^2 + |D_2\varphi|^2) + \frac{\lambda}{8}(1 - |\varphi|^2)^2$$
$$+ \text{ (total spatial divergence terms)}$$

can be more conveniently rewritten by completing its square terms.

To this purpose, we introduce the following gauge invariant Cauchy-Riemann operators:

$$D_\pm\varphi = D_1\varphi \pm iD_2\varphi \tag{13}$$

and observe that

$$|D_1\varphi|^2 + |D_2\varphi|^2 = |D_\pm\varphi|^2 \pm F_{12}|\varphi|^2 + \text{(total spatial divergence terms)}. \quad (14)$$

So we can write:

$$\mathcal{E}_{aH} = \frac{q^2}{2}F_{0j}^2 + \frac{1}{2}|D_0\varphi|^2 + \frac{1}{2}\overbrace{[q^2F_{12} \pm F_{12}(|\varphi|^2 - 1) + \frac{1}{4q^2}(|\varphi|^2 - 1)^2]}^{(q^2F_{12}\pm\frac{1}{2q}(|\varphi|^2-1))^2}$$

$$+ |D_\pm\varphi|^2 + \frac{1}{8}\left(\lambda - \frac{1}{q^2}\right)(|\varphi|^2 - 1)^2 \pm \frac{1}{2}F_{12}$$

$$+ \text{(total spatial divergence terms)}. \quad (15)$$

Hence, in the critical coupling:

$$\lambda = \frac{1}{q^2}$$

(which characterizes the selfdual regime) and at fixed magnetic flux, the minimal energy is attained by solutions of the following <u>first order selfdual</u> equations:

$$F_{0j} = 0, \ D_0\varphi = 0 \quad (16)$$

$$D_\pm\varphi = 0 \quad (17)$$

$$F_{12} = \pm\frac{1}{2q^2}(1 - |\varphi|^2) \quad (18)$$

which should be satisfied together with the Gauss law equation (5).

It is not difficult to check that solutions of the selfdual equations also satisfy the second order abelian-Higgs field equations. Moreover, equations (16)–(18) and (5) support <u>static</u> vortex configurations, since in the temporal gauge $A_0 = 0$, equation (16) just asserts that A_1, A_2 and φ are indipendent of the x^0-variable. So that (5) is automatically satisfied, and we are left to determine $\varphi = \varphi(x^1, x^2)$ and $A_j = A_j(x_1, x_2)$ $j = 1, 2$, satisfying (17) and (18), with corresponding <u>static</u> energy density:

$$\mathcal{E}_{\text{self}}^\pm = \pm\frac{1}{2}F_{12}. \quad (19)$$

We shall discuss the planar case, and we shall solve (17)–(18) in terms of <u>two</u> complex functions, namely: φ and $A_1 \pm iA_2 : \mathbb{C} \to \mathbb{C}$.

Also note that if $(\mathcal{A}, \varphi)^-$ is a solution of (17)–(18) with "minus-sign", then it can be related to a solution $(\mathcal{A}, \varphi)^+$ of (17)–(18) with the "plus-sign", by the following simple change of variables:

$$\varphi^-(x_1, x_2) = \varphi^+(-x_1, x_2)$$

$$A_1^-(x_1, x_2) = -A_1^+(-x_1, x_2),$$

$$A_2^-(x_1, x_2) = A_1^-(-x_1, x_2).$$

Thus, without loss of generality, we will fucus only to study (17)–(18) with "plus-sign".

In this case, equation (17) gives:

$$D_+\varphi = D_1\varphi + iD_2\varphi = (\partial_1\varphi + i\partial_2\varphi) - i(A_1 + iA_2)\varphi = 0. \qquad (20)$$

Now notice that the equation:

$$\partial_{\bar{z}}\psi = \frac{1}{2}(A_1 + iA_2) := A \qquad (21)$$

($\partial_{\bar{z}} = \frac{1}{2}(\partial_1 + i\partial_2)$), is locally solvable by virtue of Poincaré $\bar{\partial}$-lemma (cf. [JT]). Moreover, if $A_1, A_2 \in C^\infty$ then also the solution $\psi \in C^\infty$.

By means of (21), we can write equation (20) as follows:

$$\partial_{\bar{z}}e^{-i\psi}\varphi = 0$$

and deduce that $e^{-i\psi}\varphi$ defines an holomorphic function. So φ can only admit isolated zeroes with integral multiplicity.

We point out that the zero-set of φ represents a relevant gauge invariant quantity, which is going to characterize in an important way the corresponding vortex configuration.

Moreover, to fullfil the finite energy condition, we see from (19) and (18), that we must require the condition: $(1 - |\varphi|^2) \in L^1(\mathbb{R}^2)$. We shall see that for solutions of (17)–(18), this property is essentially equivalent to the following (apparently stronger) condition:

$$|\varphi| \to 1 \ \& \ |x|^\beta|\nabla\varphi| \to 0 \text{ as } |x| \to +\infty, \qquad (22)$$

for some $\beta > 0$ (cf. [JT]). Therefore for planar vortices, the Higgs field φ can only admit a finite number of zeroes, whose total number (counted with multiplicity) we shall call the vortex number. We shall see that the vortex number bares a topological meaning according to which we can distinguish vortices in homotopic classes.

To this purpose, let $Z = \{z_1, ..., z_N\}$ be the zero-set of φ, and let N be the vortex number, so that each point in Z (known as vortex point) is repeated according to its multiplicity. Take $R > 0$ sufficently large so that $Z \subset\subset B_R$, and let ψ be a solution of (21) in B_R. Then (by using complex notations) we find,

$$e^{-i\psi}\varphi = \prod_{j=1}^{N}(z - z_j)e^{h(x)}, \qquad (23)$$

with h holomorphic in B_R. From (23) we can derive some interesting consequences.

Firstly,

$$N = \frac{1}{2\pi i} \int\limits_{\partial B_R} \frac{\frac{d}{dz}(e^{-i\psi}\varphi)}{e^{i\psi}\varphi} dz = -\frac{1}{2\pi} \int\limits_{\partial B_R} \frac{\partial \psi}{\partial z} dz + \frac{1}{2\pi i} \int\limits_{\partial B_R} \frac{\frac{d}{dz}\varphi}{\varphi} dz.$$

By virtue of (22), we readily deduce that,

$$\left| Im \left\{ \int_{\partial B_R} \frac{\frac{d}{dz}\varphi}{\varphi} \right\} \right| \to 0 \text{ as } R \to +\infty.$$

While,

$$\int\limits_{\partial B_R} \frac{\partial \psi}{\partial z} dz = 2i \int\int_{B_R} \frac{\partial^2 \psi}{\partial \overline{z} \partial z} dxdy = 2i \int\int_{B_R} \frac{\partial A}{\partial z} dxdy,$$

so,

$$Re \left(\int \frac{\partial \psi}{\partial z} dz \right) = -2Im \left(\int\int_{B_R} \frac{\partial A}{\partial z} dxdy \right) = - \int\int_{B_R} F_{12}.$$

Hence, by letting $R \to +\infty$, we find the <u>flux quantization</u> property:

$$\frac{1}{2\pi} \int\int_{\mathbb{R}^2} F_{12} = N = \text{ winding number of } \varphi = \deg(\varphi, \mathbb{R}^2, 0). \qquad (24)$$

In other words, for each $N \in \mathbb{N}$ a selfdual solution of (17)–(18) with vortex number N defines a <u>minimizer</u> of the energy within the class of configurations for which the Higgs field admits fixed Brower degree N (topological constraint). The corresponding minimal value of the energy is exactly πN. It is possible to characterize completely the class of such (energy minimizers) N-vortex configurations.

To this purpose, we still use (23) to observe that

$$\Delta \log(|e^{-i\psi}\varphi|^2) = 4\pi \sum_{j=1}^{N} \delta_{z_j},$$

where δ_z denotes the Dirac measure with pole at z. On the other hand, from (21) we see that $\Delta\{Im\psi\} = F_{12}$ and so we deduce:

$$-\Delta \log |\varphi|^2 = 2\Delta\{Im\psi\} - 4\pi \sum_{j=1}^{N} \delta_{z_j}.$$

By means of (18) we arrive at the following elliptic equation for $u = \log |\varphi|^2$,

$$-\Delta u = \frac{1}{q^2}(1 - e^u) - 4\pi \sum_{j=1}^{N} \delta_{z_j} \qquad (25)$$

satisfying:
$$u(x) \to 0, \text{ as } |x| \to +\infty.$$

Similarly, if we analyze the solutions of (17), (18) with the "minus"-sign, then by recalling that: $\partial_z \overline{f} = \overline{\partial_{\overline{z}} f}$, we can apply the argument above to $\overline{\varphi}$ (instead of φ) and use $-\psi$ (instead of ψ given by (21)), to deduce similar properties.

In particular in this case (24) holds as follows:

$$\frac{1}{2\pi} \int_{\mathbb{R}^2} F_{12} = \text{ winding number of } \varphi = \deg(\varphi, \mathbb{R}^2, 0) = -N \in \mathbb{Z}^-,$$

where $N = \deg(\overline{\varphi}, \mathbb{R}^2, 0)$ defines the vortex number.

Furthermore, we arrive at the same elliptic problem (25) for $u = \log |\varphi|^2 = \log |\overline{\varphi}|^2$. In fact in this case we have: $\Delta Im(-\psi) = -F_{12}$, and at the same time (18) is satisfied with the "minus"-sign.

In other words, we have seen that each (static) planar selfdual configuration $(\mathcal{A}, \varphi)$, solution of (17), (18) and (22) with "plus-sign" or with "minus-sign", identifies an homotopy class of $\pi_1(S^1) = \pi_1(U(1)) = \mathbb{Z}$, given by $\pm N = \deg(\varphi, \mathbb{R}^2, 0)$, where N is the vortex number and the $\pm$ sign is chosen accordingly. Recall that N counts (with multiplicity) the vortex points (i.e. the zeroes of φ). We also know that $(\mathcal{A}, \varphi)$ corresponds to a minimum of the energy among all static configurations (satisfying (22)) with the same homotopy. While, $u = \log |\varphi|^2$ is a solution for (25), (26), with $\{z_1, \ldots, z_N\}$ the corresponding vortex points.

This provides a complete analytical description of vortices for the abelian-Higgs model in the selfdual regime, once we take into account that the following holds:

Theorem 1 *(cf. [Ta2], [JT]) Every finite energy solution $(\mathcal{A}, \varphi)$ of (4) in $\mathbb{R}^2$ with V in (2) and $\lambda = \frac{1}{q^2}$ (i.e. a planar selfdual abelian-Higgs vortex) coincides with a (static) solution of (17)–(18) satisfying (22), and the following estimate holds:*
$\forall \varepsilon > 0, \exists C_\varepsilon > 0:$

$$|F_{12}| + |D_1\varphi| + |D_2\varphi| \leq \frac{1}{q^2}(1 - |\varphi|) \leq C_\varepsilon e^{-\frac{1-\varepsilon}{q}|x|}, \text{ in } \mathbb{R}^2. \qquad (26)$$

We mention that similar exponential decay estimates also hold away from the selfdual regime, i.e. when $\lambda \neq \frac{1}{q^2}$, see [JT].

To complete the study of (17), (18) we observe:

Theorem 2 *(cf. [Ta1], [JT]) Let u be a solution (in the sense of distribution) of (25) and suppose that $(1 - e^u) \in L^1(\mathbb{R}^2)$. Using complex notations set,*

$$\varphi^\pm(z) = e^{\frac{u}{2} \pm i \sum_{j=1}^N \arg(z-z_j)}, \quad \mathcal{A} = -iA_\alpha^\pm dx^\alpha \qquad (27)$$

where

$$A_0^{\pm} = 0, \ A_1^{\pm} \pm i A_2^{\pm} = -i\partial_{\pm}(\log \varphi),$$

with $\partial_{\pm} = \partial_1 \pm i\partial_2$. Then $(\mathcal{A}, \varphi)_{\pm}$ defines a static solution for (17), (18) (with the $\pm$ sign chosen accordingly) and with vortex points $\{z_1, \ldots, z_N\}$.
Furthermore, it satisfies (26) and the following holds:

1. $|\varphi| < 1$ in $\mathbb{R}^2$,
2. $\varphi^+(z) = O((z - z_j)^{n_j}) = \overline{\varphi^-}(z)$ as $z \to z_j$, with n_j the multiplicity of z_j, for $j = 1, \ldots, N$.
3.

$$\text{Magnetic flux } \Phi^{\pm} = \int_{\mathbb{R}^2} F_{12}^{\pm} = \pm 2\pi N; \ \text{Total energy} \int_{\mathbb{R}^2} \mathcal{E}^{\pm} = \pi N \quad (28)$$

Proof. Note that (27) follows easily by direct computations, while (28) is a simple consequence of our construction. We shall give indications of (26) in the following section and we refer to [JT] for more details. ∎

Theorem 2 reduces the search of abelian-Higgs vortices to the study of the (singular) elliptic problem (25) subject to the integrability condition:

$$(1 - e^u) \in L^1(\mathbb{R}^2).$$

This task will be carried out in the following section.

Next we wish to show that it is possible to specify the strenght of the Chern-Simons field and the scalar potential in order to attain a selfdual regime also for the Chern-Simons-Higgs model (9).
To this purpose, recalling (6) and (12), we see that the Chern-Simons energy is given as follows (where as usual we have neglected the total spatial divergence terms):

$$\mathcal{E}_{MCS} = \frac{q^2}{2}|F_{0j}|^2 + \frac{q^2}{2}F_{12}^2 + \frac{1}{2}(|D_0\varphi|^2 + |D_1\varphi|^2 + |D_2\varphi|^2) + V$$

In this case, a first way to attain selfduality is described by Jackiw-Weinberg [JW] and Hong-Kim-Pac [HKP], who propose to neglect the Maxwell term in the energy (i.e. set $q = 0$), and thus consider a model whose electrodynamics is governed solely by the Chern-Simons term.
In this situation, still neglecting total spatial divergence terms and by recalling (14), we find:

$$\mathcal{E}_{CSH} = \frac{1}{2}(|D_0\varphi|^2 + |D_1\varphi|^2 + |D_2\varphi|^2) + V =$$
$$= \frac{1}{2}|D_0\varphi \pm i\varphi W(|\varphi|)|^2 + \frac{1}{2}|D_{\pm}\varphi|^2 \pm \frac{1}{2}F_{12}(|\varphi|^2 - 1) + V(|\varphi|) \pm \frac{1}{2}F_{12}$$
$$\pm \frac{i}{2}(\overline{\varphi}D_0\varphi - \varphi\overline{D_0\varphi})W(|\varphi|) - \frac{1}{2}|\varphi|^2 W^2(|\varphi|)$$

with $W(|\varphi|)$ an arbitrary real valued function to be specified later. Hence, using (11) and the Gauss law (12) (with $q = 0$) we find:

$$\mathcal{E}_{CSH} = \frac{1}{2}|D_0\varphi \pm i\varphi W(|\varphi|)|^2 + \frac{1}{2}|D_\pm\varphi|^2 \pm \frac{1}{2}F_{12}(|\varphi|^2 - 1) \pm \kappa F_{12}W(|\varphi|)$$

$$+ V(|\varphi|) - \frac{1}{2}|\varphi|^2 W^2(|\varphi|) \pm \frac{1}{2}F_{12}.$$

Hence to obtain an energy similar to (15), we need to require that,

$$\frac{1}{2}(|\varphi|^2 - 1) + \kappa W(|\varphi|) = 0 \ \& \ V(|\varphi|) = \frac{1}{2}|\varphi|^2 W^2(|\varphi|).$$

Consequently (see [JW] and [HKP]) we must take:

$$\kappa W(|\varphi|) = \frac{1}{2}(1 - |\varphi|^2),$$

and the selfdual Chern-Simons scalar potential takes the form of a "triple" well potential given as follows:

$$V_{CS} = \frac{1}{8\kappa^2}|\varphi|^2(1 - |\varphi|^2)^2.$$

Thus, for the selfdual Chern-Simons-Higgs model:

$$\mathcal{L}_{CSH} = -\frac{\kappa}{4}\varepsilon^{\alpha\beta\gamma}A_\alpha F_{\beta\gamma} + \frac{1}{2}(|D_0\varphi|^2 - |D_1\varphi|^2 - |D_2\varphi|^2) - \frac{1}{8\kappa^2}|\varphi|^2(1 - |\varphi|^2)^2 \tag{29}$$

we obtain the following set of (time-dependent) selfdual equation:

$$\begin{cases} D_0\varphi = \pm\frac{i}{\kappa}\varphi(|\varphi|^2 - 1) \\ D_\pm\varphi = 0 \end{cases} \tag{30}$$

to be satisfied together with the Gauss law constraint:

$$\kappa F_{12} = \rho = \frac{i}{2}(\overline{\varphi}D_0\varphi - \varphi\overline{D_0\varphi}). \tag{31}$$

In particular, by putting together the first equation in (30) and (31) we get,

$$F_{12} = \pm\frac{1}{\kappa^2}|\varphi|^2(1 - |\varphi|^2). \tag{32}$$

To obtain a Chern-Simons-Higgs vortex configuration, we first solve with $\varphi = \varphi(x^1, x^2)$, $A_j = A_j(x_1, x_2)$, $j = 1, 2$, the following selfdual equations:

$$\begin{cases} D_\pm\varphi = 0 \\ F_{12} = \pm\frac{1}{\kappa^2}|\varphi|^2(1 - |\varphi|^2). \end{cases} \tag{33}$$

whose structure has much in common with the selfdual equations of the abelian-Higgs model.

Then we deduce a stationary solution for (30) once that we check the following:

Proposition 3 *Let $\hat{\varphi}$ and $\hat{\mathcal{A}} = (\hat{A}_1, \hat{A}_2)_\pm$ be a solution of (33) in $\mathbb{R}^2$ and set $A_0 = \pm\frac{1}{2\kappa}(1-|\hat{\varphi}|^2)$, $A_j = \hat{A}_j + (\partial_j A_0)x^0$ and $\varphi = e^{-ix_0 A_0}\hat{\varphi}$. Then $(\mathcal{A}, \varphi)$ with $\mathcal{A} = -i\hat{A}_\alpha dx^\alpha$, defines a stationary solution for the selfdual equation (30). In particular, the corresponding electric field, magnetic field and energy density define time-independent quantities such that the following relation holds:*

$$Magnetic\ flux = \pm\frac{1}{2}\ Total\ Energy = \frac{1}{\kappa}Electric\ charge \left(= \int \hat{F}_{12}dx^1 dx^2\right)$$

As above, problem (33) involves as unknowns two complex functions, namely: φ, $A_1 \pm iA_2 : \mathbb{C} \to \mathbb{C}$, which however are defined only up to gauge transformations.

To solve (33) we can use the exact same argument presented for the abelian-Higgs model, and show that via (27) we are reduced to find solutions for the elliptic problem:

$$-\Delta u = \frac{1}{\kappa^2}e^u(1 - e^u) - 4\pi\sum_{j=1}^{N}\delta_{z_j}\ in\ \mathbb{R}^2, \tag{34}$$

where $u = \log|\varphi|^2$, N is the vortex number and $z_1, \ldots, z_N$ are the corresponding vortex points (repeated with multiplicity).

The finite energy condition now requires that:

$$\int_{\mathbb{R}^2} e^u(1 - e^u) < +\infty. \tag{35}$$

In principle, property (35) allows for two type of admissible boundary conditions, namely:

$$u \to 0\ as\ |x| \to +\infty\ \ or\ \ u \to -\infty\ as\ |x| \to +\infty.$$

To have physically meaningful solutions we also need that,

$$u < 0\quad (i.e.\ |\varphi| < 1)\ in\ \mathbb{R}^2. \tag{36}$$

So, now it makes sense to consider the following two boundary value problems:

$$(I)\quad \begin{cases} -\Delta u = \frac{1}{\kappa^2}e^u(1 - e^u) - 4\pi\sum_{j=1}^{N}\delta_{z_j} \\ u < 0 \\ \int_{\mathbb{R}^2}(1 - e^u) < +\infty \end{cases}$$

and

$$(II) \quad \begin{cases} -\Delta u = \frac{1}{\kappa^2} e^u (1 - e^u) - 4\pi \sum_{j=1}^{N} \delta_{z_j} \\ u < 0 \\ \int_{\mathbb{R}^2} e^u < +\infty \end{cases}$$

We shall see that problem (I) shares many features with the elliptic problem corresponding to the abelian-Higgs model; while problem (II) admits features common to the singular Liouville equations (cf. [Lio]) and it will be characterized by other interesting analytical aspects.

However, contrary to the abelian-Higgs model (see Theorem 1), it is still open the question of whether Chern-Simons vortex configurations are fully described (via Proposition 3) by (stationary) solutions of the selfdual equation (30) with finite energy, or, on the contrary, non-selfdual solutions exist, as it occurs for Yang-Mills fields (cf. [Ta3]).

For the moment (still using complex notations) we can easily check the following:

Proposition 4 *For given* $\{z_1, \ldots, z_N\} \subset \mathbb{C}$*, let* u *be a solution of (34), (35) and (36). Then*

$$\varphi(z) = e^{\frac{u}{2} \pm i \sum_{j=1}^{N} \arg(z - z_j)} \quad \& \quad A_1 \pm i A_2 = -i \partial_\pm \log \varphi$$

defines a solution for the selfdual equations (33), with $\pm$ *sign chosen accordingly, such that:*

(a) $|\varphi| < 1$*, and* φ *vanishes exactly at* $\{z_1, \ldots, z_N\}$*;*
(b)

$$\int_{\mathbb{R}^2} F_{12} = \pm 2\pi N.$$

In concluding this section, we mention that after [JW] and [HKP], many other selfdual Chern-Simons models have been introduced, including the Maxwell term and in the non-abelian setting, see e.g. [D1], [D2] [LLM], [CaL], [KiKi], [Va] and references therein.

For those models it has been possible to use an analogous approach to deduce results with a similar flavor about their vortex configurations.

In this respect, we bring to the reader's attention the following work: [Ch3], [ChCh1], [ChK1], [ChK2], [ChNa], [DJLW2], [Ha1], [Ha2], [Ha3], [JoLW], [JoW1], [JoW2], [LN], [NT1], [NT3], [Ol], [Ri1], [Ri2], [Ri3], [RT1], [SY1], [T1], [T2], [T3], [Wa] and [Y3].

4 On the Elliptic Problem Concerning Abelian-Higgs Vortices

In the previous section we have seen how the selfdual vortex problem for both abelian-Higgs and Chern-Simons-Higgs model can be reduced to the study of elliptic problems involving exponential nonlinearities and including Dirac measures supported at the vortex points.

In this and the following section, we show how to approach such elliptic problems by means of variational techniques. To this purpose, it will be convenient to use complex notations by the usual identification: $\mathbb{R}^2 \simeq \mathbb{C}$.

We shall focus mainly on the planar case, and study vortex configurations in $\mathbb{R}^2$. See [T2], [T3], [T8] and references therein, for the study of periodic vortices (as motivated by [Ab], [Ol], ['tH1] and ['tH2]) or more generally vortices defined on compact surfaces (as motivated in [Bra1], [Ga1] [Hi] and [KiKi]).

We start in this section to analize the elliptic problem relative to abelian-Higgs vortices.

Hence, for a given $N \in \mathbb{N}$ and $\{z_1, \dots z_N\} \subset \mathbb{C}$ a set of N-points (not necessarily distinct), we seek solutions $u \in C^2(\mathbb{R}^2 \setminus \{z_1, \dots z_N\})$ for the problem:

$$-\Delta u = \frac{1}{q^2}(1 - e^u) - 4\pi \sum_{j=1}^{N} \delta_{z_j} \text{ in } \mathbb{R}^2 \tag{37}$$

$$(1 - e^u) \in L^1(\mathbb{R}^2) \tag{38}$$

where (37) is satisfied in the sense of distributions.

Clearly,

$$u(z) = O(2n_j \log(|z - z_j|)) \text{ as } z \to z_j,$$

(n_j is the multiplicity of z_j), $j = 1, \dots, N$.

So $e^u \in C^2(\mathbb{R}^2)$, and it vanishes exactly at each z_j with multiplicity $2n_j$, for every $j = 1, \dots, N$.

The following holds:

Proposition 5 *Let $u \in C^2(\mathbb{R}^2 \setminus \{z_1, \dots z_N\})$ be a solution (in the sense of distributions) of (37), (38). Then,*

i)

$$u < 0, \tag{39}$$

ii)

$$|\nabla u(z)| + |u(z)| \to 0 \text{ as } |z| \to +\infty \tag{40}$$

iii)

$$\int_{\mathbb{R}^2} (1 - e^u) = 4\pi N. \tag{41}$$

Proof. To establish i) we start by showing that u admits positive part $u_+ \equiv 0$. To this purpose, we argue by contradiction and assume that $\Omega_+ = \{z : u(z) > 0\}$ is non-empty. In other words, Ω_+ defines a non-empty open set disjoint from $\{z_1, \ldots, z_N\}$. Since $0 \leq u_+ \leq (e^u - 1) \in L^1(\mathbb{R}^2)$, we see that $u_+ \in L^1(\mathbb{R}^2)$ and it satisfies:

$$-\Delta u_+ = \frac{1}{q^2}(1 - e^{u_+}) \leq 0 \text{ in } \Omega_+. \tag{42}$$

So u_+ defines a (smooth) superharmonic function in Ω_+, and we can use the mean value theorem to check that $u_+ \to 0$ as $|x| \to +\infty$. In particular, u_+ is bounded from above in Ω_+, where it attains its maximum point, say at $z_0 \in \Omega_+$. Hence $u_+(z_0) > 0$ and $\Delta u_+(z_0) \leq 0$. But this is impossible since by (42) we find:

$$0 \geq \Delta u_+(z_0) = \frac{1}{q^2}(e^{u_+} - 1) > 0,$$

an absurd. Hence, $u_+ \equiv 0$, and so $u \leq 0$. But at this point we can use the strong maximum principle to conclude that $u < 0$ as claimed.

To establish (40), we argue again by contradiction and assume there exist: $l > 0$ and a sequence $\{y_n\} \subset \mathbb{C}$ such that, $|y_n| \to +\infty$ and $u(y_n) \leq -l$, $\forall n \in \mathbb{N}$. Without loss of generality we can further assume that $\{z_1, \ldots, z_N\} \cap B_1(y_n) = \emptyset$, $\forall n \in \mathbb{N}$ and $B_1(y_n) \cap B_1(y_m) = \emptyset$, for $n \neq m$.

We can use Harnack estimates to find two universal constants $c_1 > 0$ and $c_2 > 0$ such that,

$$\sup_{B_1(y_n)} u \leq c_1 \inf_{B_1(y_n)} u + c_2 \|1 - e^u\|_{L^2(B_2(y_n))}, \quad \forall n \in \mathbb{N}.$$

Since

$$\|1 - e^u\|_{L^2(B_2(y_n))} \leq \|1 - e^u\|_{L^1(B_2(y_n))} \to 0, \text{ as } n \to \infty,$$

from the inequality above, we deduce that

$$\sup_{B_1(y_n)} u \leq -c_1 l + o(1), \text{ as } n \to \infty.$$

Consequently, we find $n_0 \in \mathbb{N}$ and a suitable constant $\beta > 0$, such that

$$\inf_{B_1(y_n)} (1 - e^u) \geq \beta, \forall n \geq n_0.$$

Since the balls $B_1(y_n)$ are mutually disjoint, this clearly violates the interability of $(1 - e^u)$. Thus, we must have that necessarily: $\limsup_{|z| \to +\infty} u = \liminf_{|z| \to +\infty} u = 0$, that is, $u \to 0$ as $|z| \to +\infty$.

To establish ii) we only need to use the following well known gradient estimates for Poisson's equation (cf. [GT]),

$$|\nabla u(y)| \leq C \left(\sup_{B_1(y)} |u| + \sup_{B_1(y)} |\Delta u| \right) \tag{43}$$

which holds with an universal constant $C > 0$, provided $B_1(y)$ does not contain any of the vortex points z_j, for every $j = 1, \ldots, N$. Thus, for $R > \max_{j=1,\ldots,N}(|z_j| + 2)$, we get,

$$\sup_{|y| \geq R} |\nabla u| \leq C \left(\sup_{|y| \geq R-1} |u| + \sup_{|y| \geq R-1} \frac{1}{q^2}(1 - e^u) \right) \tag{44}$$

which allow us to complete the proof of ii). Finally, to obtain (41), we integrate (37) over the ball B_R. Then by letting $R \to +\infty$, we deduce the desired conclusion by Green-Gauss theorem and by virtue of ii).   ∎

The "quantized" value of the integral (41) is consistent with the discussion of the previous section.

We show next that the convergence in (40) actually holds exponentially fast.

Proposition 6 *Let* $u \in C^2(\mathbb{R}^2 \setminus \{z_1, \ldots, z_N\})$ *satisfy (in the sense of distributions) (37) and assume that (38) holds. Then, for every* $0 < \varepsilon < 1$, *there exists a constsant* $C_\varepsilon > 0$:

$$0 < 1 - e^u \leq C_\varepsilon e^{-\frac{1-\varepsilon}{q}|x|}, \quad \forall x \in \mathbb{R}^2. \tag{45}$$

Proof. By virtue of (40) we can fix $R_\varepsilon > 0$ sufficiently large, so that:

$$q|\nabla u(z)| + (1 - e^{u(z)}) < \varepsilon, \quad \forall z : |z| \geq R_\varepsilon. \tag{46}$$

For $R \geq R_\varepsilon$, we introduce the function:

$$\psi(z) = e^{\frac{1-\varepsilon}{q}(R-|z|)} - (1 - e^{u(z)}).$$

It satisfies:

$$\begin{cases} \Delta\psi = \frac{1}{q^2}e^{\frac{1-\varepsilon}{q}(R-|z|)}\left((1-\varepsilon)^2 - \frac{(1-\varepsilon)q}{|z|} - e^u\right) + \frac{1}{q^2}e^u\psi + e^u|\nabla u|^2 \\ \psi_{|\partial B_R} > 0 \ \& \ \psi \to 0 \text{ as } |z| \to +\infty. \end{cases}$$

We claim that,

$$\psi \geq 0, \forall |z| \geq R.$$

Indeed, if on the contrary we suppose $\inf_{|z| \geq R} \psi < 0$, then ψ would assume its minimum value at a point, say z_0, where we have:

$$\psi(z_0) < 0 \leq \Delta\psi(z_0) \ \& \ \nabla\psi(z_0) = -\frac{1-\varepsilon}{q}e^{\frac{1-\varepsilon}{q}(R-|z_0|)}\frac{z_0}{|z_0|} + e^{u(z_0)}\nabla u(z_0) = 0$$

Consequently,

$$0 \le \Delta\psi(z_0) =$$
$$\frac{1}{q^2}e^{\frac{1-\varepsilon}{q}(R-|z_0|)}\left((1-\varepsilon)^2 - \frac{(1-\varepsilon)q}{|z_0|} - e^{u(z_0)} + (1-\varepsilon)q\frac{z_0}{|z_0|}\cdot\nabla u(z_0)\right)$$
$$+\frac{1}{q^2}e^{u(z_0)}\psi(z_0) \le \frac{1}{q^2}e^{\frac{1-\varepsilon}{q}(R-|z_0|)}(-2\varepsilon + \varepsilon^2 + q|\nabla u(z_0)| + (1 - e^{u(z_0)})) < 0,$$

a contradiction. Thus we have established that:

$$\forall \varepsilon > 0, \exists R_\varepsilon > 0 : \forall R \ge R_\varepsilon, 0 < 1 - e^{u(z)} \le e^{\frac{1-\varepsilon}{q}(R-|z|)}, \qquad (47)$$

and clearly (45) holds with $C_\varepsilon = e^{\frac{1-\varepsilon}{q}R_\varepsilon}$. ∎

By combining (43) and (45), we conclude:

Corollary 7 Let $u \in C^2(\mathbb{R}^2 \setminus \{z_1,\ldots,z_N\})$ satisfy (37) (in the sense of distributions) and assume that (38) holds. For every $\varepsilon \in (0,1)$ and $\delta > 0$ there exists a constant $C_{\varepsilon,\delta} > 0$ such that,

$$|\nabla u| + |u| \le C_{\varepsilon,\delta}e^{-\frac{1-\varepsilon}{q}|x|}, \ \forall x \in \mathbb{R}^2 \setminus \bigcup_{j=1}^{N} B_\delta(z_j).$$

By the way we have obtained the decay estimates above, it may seem that all constants there involved depend on the given solution. As a matter of fact, it is worth to note that the above estimates hold <u>uniformly</u> with respect to the solution and the given data (namely, the vortex points and the parameter $q > 0$) as follows:

Lemma 8 Given $\varepsilon \in (0,1)$, $N \in \mathbb{N}$ and $\rho > 0$, there exist $R_\varepsilon = R_\varepsilon(N,\rho) > 0$ such that every solution u of (37), (38) with $\max_{j,k=1,\ldots,N}|z_j - z_k| \le \rho$ and $0 < q \le 1$ satisfies:

$$0 < 1 - e^{u(z)} < \varepsilon, \ \forall z \in \mathbb{R}^2 \setminus \left(\bigcup_{j=1}^{N} B_\delta(z_j)\right)$$

for every $\delta \ge R_\varepsilon$.

Proof. By means of a translation, we can always be reduced to show the following:
<u>Claim:</u> if $|z_j| \le \rho$ for every $j = 1,\ldots,N$, then

$$0 < 1 - e^{u(z)} < \varepsilon, \ \forall|z| \ge R_\varepsilon.$$

To establish the Claim, we argue by contradiction and assume there exist $\varepsilon_0 \in (0,1)$, a solution u_n of (37) and (38) with $z_j = z_j^n$ and $q = q_n$, satisfying:

$0 < q_n \leq 1$ and $|z_j^n| < \rho$ for every $j = 1, \ldots, N$ and $\forall n \in \mathbb{N}$; such that, for suitable $R_n \to +\infty$ and $y_n \in \mathbb{R}^2 \setminus B_{2R_n}$ we have,

$$(1 - e^{u_n(y_n)}) \geq \varepsilon_0.$$

Since $e^{u_n(x)} \to 1$, as $|x| \to +\infty$, we can actually claim that for every $0 < \varepsilon \leq \varepsilon_0$, there exist a sequence $x_n \in \mathbb{R}^2 \setminus B_{2R_n}$ such that: $e^{u_n(x_n)} = 1 - \varepsilon$. Observe that $\frac{x_n - z_j^n}{q_n} \notin B_{R_n}(0)$, $\forall j = 1, \ldots, N$ and $\forall n \in \mathbb{N}$ sufficiently large. Thus, setting $U_n(x) = u_n(x_n + q_n x)$, $x \in B_{R_n}(0)$; we see that, $\forall n \in \mathbb{N}$ sufficiently large, U_n satisfies:

$$\begin{cases} -\Delta U_n = 1 - e^{U_n}, \text{ in } B_{R_n}(0) \\ U_n(0) = \log(1 - \varepsilon) \\ U_n < 0; \quad \int\limits_{B_{R_n}(0)} (1 - e^{U_n}) \leq 4\pi N. \end{cases} \tag{48}$$

By Harnack estimates, we deduce that U_n is uniformly bounded on compact sets of $\mathbb{R}^2$. Hence by elliptic regularity theory and a diagonalization procedure, we find a subsequence (denoted the same way), such that $U_n \to U$ in $C_{\text{loc}}^{2,\alpha}(\mathbb{R}^2)$ (for some $\alpha \in (0,1)$) and U satisfies:

$$\begin{cases} -\Delta U = 1 - e^U \text{ in } \mathbb{R}^2 \\ U(0) = \log(1 - \varepsilon), \ U < 0 \\ \int\limits_{\mathbb{R}^2} (1 - e^U) \leq 4\pi N \end{cases} \tag{49}$$

As in Proposition 5 we check that $U(x) \to 0$ as $|x| \to +\infty$, and $U \not\equiv 0$. So (being U regular) it must attain its minimum value at a point $x_0 \in \mathbb{R}^2$ and there holds: $U(x_0) < 0 \leq \Delta U(x_0) = e^{U(x_0)} - 1 < 0$, an absurd. ∎

Remark 9 *Notice that, by the analysis above, we can also treat the case $q > 1$ via the transformations:*

$$u(z) \to u(qz); \ z_j \to \frac{z_j}{q} \ j = 1, \ldots, N, \tag{50}$$

which allows one to reduce the case $q > 1$ to the case $q = 1$.

Since $R_\varepsilon > 0$ in Lemma 8 is indipendent of q, we can argue as in Proposition 6 and (by means of (44)) obtain the following uniform exponential decay estimates:

Corollary 10 *Given $\varepsilon \in (0,1)$, $N \in \mathbb{N}$ and $\rho > 0$, there exist $\hat{R}_\varepsilon = \hat{R}_\varepsilon(N, \rho) > 0$ and a constant $C_\varepsilon = C_\varepsilon(N, \rho) > 0$ such that every solution u of (37), (38) with $\max\limits_{j=1,\ldots,N} |z_j| \leq \rho$ and $0 < q \leq 1$ satisfies:*

$$0 < 1 - e^u \leq C_\varepsilon e^{-\frac{1-\varepsilon}{q}|x|}, \quad \forall |x| \geq \hat{R}_\varepsilon; \tag{51}$$

$$|\nabla u| + |u| \leq C_\varepsilon e^{-\frac{1-\varepsilon}{q}|x|}, \quad \forall |x| \geq \hat{R}_\varepsilon. \tag{52}$$

Next, we direct our attention to show that problem (37), (38) is uniquely solvable. To this purpose, by scaling:

$$u(z) \to u(qz); \; z_j \to \frac{z_j}{q} \; j = 1, \ldots, N, \tag{53}$$

we can always assume that,

$$q = 1.$$

For fixed $\mu > 0$ sufficiently large (to be specified below), we consider the function:

$$u_0(z) = \sum_{j=1}^{N} \log \frac{|z - z_j|^2}{\mu + |z - z_j|^2} \tag{54}$$

Hence,

$$u_0 < 0 \text{ in } \mathbb{R}^2, and \; u_0(z) = O\left(\frac{1}{|z|^2}\right) \text{ as } |z| \to +\infty.$$

So, $u_0 \in L^p(\mathbb{R}^2)$, $\forall 1 < p < +\infty$; while $u_0 \in L^1_{\text{loc}}(\mathbb{R}^2)$ and it satisfies (in the sense of distributions):

$$-\Delta u_0 = g_0 - 4\pi \sum_{j=1}^{N} \delta_{z_j} \tag{55}$$

with

$$g_0(z) = 4\mu \sum_{j=1}^{N} \frac{1}{(\mu + |z - z_j|^2)^2} \in L^1(\mathbb{R}^2) \cap L^\infty(\mathbb{R}^2). \tag{56}$$

Since $\|g_0\|_{L^2(\mathbb{R}^2)} \to 0$ as $\mu \to +\infty$, from now on we shall fix $\mu > 0$ so that,

$$\|g_0\|_{L^2(\mathbb{R}^2)} < \frac{1}{2}. \tag{57}$$

Set

$$u = u_0 + v. \tag{58}$$

We have:

Proposition 11 *According to the decomposition (58), $u \in C^2(\mathbb{R}^2 \setminus \{z_1, \ldots, z_N\})$ satisfies:*

$$\begin{cases} -\Delta u = 1 - e^u - 4\pi \sum_{j=1}^{N} \delta_{z_j}, & in \; \mathbb{R}^2 \\ (1 - e^u) \in L^1(\mathbb{R}^2) \end{cases} \tag{59}$$

if and only if v satisfies:

$$\begin{cases} -\Delta v = 1 - e^u - g_0, & in \ \mathbb{R}^2 \\ v \in H^1(\mathbb{R}^2). \end{cases} \tag{60}$$

Proof. If $u \in C^2(\mathbb{R}^2 \setminus \{z_1, \ldots, z_N\})$ satisfies (59), then v in (58) defines a smooth solution of (60) where both functions e^{u_0} and $g_0 \in C^\infty(\mathbb{R}^2)$. Furthermore, $v \to 0$ as $|z| \to +\infty$, and $v \in L^p(\mathbb{R}^2)$, $\forall p > 1$. In particular $v \in L^2(\mathbb{R}^2)$.

Denote by $\chi \in C_0^\infty(\mathbb{R}^2)$ a standard cut off function, so that $0 \leq \chi \leq 1$, $\chi_{|B_1(0)} \equiv 1$ and $\chi \equiv 0$ in $\mathbb{R}^2 \setminus B_2(0)$. For $R > 0$, set $\chi_R(z) = \chi\left(\frac{z}{R}\right)$. We use $v\chi_R$ as a test function in the equation (60) to obtain,

$$-\int_{R^2} (\Delta v)v\chi_R = \int_{R^2} (1 - e^u)v\chi_R - \int_{R^2} g_0 v\chi_R \to \int_{R^2} (1 - e^u)v - \int_{R^2} g_0 v,$$

as $R \to +\infty$, which follows by Lebesgue dominated convergence theorem.

On the other hand,

$$-\int_{R^2} (\Delta v)v\chi_R = \int_{R^2} |\nabla v|^2 \chi_R + \int_{R^2} v\nabla v \cdot \nabla \chi_R,$$

and

$$\left| \int_{R^2} v\nabla v \cdot \nabla \chi_R \right| = \frac{1}{2} \left| \int_{R^2} \nabla v^2 \cdot \nabla \chi_R \right| \leq \frac{1}{2} \int_{R \leq |x| \leq 2R} v^2 |\Delta \chi_R| =$$

$$= \frac{1}{2} \max_{R \leq |x| \leq 2R} v^2 \int_{1 \leq |x| \leq 2} |\Delta \chi| \to 0 \text{ as } R \to \infty.$$

Consequently, $|\nabla v| \in L^2(\mathbb{R}^2)$ and so $v \in H^1(\mathbb{R}^2)$, as claimed.

Viceversa, assume that (60) holds, namely $v \in H^1(\mathbb{R}^2)$ and

$$\int_{\mathbb{R}^2} \nabla v \cdot \nabla \varphi = \int_{\mathbb{R}^2} (1 - e^u)\varphi - \int_{\mathbb{R}^2} g_0 \varphi, \ \forall \varphi \in H^1(\mathbb{R}^2). \tag{61}$$

By well known elliptic estimates, it follows that $v \in C^\infty(\mathbb{R}^2)$ and $v \to 0$ as $|z| \to +\infty$. Consequently, $u \in C^\infty(\mathbb{R}^2 \setminus \{z_1, ..., z_N\})$ and $u(z) \to 0$, as $|z| \to +\infty$. So (as in Proposition 5), we see that $u < 0$. Furthermore, if we use (61) with $\varphi = \chi_R$, arguing as above, we find that $(1 - e^u) \in L^1(\mathbb{R}^2)$ as claimed. ∎

The advantage of introducing the equivalent problem (60) is that it admits a variational formulation.

To this purpose, recall that if $v \in H^1(\mathbb{R}^2)$, then $(e^v - 1 - v) \in L^1(\mathbb{R}^2)$ and $(e^v - 1) \in L^p(\mathbb{R}^2)$, $\forall p \geq 2$. The proof of those integrability properties can be found in [JT], [Y1] and [T8], and follows by an approach of Trudinger [Tr] introduced to show similar integrability properties on bounded domains via an inequality that subsequently has been re-derived in its sharp form by Moser in [Mo].

Remark 12 *Incidentally, let us mention that many other sharp versions of the Moser-Trudinger inequality are now available over manifolds and in the context of systems, (see e.g. [Ad], [Au], [Ban], [Be], [ChY3], [Che], [CSW], [DET], [Fo], [JoW1], [JoLW], [NT2], [On], [SW1], [SW2] and [W]). All those inequalities have proved extremely usefull in the study of related variational problems.*

Observing that $0 \leq e^{u_0} < 1$ and that $(1 - e^{u_0}) \in L^2(\mathbb{R}^2)$, we can consider the functional:

$$J(v) = \frac{1}{2} \int_{\mathbb{R}^2} |\nabla v|^2 + \int_{\mathbb{R}^2} e^{u_0}(e^v - 1 - v) + \int_{\mathbb{R}^2} (e^{u_0} - 1 + g_0)v, \qquad (62)$$

$\forall v \in H^1(\mathbb{R}^2)$.

It is easy to check that $J \in C^1(H^1(\mathbb{R}^2))$, and every critical point of J satisfies (61). Therefore, by well known elliptic regularity theory, critical points of J define solutions for (60).

To obtain critical points for J we prove the following:

Proposition 13 *The functional J is (strictly) convex and satisfies:*

$$J(v) \geq a\|v\| - b, \ \forall v \in H^1(\mathbb{R}^2), \qquad (63)$$

for suitable $a, b > 0$.

Proof. The convexity of J easily follows once we observe that

$$J(v) = \frac{1}{2} \int_{\mathbb{R}^2} |\nabla v|^2 + \int_{\mathbb{R}^2} e^{u_0}(e^v - 1) + \int_{\mathbb{R}^2} (g_0 - 1)v.$$

To establish (63), we use the inequality:

$$\|v\|^2_{L^2(\mathbb{R}^2)} \leq 2(1 + 2\|\nabla v\|^2_{L^2(\mathbb{R}^2)}) \int_{\mathbb{R}^2} \frac{v^2}{(1 + |v|)^2}, \ \forall v \in H^1(\mathbb{R}^2) \qquad (64)$$

which can be checked as follows:

$$\|v\|_{L^2(\mathbb{R}^2)}^2 = \int_{\mathbb{R}^2} \frac{v}{1+|v|} v(1+|v|) \leq \left(\int_{\mathbb{R}^2} \frac{v^2}{(1+|v|)^2} \right)^{\frac{1}{2}} \left(\int_{\mathbb{R}^2} v^2(1+|v|)^2 \right)^{\frac{1}{2}} \leq$$

$$\leq \left(\int_{\mathbb{R}^2} \frac{v^2}{(1+|v|)^2} \right)^{\frac{1}{2}} \left(2 \int_{\mathbb{R}^2} (v^2+v^4) \right)^{\frac{1}{2}} \leq$$

$$\leq \left(\int_{\mathbb{R}^2} \frac{v^2}{(1+|v|)^2} \right)^{\frac{1}{2}} \sqrt{2} \|v\|_{L^2(\mathbb{R}^2)} (1+2\|\nabla v\|_{L^2(\mathbb{R}^2)}^2)^{\frac{1}{2}},$$

where we have used the well known Sobolev inequality:

$$\int_{\mathbb{R}^2} v^4 \leq 2\|v\|_{L^2(\mathbb{R}^2)}^2 \|\nabla v\|_{L^2(\mathbb{R}^2)}^2.$$

Furtheremore, if we write $v = v_+ - v_-$ ($v_\pm$ respectively the positive and negative part of v) then $v_\pm \in H^1(\mathbb{R}^2)$ and $J(v) = J(v_+) + J(-v_-)$. So it suffices to check (63) under the assumption that v does not change sign.

Case 1. $v \leq 0$ a.e. in $\mathbb{R}^2$.

We have:

$$J(v) = \frac{1}{2} \int_{\mathbb{R}^2} |\nabla v|^2 + \int_{\mathbb{R}^2} (e^{u_0}-1-u_0)(e^v-1) + \int_{\mathbb{R}^2} (e^v-1-v) + \int_{\mathbb{R}^2} u_0(e^v-1) + \int_{\mathbb{R}^2} g_0 v.$$

Observing that $(e^{u_0} - 1 - u_0) \geq 0$, we use the inequality: $\frac{|t|}{1+|t|} \leq |e^t - 1|$ to estimate:

$$J(v) \geq \frac{1}{2} \int_{\mathbb{R}^2} |\nabla v|^2 - \int_{\mathbb{R}^2} \frac{|v|}{1+|v|} (e^{u_0}-1-u_0) + \int_{\mathbb{R}^2} \frac{v^2}{1+|v|} + \int_{\mathbb{R}^2} g_0 v \geq$$

$$\geq \frac{1}{2} \int_{\mathbb{R}^2} |\nabla v|^2 - \left(\int_{\mathbb{R}^2} \frac{v^2}{(1+|v|)^2} \right)^{\frac{1}{2}} \left(\int_{\mathbb{R}^2} (e^{u_0}-1-u_0)^2 \right)^{\frac{1}{2}} + \int_{\mathbb{R}^2} \frac{v^2}{(1+|v|)^2} + \int_{\mathbb{R}^2} g_0 v \geq$$

$$\geq \frac{1}{2} \int_{\mathbb{R}^2} |\nabla v|^2 - \frac{1}{2} \int_{\mathbb{R}^2} \frac{v^2}{(1+|v|)^2} - \frac{1}{2} \int_{\mathbb{R}^2} (e^{u_0}-1-u_0)^2 + \int_{\mathbb{R}^2} \frac{v^2}{(1+|v|)^2} + \int_{\mathbb{R}^2} g_0 v \geq$$

$$\geq \frac{1}{2} \int_{\mathbb{R}^2} |\nabla v|^2 + \frac{1}{2} \int_{\mathbb{R}^2} \frac{v^2}{(1+|v|)^2} + \int_{\mathbb{R}^2} g_0 v - C_1,$$

with $C_1 > 0$ a suitable constant. Thus, by using (64) we find:

$$J(v) \geq \frac{1}{2} \int_{\mathbb{R}^2} |\nabla v|^2 + \frac{\|v\|_{L^2(\mathbb{R}^2)}^2}{4(1 + 2\|\nabla v\|_{L^2(\mathbb{R}^2)}^2)} + \int_{\mathbb{R}^2} g_0 v - C_1 \geq$$

$$\geq \frac{1}{2}\left(\|\nabla v\|_{L^2(\mathbb{R}^2)}^2 + \frac{\|v\|_{H^1(\mathbb{R}^2)}^2}{2(1+2\|\nabla v\|_{L^2(\mathbb{R}^2)}^2)}\right) - \frac{\|\nabla v\|_{L^2(\mathbb{R}^2)}^2}{4(1+2\|\nabla v\|_{L^2(\mathbb{R}^2)}^2)} + \int_{\mathbb{R}^2} g_0 v - C_1 \geq$$

$$\geq \frac{1}{2}\left(\|\nabla v\|_{L^2(\mathbb{R}^2)}^2 + \frac{\|v\|_{H^1(\mathbb{R}^2)}^2}{2(1 + 2\|\nabla v\|_{L^2(\mathbb{R}^2)}^2)}\right) - \|g_0\|_{L^2(\mathbb{R}^2)}\|v\|_{L^2(\mathbb{R}^2)} - \frac{1}{8} - C_1$$

$$\geq \frac{1}{2}\|v\|_{H^1(\mathbb{R}^2)} - \|g_0\|_{L^2(\mathbb{R}^2)}\|v\|_{H^1(\mathbb{R}^2)} - C,$$

with $C > 0$ a suitable constant. Recalling (57), we conclude that (63) holds in this case.

Case 2. $v \geq 0$ a.e. in $\mathbb{R}^2$.

We have:

$$J(v) = \frac{1}{2} \int_{\mathbb{R}^2} |\nabla v|^2 + \int_{\mathbb{R}^2} e^{u_0}(e^v - 1 - v) - \int_{\mathbb{R}^2}(1 - e^{u_0})v + \int_{\mathbb{R}^2} g_0 v \geq$$

$$\geq \frac{1}{2} \int_{\mathbb{R}^2} |\nabla v|^2 + \frac{1}{2} \int_{\mathbb{R}^2} e^{u_0} v^2 - \int_{\mathbb{R}^2}(1 - e^{u_0})v + \int_{\mathbb{R}^2} g_0 v.$$

Since $u_0 \to 0$ as $|x| \to \infty$, we fix $R \geq 1$ large enough so that $e^{u_0} \geq \frac{1}{2}$ for $x \in B_R^c = \mathbb{R}^2 \setminus B_R$. Set:

$$M := \frac{1}{\pi R^2} \int_{B_R} v\,dx \geq 0$$

the average of v over B_R. We estimate,

$$J(v) \geq \frac{1}{2}\int_{B_R} |\nabla v|^2 + \frac{1}{2} \int_{B_R} e^{u_0}(v - M + M)^2 - \int_{B_R}(v - M)(1 - e^{u_0}) - M\int_{B_R}(1 - e^{u_0}) +$$

$$+ \frac{1}{2} \int_{B_R^c} |\nabla v|^2 + \frac{1}{4} \int_{B_R^c} v^2 - \|1 - e^{u_0}\|_{L^2(B_R^c)}\|v\|_{L^2(B_R^c)} + \int_{\mathbb{R}^2} g_0 v \geq$$

$$\geq \frac{1}{2} \int_{B_R} |\nabla v|^2 + \frac{1}{2} \int_{B_R} e^{u_0}(v-M)^2 + M \int_{B_R} e^{u_0}(v-M) + \frac{M^2}{2} \int_{B_R} e^{u_0} -$$

$$- \|v-M\|_{L^2(B_R)} \|1 - e^{u_0}\|_{L^2(B_R)} - M \int_{B_R} (1-e^{u_0}) +$$

$$+ \frac{1}{2} \int_{B_R^c} |\nabla v|^2 + \frac{1}{8} \int_{B_R^c} v^2 - 2\|1 - e^{u_0}\|_{L^2(B_R^c)}^2 + \int_{\mathbb{R}^2} g_0 v.$$

By recalling the Poincaré inequality,

$$\left(\int_{B_R} \left(v - \frac{1}{\pi R^2} \int_{B_R} v \right)^2 \right)^{\frac{1}{2}} = \left(\int_{B_R} (v-M)^2 \right)^{\frac{1}{2}} \leq 4R \left(\int_{B_R} |\nabla v|^2 \right)^{\frac{1}{2}}$$
(65)

we get:

$$J(v) \geq \frac{1}{2} \int_{B_R} |\nabla v|^2 + \frac{1}{2} \int_{B_R} e^{u_0}(v-M)^2 + M \int_{B_R} e^{u_0}(v-M) + \frac{M^2}{2} \int_{B_R} e^{u_0} -$$

$$-4R\|1 - e^{u_0}\|_{L^2(B_R)} \|\nabla v\|_{L^2(B_R)} - M \int_{B_R} (1-e^{u_0}) +$$

$$+ \frac{1}{2} \int_{B_R^c} |\nabla v|^2 + \frac{1}{8} \int_{B_R^c} v^2 - 2\|1 - e^{u_0}\|_{L^2(B_R^c)}^2 + \int_{\mathbb{R}^2} g_0 v. \qquad (66)$$

Minimizing the expression:

$$t \left[\int_{B_R} e^{u_0}(v-M) - \int_{B_R} (1-e^{u_0}) \right] + \frac{t^2}{2} \int_{B_R} e^{u_0},$$

with respect to $t \in \mathbb{R}$, from (66) we obtain:

$$J(v) \geq \frac{1}{2} \int_{B_R} |\nabla v|^2 + \frac{1}{2} \int_{B_R} e^{u_0}(v-M)^2$$

$$- \frac{1}{2} \frac{\left[\int_{B_R} e^{u_0}(v-M) - \int_{B_R}(1-e^{u_0}) \right]^2}{\int_{B_R} e^{u_0}} -$$

$$- 4R\|1 - e^{u_0}\|_{L^2(B_R)} \|\nabla v\|_{L^2(B_R)} + \frac{1}{2} \int_{B_R^c} |\nabla v|^2 + \frac{1}{8} \int_{B_R^c} v^2 -$$

$$- 2\|1 - e^{u_0}\|_{L^2(B_R^c)}^2 + \int_{\mathbb{R}^2} g_0 v \geq$$

$$\geq \frac{1}{4}\int\limits_{B_R}|\nabla v|^2 + \frac{1}{2}\int\limits_{B_R}e^{u_0}(v-M)^2 - \frac{1}{2}\frac{\left(\int_{B_R}e^{u_0}(v-M)\right)^2}{\int_{B_R}e^{u_0}} -$$

$$-\frac{1}{2}\frac{\left(\int_{B_R}(1-e^{u_0})\right)^2}{\int_{B_R}e^{u_0}} + \frac{\int_{B_R}e^{u_0}(v-M)\int_{B_R}(1-e^{u_0})}{\int_{B_R}e^{u_0}} +$$

$$+\frac{1}{2}\int\limits_{B_R^c}|\nabla v|^2 + \frac{1}{8}\int\limits_{B_R^c}v^2 + \int\limits_{\mathbb{R}^2}g_0 v - C, \tag{67}$$

with a suitable constant $C = C(R) > 0$. On the other hand, we observe that:

$$\int\limits_{B_R}e^{u_0}(v-M)^2 - \frac{\left(\int_{B_R}e^{u_0}(v-M)\right)^2}{\int_{B_R}e^{u_0}} \geq 0,$$

while, via Poincaré inequality, we estimate,

$$\left|\frac{\int_{B_R}e^{u_0}(v-M)\int_{B_R}(1-e^{u_0})}{\int_{B_R}e^{u_0}}\right| \leq$$

$$\frac{\left(\int_{B_R}e^{2u_0}\right)^{\frac{1}{2}}\int_{B_R}(1-e^{u_0})}{\int_{B_R}e^{u_0}}\left(\int\limits_{B_R}(v-M)^2\right)^{\frac{1}{2}} \leq \frac{1}{8}\int\limits_{B_R}|\nabla v|^2 + C,$$

with another suitable constant $C = C(R) > 0$. Thus we have shown that if $v \in H^1(\mathbb{R}^2)$ and $v \geq 0$ a.e. in $\mathbb{R}^2$ then

$$J(v) \geq \frac{1}{8}\int\limits_{B_R}|\nabla v|^2 + \frac{1}{2}\int\limits_{B_R^c}|\nabla v|^2 + \frac{1}{8}\int\limits_{B_R^c}v^2 + \int\limits_{\mathbb{R}^2}g_0 v - C$$

$$\geq \frac{1}{8}\int\limits_{\mathbb{R}^2}|\nabla v|^2 + \frac{1}{8}\int\limits_{B_R^c}v^2 + \int\limits_{\mathbb{R}^2}g_0 v - C,$$

with a suitable constant $C > 0$. At this point, to verify (63), we observe that both g_0 and v are non-negative, so that

$$\int\limits_{B_R}g_0 v \geq \left(\inf_{B_R} g_0\right)\int\limits_{B_R}v.$$

By the definition of g_0, we see that $\inf\limits_{B_R} g_0 \geq \frac{c_0}{R^4}$ with $c_0 > 0$ a suitable constant indipendent of R. So, using Poincaré inequality, we get

$$\int\limits_{B_R}g_0 v \geq \frac{c_0}{R^3}\left(\|v\|_{L^2(B_R)} - 4R\|\nabla v\|_{L^2(B_R)}\right).$$

By inserting this inequality in the estimate above, we readily check that inequality (63) holds with suitable constants $a > 0$ and $b > 0$.   ∎

As an immediate consequence of Proposition 13 we have:

Corollary 14 *The functional J is bounded from below and admits a unique critical point given by its minimum point. In particular, for every $\{z_1, \ldots, z_N\}$, problem (59) admits a <u>unique</u> solution, which we denote by $u = u_{(z_1, \ldots, z_N)}$.*

Proof. The strict convexity implies that J is weakly lower semicontinuos anb admits at most one critical point. On the other hand, by (63), we also know that J is coercive and bounded from below. Hence J attains its minimum value at a point that gives its <u>only</u> critical point, and defines the unique solution of (59).   ∎

Observe that the operator $L : -\Delta + e^{u_{(z_1, \ldots, z_N)}} : L^2(\mathbb{R}^2) \to L^2(\mathbb{R}^2)$ is strictly positive defined, as we have:

$$< L\varphi, \varphi >_{L^2(\mathbb{R}^2)} = \|\nabla\varphi\|^2_{L^2(\mathbb{R}^2)} + \int_{\mathbb{R}^2} e^{u_{(z_1, \ldots, z_N)}} \varphi^2 > 0, \text{ for } \varphi \neq 0.$$

Furthermore, due to decay rates established in Proposition 6, we see that L defines a Fredholm operator of index zero. So L is <u>invertible</u>, and we can use the Implicit Function Theorem, to see that, for every $\{z_1, \ldots, z_N\} \in \mathbb{R}^{2N}/\Gamma_N$ (Γ_N = group of permutation of N-elements) the map:

$$\{z_1, \ldots, z_N\} \to u_{(z_1, \ldots, z_N)}$$

is <u>smoothly</u> defined. Moreover, we can use Corollary 10, to deduce the following uniform decay estimates:

Corollary 15 $\forall \varepsilon > 0$, $N \in \mathbb{N}$ and $\forall \rho > 0$, there exist $R_\varepsilon = R_\varepsilon(N, \rho) > 0$ and·a constant $C_\varepsilon = C_\varepsilon(N, \rho) > 0$ such that, for every $\{z_1, \ldots z_N\} \subset B_\rho$, we have:

$$0 < 1 - e^{u_{(z_1, \ldots, z_N)}} \leq C_\varepsilon e^{-\frac{1-\varepsilon}{q}|x|} \quad in \ \mathbb{R}^2 \setminus B_{R_\varepsilon}, \tag{68}$$

$$|\nabla u_{(z_1, \ldots, z_N)}| + |u_{(z_1, \ldots, z_N)}| \leq C_\varepsilon e^{-\frac{1-\varepsilon}{q}|x|}, \quad \forall |x| \geq R_\varepsilon. \tag{69}$$

Combining all the results above, we can conclude the following about selfdual abelian-Higgs vortices.

Theorem 16 *Every selfdual abelian-Higgs vortex (A, φ) (i.e. a smooth (static) solution of (4) with V in (2) and $\lambda = \frac{1}{q^2}$, having finite energy), admits a finite number of vortex points (i.e. zeroes of the Higgs field) with integral multiplicity.*

Moreover, for any integer $N \in \mathbb{N}$, the family of selfdual abelian-Higgs N-vortices, having N as the total number of vortex points (counted with multeplicity), is described by a family-pair of solutions $\{(A, \varphi)_\pm\}$ parametrized by the manifold $\mathbb{R}^{2N}/\Gamma_N$ and satisfying:

(a) $|\varphi| < 1$ in $\mathbb{R}^2$, and $|\varphi|$ satisfies the exponential decay estimates (26);
(b)

$$\deg(\varphi_+, \mathbb{R}^2, 0) = N = \deg(\overline{\varphi}_-, \mathbb{R}^2, 0)$$

and $(A, \varphi)_+$ (respectively $(A, \varphi)_-$) minimizes the energy among all configurations for which φ (resp. $\overline{\varphi}$) admits fixed Brower degree N.
(c) the magnetic flux $\Phi = \pm 2\pi N$, the total energy $\mathcal{E} = \pi N$.

More precisely the following holds:

for every $\{z_1, \ldots, z_N\} \subset R^{2N}/\Gamma_N$ there exists a <u>unique</u> pair of N-vortices $(A, \varphi)_\pm$ (up to gauge transformations) such that,

$$\varphi_+(z) = O((z - z_j)^{n_j}) = \overline{\varphi}_-(z), \quad as \ z \to z_j; \tag{70}$$

with n_j the multiplicity of z_j, $j = 1, \ldots N$. In particular, $\varphi_\pm$ vanishes exactly at $\{z_1, \ldots z_N\}$ (with the given multiplicity) and depends smoothly on those points.

We shall see next that the Chern-Simons-Higgs model described above carries a much richer vortex structure for which such a complete description is not yet available.

5 On the Elliptic Problem of Selfdual Chern-Simons Vortices

In this section we are concerned with the elliptic problem that describes planar selfdual Chern-Simons vortices. Hence for given $N \in \mathbb{N}$ and $\{z_1, \ldots, z_N\} \subset \mathbb{R}^2$, (not necessarily distinct) we aim to describe solutions of the following:

$$-\Delta u = \frac{1}{\kappa^2} e^u (1 - e^u) - 4\pi \sum_{j=1}^{N} \delta_{z_j} \tag{71}$$

$$e^u(1 - e^u) \in L^1(\mathbb{R}^2) \tag{72}$$

where $\kappa > 0$ and $u \in C^2(\mathbb{R}^2 \setminus \{z_1, \ldots, z_N\})$ satisfies (71) in the sense of distributions.

To emphasize right away the difference with problem (37), (38) we have:

Proposition 17 *Let $u \in C^2(\mathbb{R}^2 \setminus \{z_1, \ldots, z_N\})$ be a solution (in the sense of distributions) of (71), (72). Then,*

i) $u < 0$ in $\mathbb{R}^2$,
ii) either $u(x) \to 0$ as $|x| \to +\infty$, or $u(x) \to -\infty$ as $|x| \to +\infty$.

Proof. Property i) follows essentially as in Proposition 5. In fact, since: $0 \leq u_+ \leq e^{u_+} - 1 \leq e^{u_+}(e^{u_+} - 1)$, we see that $u_+ \in L^1(\mathbb{R}^2)$. Consequently, we can show that, $u_+ \to 0$, as $|x| \to +\infty$. Thus, if $u_+ \not\equiv 0$, then it would admit a maximum point, say at x_0, with $u_+(x_0) > 0$, and,

$$0 \geq \Delta u_+(x_0) = \frac{1}{\kappa^2} e^{u_+(x_0)}(e^{u_+(x_0)} - 1) > 0,$$

which is clearly impossibile.

To establish ii) observe that if

$$\liminf_{|x| \to \infty} u(x) = 0,$$

then the first alternative would hold, since $u < 0$, and so necessarily also: $\limsup_{|x| \to \infty} u(x) = 0$. Hence, assume that,

$$\liminf_{|x| \to \infty} u(x) < 0.$$

We prove that in this case we have: $\limsup_{|y| \to \infty} u(y) = \liminf_{|y| \to \infty} u(y) = -\infty$. Argue by contradiction and suppose there exists a sequence $\{y_n\}$ such that $|y_n| \to \infty$ and $u(y_n) \to -l$, with $l > 0$. By taking a subsequence if necessary, we can always assume that:

$$B_1(y_h) \cap B_1(y_k) = \emptyset, \tag{73}$$

for $h \neq k$. We use Harnack inequality to find suitable universal constants $\alpha > 0$ and $\beta > 0$ such that (for n sufficiently large) we have:

$$\sup_{B_1(y_h)} u \leq \alpha \inf_{B_1(y_h)} u + \beta \|e^u(1 - e^u)\|_{L^2(B_2(y_h))}.$$

Consequently:

$$\inf_{B_1(y_h)} u \geq -\frac{l}{\alpha} + o(1),$$

$$\sup_{B_1(y_h)} u \leq -l\alpha + o(1),$$

as $h \to \infty$. Hence, for h sufficiently large, we would find a suitable constant $c_0 > 0$ such that:

$$\int\limits_{B_1(y_h)} e^u(1 - e^u) \geq c_0,$$

which is impossible, since in view of (73), this would contradict the integrability of $e^u(1 - e^u)$ in $\mathbb{R}^2$. ∎

Thus, it makes good sense to consider two types of solutions, namely:

(i) the topological solutions which satisfy (71)–(72) subject to the boundary conditions:
$$u \to 0 \text{ as } |x| \to +\infty;$$

(ii) the non-topological solutions which satisfy (71)–(72) under the boundary conditions:
$$u \to -\infty \text{ as } |x| \to +\infty.$$

The terminology topological/non-topological solution is justified by the discussion of the previous section, since only in the first case we can assign to the vortex number N a topological meaning, that will distinguish Chern-Simons vortex configurations into homotopic classes.

Actually, as we shall see, topological Chern-Simons vortices will have much more in common with abelian-Higgs vortices.

5.1 Topological Chern-Simons Voritces

Let $\lambda = \frac{1}{\kappa^2}$, we devote this section to the solvability of the problem:

$$\begin{cases} -\Delta u = \lambda e^u(1 - e^u) - 4\pi \sum_{j=i}^{N} \delta_{z_j} \\ u \to 0 \text{ as } |x| \to +\infty \end{cases} \tag{74}$$

Arguing as in Section 1.3, we easily check that, any solution $u \in C^2(\mathbb{R}^2 \setminus \{z_1, \ldots, z_N\})$ of (74) satisfies: $u < 0$ in $\mathbb{R}^2$, $e^u(1-e^u) \in L^1(\mathbb{R}^2)$ and $\lambda \int_{\mathbb{R}^2} e^u(1 - e^u) = 4\pi N$. In fact most of the properties discussed for the abelian-Higgs vortices remain valid here. As for example the following uniform estimates (to be compared with Lemma 8):

Lemma 18 *Given* $\varepsilon \in (0,1)$, $N \in \mathbb{N}$ *and* $\rho > 0$, *there exist* $R_\varepsilon = R_\varepsilon(N, \rho) > 0$, *such that every solution* u *of (74) with* $\lambda \geq 1$ *and* $\max\limits_{j,k=1,\ldots,N} |z_j - z_k| \leq \rho$, *satisfies:*

$$0 < 1 - e^{u(z)} < \varepsilon, \ \forall z \in \mathbb{R}^2 \setminus \left(\bigcup_{j=1}^{N} B_\delta(z_j) \right) \tag{75}$$

for every $\delta \geq R_\varepsilon$.

Proof. Exactly as in the proof of Lemma 8, we use an argument by contradiction to find $\varepsilon_0 \in (0,1)$ such that, for every $0 < \varepsilon \le \varepsilon_0$, we obtain a solution $U = U_\varepsilon$ of the following problem:

$$\begin{cases} -\Delta U = e^U(1 - e^U), \text{ in } \mathbb{R}^2 \\ U(0) = \log(1 - \varepsilon), \ U < 0 \\ \int\limits_{\mathbb{R}^2} e^U(1 - e^U) \le 4\pi N \end{cases} \tag{76}$$

Hence, we can argue as in Proposition 17, to check that either $U(x) \to 0$ or $U(z) \to -\infty$, as $|z| \to +\infty$. In the first case, we easily reach a contradiction, since we find (as in Lemma 8), that the only solution of (76) such that: $U \to 0$ at infinity, is given by the trivial solution $U \equiv 0$. Thus , $U \to -\infty$, as $|z| \to +\infty$. According to a result of Spruck-Yang, (cf. [SY1]), U must be radially symmetric about a point, say $x_0 \in \mathbb{R}^2$, corresponding to its unique maximum point. Hence for $V(z) = U(z + x_0)$, we see that $V = V(|x|)$ is radially symmetric about the origin and satisfies:

$$\begin{cases} -(V_{rr} + \frac{1}{r}V_r) = e^V(1 - e^V), \ r > 0 \\ 0 > V(0) = \max_{\mathbb{R}^2} V := s_\varepsilon \ge \log(1 - \varepsilon), \ V(r) \to -\infty \text{ as } r \to +\infty \\ \int\limits_0^\infty e^{V(r)}(1 - e^{V(r)})rdr \le 2N. \end{cases} \tag{77}$$

A detailed analysis of such radial problem has been carried out by Chan-Fu-Lin in [CFL]. Thus, according to [CFL], we know that for $s < 0$, the solution $v(r, s)$ of the differential equation in (77) satisfying: $v(0, s) = s$ and $v_r(0, s) = 0$, is unique and globally defined $\forall r > 0$. Moreover:

$$v(r, s) < 0 \ \& \ v(r, s) \to -\infty \text{ as } r \to +\infty.$$

Furthermore, $e^{v(r,s)}(1 - e^{v(r,s)})r \in L^1(\mathbb{R}^+)$ and letting:

$$\beta(s) = \int_0^\infty e^{v(r,s)}(1 - e^{v(r,s)})rdr. \tag{78}$$

then,

$$\beta(s) \to +\infty \text{ as } s \to 0^-. \tag{79}$$

Consequently, $V(r) = v(r, s_\varepsilon)$ and so,

$$\beta(s_\varepsilon) = \int_0^\infty e^{V(r)}(1 - e^{V(r)})rdr \le 2N.$$

But since: $s_\varepsilon \to 0^-$, as $\varepsilon \to 0^+$, this is in contradiction with (79). ∎

Similarly to the previous section, from Lemma 18 we can deduce the following uniform decay estimates:

Corollary 19 *For every* $\varepsilon \in (0,1)$, $N \in \mathbb{N}$ *and for every* $\rho > 0$, *there exist* $R_\varepsilon = R_\varepsilon(\rho, N) > 0$ *and* $C_\varepsilon = C_\varepsilon(\rho, N) > 0$, *such that* <u>*every*</u> *solution* u *of* (74) *with* $\lambda \geq 1$ *and* $\max\limits_{j=1,\ldots,N} |z_j| \leq \rho$, *satisfies:*

$$0 < 1 - e^u < C_\varepsilon e^{-\sqrt{\lambda}(1-\varepsilon)|x|}, \quad \forall x : |x| \geq R_\varepsilon. \tag{80}$$

$$|u(x)| + |\nabla u(x)| \leq C_\varepsilon e^{-\sqrt{\lambda}(1-\varepsilon)|x|}, \quad \forall x : |x| \geq R_\varepsilon. \tag{81}$$

In particular, for fixed $\{z_1, \ldots, z_N\}$ *all corresponding solutions admits* <u>*uniform*</u> *exponential decay.*

At this point we turn to discuss the solvability of (74). We proceed as for the abelian-Higgs case and let,

$$u = u_0 + v \tag{82}$$

where u_0 is defined in (54), satisfies (55) with g_0 given in (56).

In analogy to Proposition 11 we find:

Proposition 20 *According to the decomposition (82), we have that* $u \in C^2(\mathbb{R}^2 \setminus \{z_1, \ldots, z_N\})$ *satisfies (74) (in the sense of distributions) if and only if* v *satisfies (in the classical sense)*

$$\begin{cases} -\Delta v = \lambda e^{u_0+v}(1 - e^{u_0+v}) - g_0 \\ v \in H^1(\mathbb{R}^2) \end{cases} \tag{83}$$

and,

$$\lambda \int_{\mathbb{R}^2} e^{u_0+v}(1 - e^{u_0+v}) = 4\pi N. \tag{84}$$

Since Proposition 20 follows exactly by the same arguments of Proposition 11, we leave the details of its proof to the interested reader.

Again, the advantage of working with problem (83) is that it admits a variational formulation, in the sense that (weak) solutions of (83) correspond to critical points for the functional

$$I_\lambda(v) = \frac{1}{2} \int_{\mathbb{R}^2} |\nabla v|^2 + \frac{\lambda}{2} \int_{\mathbb{R}^2} (e^{u_0+v} - 1)^2 - \int_{\mathbb{R}^2} g_0 v, \; v \in H^1(\mathbb{R}^2).$$

Indeed, by recalling the integrability properties of $(1 - e^{u_0})$, and of $(1 - e^v)$ for $v \in H^1(\mathbb{R}^2)$, (cf. [JT], [Y1]and [T8]), we check that $I_\lambda \in C^1(H^1(\mathbb{R}^2))$ and

$$< I_\lambda'(v), \varphi > = \int_{\mathbb{R}^2} \nabla v \cdot \nabla \varphi + \lambda \int_{\mathbb{R}^2} (e^{u_0+v} - 1)e^{u_0+v}\varphi - \int_{\mathbb{R}^2} g_0 \varphi, \; \forall \varphi \in H^1(\mathbb{R}^2).$$

Thus, by means of elliptic regularity theory, we find that

$$v \in H^1(\mathbb{R}^2) : I'_\lambda(v) = 0 \iff v \in C^\infty(\mathbb{R}^2) \text{ satisfies (83)}.$$

In complete analogy to the abelian-Higgs case we have:

Proposition 21 *Let $\lambda > 0$. There exists $a, b > 0$ such that*

$$I_\lambda(v) \geq a\|v\|_{H^1(\mathbb{R}^2)} - b, \ \forall v \in H^1(\mathbb{R}^2). \tag{85}$$

Proof. Using again the estimate $\frac{|t|}{1+|t|} \leq |e^t - 1|, \forall t \in \mathbb{R}$, we find that

$$\int_{\mathbb{R}^2} (1 - e^{v+u_0})^2 dx \geq \frac{1}{2} \int_{\mathbb{R}^2} \frac{v^2}{(1+|v|)^2} dx - 4\|u_0\|^2_{L^2(\mathbb{R}^2)}.$$

Consequently, by using (64), for any given $\varepsilon > 0$, we have:

$$I_\lambda(v) \geq \frac{1}{2}\|\nabla v\|^2_{L^2(\mathbb{R}^2)} + \frac{\lambda}{4}\int_{\mathbb{R}^2} \frac{v^2}{(1+|v|)^2} - 2\lambda\|u_0\|^2_{L^2(\mathbb{R}^2)} - \|g_0\|_{L^2(\mathbb{R}^2)}\|v\|_{L^2(\mathbb{R}^2)} \geq$$

$$\geq \frac{1}{2}\|\nabla v\|^2_{L^2(\mathbb{R}^2)} + \frac{\lambda}{4}\int_{\mathbb{R}^2} \frac{v^2}{(1+|v|)^2} - 2\lambda\|u_0\|^2_{L^2(\mathbb{R}^2)} -$$

$$- \|g_0\|_{L^2(\mathbb{R}^2)} \left(2\int_{\mathbb{R}^2} \frac{v^2}{(1+|v|)^2}\right)^{\frac{1}{2}} (1 + 2\|\nabla v\|^2_{L^2(\mathbb{R}^2)})^{\frac{1}{2}} \geq$$

$$\geq \frac{1}{2}\|\nabla v\|^2_{L^2(\mathbb{R}^2)} + \frac{\lambda}{4}\int_{\mathbb{R}^2} \frac{v^2}{(1+|v|)^2} - 2\lambda\|u_0\|^2_{L^2(\mathbb{R}^2)} -$$

$$- \frac{1}{\varepsilon}\|g_0\|^2_{L^2(\mathbb{R}^2)} \int_{\mathbb{R}^2} \frac{v^2}{(1+|v|)^2} - \frac{\varepsilon}{2}(1 + 2\|\nabla v\|^2_{L^2(\mathbb{R}^2)}) \geq$$

$$\geq \left(\frac{1}{2} - \varepsilon\right)\|\nabla v\|^2_{L^2(\mathbb{R}^2)} + \left(\frac{\lambda}{4} - \frac{1}{\varepsilon}\|g_0\|^2_{L^2(\mathbb{R}^2)}\right)\int_{\mathbb{R}^2} \frac{v^2}{(1+|v|)^2}$$

$$- 2\lambda\|u_0\|^2_{L^2(\mathbb{R}^2)} - \frac{\varepsilon}{2}.$$

Recalling that $\|g_0\|_{L^2(\mathbb{R}^2)}$ converges to zero as $\mu \to \infty$, we can choose $\mu = \mu_\lambda$ large enough to ensure that $\|g_0\|^2_{L^2(\mathbb{R}^2)} < \frac{\lambda}{16}$. Hence, we can fix $\varepsilon = \frac{1}{4}$ and obtain two suitable constants $C_1 > 0$ and $C_2 > 0$ such that:

$$I_\lambda(v) \geq C_1 \left(\|\nabla v\|^2_{L^2(\mathbb{R}^2)} + \int_{\mathbb{R}^2} \frac{v^2}{(1+|v|)^2}\right) - C_2 \geq$$

$$C_1 \left(\|\nabla v\|^2_{L^2(\mathbb{R}^2)} + \frac{\|v\|^2_{L^2(\mathbb{R}^2)}}{2(1 + 2\|\nabla v\|^2_{L^2(\mathbb{R}^2)})} \right) - C_2.$$

To complete the proof, let $\sigma \in [0, 1]$ be such that: $\|\nabla v\|^2_{L^2(\mathbb{R}^2)} = \sigma\|v\|^2_{H^1(\mathbb{R}^2)}$ and $\|v\|^2_{L^2(\mathbb{R}^2)} = (1-\sigma)\|v\|^2_{H^1(\mathbb{R}^2)}$. Then the inequality above reads as follows:

$$I_\lambda(v) \geq C_1 \left(\sigma\|v\|^2_{H^1(\mathbb{R}^2)} + \frac{(1-\sigma)\|v\|^2_{H^1(\mathbb{R}^2)}}{2(1 + 2\sigma\|v\|^2_{H^1(\mathbb{R}^2)})} \right) - C_2,$$

and by minimizing the inequality above with respect to σ, we deduce the desired estimate. ∎

Corollary 22 *For any $\lambda > 0$, I_λ attains its minumum value at $v \in H^1(\mathbb{R}^2)$ which defines a solution of problem (83).*

Proof. From (85), we know that I_λ is coercieve and bounded from below . Clearly, it is also weakly lower semicontinuous, and so it attains its infimun at a critical point, that provides the desired solution. ∎

Consequently, for every set of N-(vortex) points $\{z_1, \ldots, z_N\}$ (not necessarily distinct) and $\lambda > 0$, we have established that problem (74) admits a solution.

However, contrary to what we have seen for the abelian-Higgs model, now it is not clear whether ot not other topological solutions exit.

A first contribute to this question is given in case all vortex points coincide, i.e. $z_1 = z_2 = \ldots = z_N$. There holds,

Proposition 23 *(cf. [CHMcLY] and [Ha3]) For every $N \in \mathbb{N}$, $\lambda > 0$ and $z_0 \in \mathbb{R}^2$ the problem:*

$$\begin{cases} -\Delta u = \lambda e^u(1 - e^u) - 4\pi N\delta_{z_0}, & in \ \mathbb{R}^2 \\ u \to 0 \ as \ |z| \to \infty \end{cases}$$

admits a <u>unique</u> solution u_{z_0} which is radially symmetric about the point z_0.

As a matter of fact, the linearized operator:

$$L = -\Delta + \lambda e^{u_{z_0}}(2e^{u_{z_0}} - 1) : L^2(\mathbb{R}^2) \to L^2(\mathbb{R}^2)$$

is strictly positive definite (hence invertible), see [CFL] and [T8] for details.

Recently this result has been extended to hold also when the vortex points may be distinct, provided the parameter $\lambda > 0$ is taken large enough. More precisely the following holds:

Theorem 24 *(cf. [T8]) For every $N \in \mathbb{N}$ and every assigned set of points $\{z_1, \ldots, z_N\} \subset \mathbb{R}^2$, there exists $\lambda_* > 0$ and $\mu_* > 0$, such that for every solution u of (74) with $\lambda > \lambda_*$, we have:*

$$\inf_{\varphi \in H^1(\mathbb{R}^2),\, \|\varphi\|_{H^1(\mathbb{R}^2)}=1} \|\nabla\varphi\|_{L^2(\mathbb{R}^2)}^2 + \int_{\mathbb{R}^2} e^u(2e^u - 1)\varphi^2 \geq \mu_*$$

As an important consequence of Theorem 24 we have:

Theorem 25 *(cf. [Cho1], [T8]) For every $N \in \mathbb{N}$ and $\{z_1, \ldots, z_N\} \subset \mathbb{R}^2$ there exists $\lambda_* > 0$ such that, for every $\lambda > \lambda_*$, problem (74) admits a <u>unique</u> solution.*

Proof. By virtue of Theorem 24, if $u = u_0 + v$ is a solution of (74) with $\lambda > \lambda_*$ then v defines a <u>strict</u> local mimimum for I_λ. Suppose by contradiction that there exist <u>two</u> distinct solutions $u_1 \neq u_2$ for (74) with $\lambda > \lambda_*$. Setting $u_j = u_0 + v_j$, $j = 1, 2$ and observing that $\hat{v}(x) = \max\{v_1(x), v_2(x)\}$ defines a subsolution for (83), we can consider the minimization problem:

$$\inf\{I_\lambda(v),\, \forall v \in H^1(\mathbb{R}^2): v(x) \geq \hat{v}(x) \text{ a.e. } x \in \mathbb{R}^2\}$$

and show that the infimun is attained at a critical point for I_λ. Hence, we obtain <u>another</u> solution for (83) which is pointwise greater than v_1 and v_2. In other words, we can always assume (by contradiction) to have two ordered solutions for (83), say $v_1 < v_2$, which define two distinct strict local minima for I_λ. Then we can use a "mountain pass" constuction between v_1 and v_2, (see Theorem 12.8 in [St] for details) to obtain a third solution v_3 for (83), such that $v_1(x) < v_3(x) < v_2(x)$, $\forall x \in \mathbb{R}^2$ and v_3 is <u>not</u> a local minimum for I_λ. But this is in contradiction with Theorem 24.   ∎

Generalizations of this uniqueness result can be found in [T7], [ChoN], and [MaN].

<u>Open question</u>: It remains as a challenging open question to know whether uniqueness holds for problem (74), without any restriction on $\lambda > 0$.

All the results above show strong analogies between abelian-Higgs vortices and topological Chern-Simons vortices. It should be noticed however, that in fact their profiles around the vortex points is quite different, as we see from the following result:

Theorem 26 *For every $N \in \mathbb{N}$, $\{z_1, \ldots, z_N\} \subset \mathbb{R}^2$ and $\lambda > 0$ sufficiently large, denote by u_{aH}^λ the unique solution for (37)–(38) (with $\lambda = \frac{1}{q^2}$) and by u_{CS}^λ the unique (topological) solution for (74) (with $\lambda = \frac{1}{\kappa^2}$). Then, as $\lambda \to +\infty$:*

(i)

$$1 - e^{u_{aH}^\lambda} \rightharpoonup 4\pi \sum_{j=1}^N \delta_{z_j}$$

(ii)

$$1 - e^{u^\lambda_{CS}} \rightharpoonup 4\pi \left(\sum_{j=1}^{N} \delta_{z_j} + \sum_{j=1}^{N} n_j \delta_{z_j} \right)$$

weakly in the sense of measure. Here $n_j \in \mathbb{N}$ defines the multiplicity of z_j, $j = 1, \ldots, N$.

See [T8] for details.

In other words, in terms of the corresponding vortex configurations, we see that, the function: $(1 - |\varphi|^2)$ concentrates around the vortex points for both the abelian-Higgs and topological Chern-Simons-Higgs model, (respectively, as $q \to 0$, and $\kappa \to 0$). But the "local" behaviour around those points is quite different, as a result of the different scalar potential.

In terms of Chern-Simons vortices, the results established above can be summarized as follows (cf. [W], [Y1] and [T8]):

Theorem 27 *For any given $\kappa > 0$, $N \in \mathbb{N}$, and any assigned set of (vortex) points $Z = \{z_1, \ldots, z_N\} \subset \mathbb{R}^2$ (repeated according to their multiplicity) there exists $(\mathcal{A}, \varphi)_\pm$ a smooth solution to the selfdual equation (33) in $\mathbb{R}^2$ (with the $\pm$ sign choosen accordingly) such that:*

i) *$|\varphi_\pm| < 1$ in $\mathbb{R}^2$, $\varphi_\pm$ vanishes exactly at the set Z and if $n_j \in \mathbb{Z}$ is the multiplicity of $z_j \in Z$ $j = 1, \ldots, N$, then:*

$$\varphi_+(z) \ \& \ \overline{\varphi}_-(z) = O((z - z_j)^{n_j}), \text{ as } z \to z_j. \tag{86}$$

ii) *For every $\varepsilon \in (0,1)$ there exist a constant $R_\varepsilon > 0$ such that, for every $0 < \kappa \leq 1$, the following estimate holds:*

$$|D_1\varphi_\pm| + |D_2\varphi_\pm| + |F_{12}| \leq c_\varepsilon (1 - |\varphi_\pm|) \leq C_\varepsilon e^{-\frac{1}{\kappa}(1-\varepsilon)|z|}, \ \forall |z| \geq R_\varepsilon, \tag{87}$$

with $c_\varepsilon > 0$ and $C_\varepsilon > 0$ suitable constants (indipendent of κ). In particular $|\varphi_\pm| \to 1$, as $|z| \to +\infty$.
Furthermore, if $\kappa > 0$ is sufficiently small, then $(\mathcal{A}, \varphi)_\pm$ is the <u>only</u> solution (up to gauge transformation) satisfying (86) and (87).

iii) *The following holds respectively for the magnetic flux, electric charge and total energy:*

$$\begin{aligned} &\text{Magnetic flux } \Phi = \int_{\mathbb{R}^2} (F_{12})_\pm = \pm 2\pi N; \\ &\text{Electric charge } Q = \int_{\mathbb{R}^2} (J^0)_\pm = \pm 2\pi k N; \\ &\text{Total energy } E = \int_{\mathbb{R}^2} \mathcal{E}_\pm = \pi N. \end{aligned} \tag{88}$$

In concluding this section, we mention that it is possible to deduce the following asymptotic behaviour for the vortex solution:

Theorem 28 *(cf. [T8]) As $\kappa \to 0^+$, we have:*

$$(A_0)_\pm \to 0, \ (J^0)_\pm \to 0, in L^1(\mathbb{R}^2);$$

while,

$$(F_{12})_\pm \to \pm 2\pi \sum_{j=1}^{N} \delta_{z_j}, \tag{89}$$

$$(A_0)_\pm^2 \to \pm\pi \sum_{j=1}^{N} n_j \delta_{z_j} \ \& \ \frac{1}{k}(A_0)_\pm \to \pm\pi \sum_{j=1}^{N} (n_j + 1)\delta_{z_j},$$

weakly in the sense of measure in $\mathbb{R}^2$.

We observe that the abelian-Higgs vortices of Theorem 16 also satisfy the "concentration" property (89) as $q \to 0$, as shown in [HJS].

5.2 Non Topological Chern-Simons Vortices

We now explore the possibility to establish the existence of non-topological Chern-Simons vortices, by analyzing the elliptic problem:

$$\begin{cases} -\Delta u = \lambda e^u (1 - e^u) - 4\pi \sum_{j=1}^{N} \delta_{z_j} \\ u \to -\infty, \ \text{as } |x| \to +\infty. \end{cases} \tag{90}$$

To be consistent with the physical applications, we need to solve (90) together with the properties:

$$u < 0 \text{ and } e^u(1 - e^u) \in L^1(\mathbb{R}^2). \tag{91}$$

By virtue of our discussion at the biginning of this section, it is not difficult to check that problem (90) and (91) can be equivalently formulated as follows,

$$\begin{cases} -\Delta u = \lambda e^u (1 - e^u) - 4\pi \sum_{j=1}^{N} \delta_{z_j} \\ e^u \in L^1(\mathbb{R}^2). \end{cases} \tag{92}$$

The advantage of (92) is that it may be viewed as a perturbation of the (singular) Liouville problem (see (93) below) for which we have explicit solutions available. More precisely, if u solves (92) then for every $\varepsilon > 0$, the scaled function

$$u_\varepsilon(z) = u(\frac{z}{\varepsilon}) + 2\log(\frac{1}{\varepsilon}),$$

satisfies:

$$-\Delta u_\varepsilon = \lambda e^{u_\varepsilon} - \lambda \varepsilon^2 e^{2u_\varepsilon} - 4\pi \sum_{j=1}^{N} \delta_{\varepsilon z_j} \text{ in } \mathbb{R}^2,$$

which, we may regard as a perturbation of the "singular" Liouville problem:

$$\begin{cases} -\Delta u = \lambda e^u - 4\pi \sum_{j=1}^{N} \delta_{\varepsilon z_j} \\ \int_{\mathbb{R}^2} e^u < \infty. \end{cases} \tag{93}$$

We can exhibit an explicit solution for (93). To this purpose, let

$$f(z) = (N+1) \prod_{j=1}^{N} (z - z_j) \text{ and } F(z) = \int_0^z f(\xi) d\xi$$

and set,

$$f_\varepsilon(z) = (N+1) \prod_{j=1}^{N} (z - \varepsilon z_j) \text{ and } F_\varepsilon(z) = \int_0^z f_\varepsilon(\xi) \, d\xi.$$

By Liouville formula (cf. [Lio]) we know that,

$$u_{\varepsilon,a}^0(z) = \log \frac{8|f_\varepsilon(z)|^2}{\lambda(1 + |F_\varepsilon(z) + a|^2)^2} \tag{94}$$

satisfies (93) for any $\varepsilon \in \mathbb{R}$ and $a \in \mathbb{C}$. Thus, it is reasonable to search for solution of problem (92) in the form

$$u(z) = u_{\varepsilon,a}^0(\varepsilon z) + \log \varepsilon^2 + \varepsilon^2 w(\varepsilon z) \tag{95}$$

with w a suitable error term that satisfies:

$$-\Delta w = \lambda e^{u_{\varepsilon,a}^0} \left(\frac{e^{\varepsilon^2 w} - 1}{\varepsilon^2} \right) - \lambda e^{2(u_{\varepsilon,a}^0 + \varepsilon^2 w)}. \tag{96}$$

We consider the free parameters ε and a as part of our unknowns and concentrate around the values $\varepsilon = 0$ and $a = 0$ where (96) reduces to:

$$\Delta w + \rho w = \frac{1}{\lambda} \rho^2 \tag{97}$$

with $\rho = \lambda e^{u_{\varepsilon=0,a=0}^0}$, the radial function given as follows:

$$\rho(r) = \frac{8(N+1)^2 r^{2N}}{(1 + r^{2(N+1)})^2}, \quad r = |z|. \tag{98}$$

We will solve (97) within the class of radial functions, and find an explicit solution $w_0 = w_0(r)$ given in Lemma 31 below. Thus, using the decomposition:

$$w(z) = w_0(|z|) + u_1(z) \tag{99}$$

we need to solve for u_1 the following equation:

$$P(u_1, a, \varepsilon) :=$$

$$\Delta u_1 + \lambda e^{u^0_{\varepsilon,a}} \left(\frac{e^{\varepsilon^2 w_0 + \varepsilon^2 u_1} - 1}{\varepsilon^2} \right) - \rho w_0 + \lambda e^{2(u^0_{\varepsilon,a} + \varepsilon^2 w_0 + \varepsilon^2 u_1)} - \frac{1}{\lambda}\rho^2 = 0. \tag{100}$$

To this end, we aim to apply the Implicit Function Theorem (cf. [Nir]) to the operator P acting between suitable functional spaces where it extends smoothly at $\varepsilon = 0$ to satisfy $P(0,0,0) = 0$. To this purpose, Chae-Imanuvilov in [ChI1] have introduced the following spaces:

$$X_\alpha = \left\{ u \in L^2_{\text{loc}}(\mathbb{R}^2) : (1 + |z|^{2+\alpha})u^2 \in L^1(\mathbb{R}^2) \right\}, \ \alpha > 0$$
$$Y_\alpha = \left\{ u \in W^{2,2}_{\text{loc}}(\mathbb{R}^2) : \Delta u \in X_\alpha, \ \frac{u}{(1+|z|)^{1+\frac{\alpha}{2}}} \in L^2(\mathbb{R}^2) \right\}, \ \alpha > 0 \tag{101}$$

equipped respectively with scalar product:

$$(u,v)_{X_\alpha} = \int_{\mathbb{R}^2} (1 + |z|^{2+\alpha})uv \ \text{and} \ (u,v)_{Y_\alpha} = (\Delta u, \Delta v)_{X_\alpha} + \int_{\mathbb{R}^2} \frac{uv}{(1+|z|)^{2+\alpha}}$$

and relative norms denoted by $\| \cdot \|_{X_\alpha}$ and $\| \cdot \|_{Y_\alpha}$ respectively.

For any $\alpha > 0$, the following continuous embedding properties hold:
$$X_\alpha \hookrightarrow L^q(\mathbb{R}^2), \ \forall q \in [1,2);$$
$$Y_\alpha \hookrightarrow C^0_{\text{loc}}(\mathbb{R}^2).$$

Furthermore, we have:

Lemma 29 *Let $\alpha \in (0,1)$ and $v \in Y_\alpha$,*

(a) if v is harmonic then v is a constant;
(b) the following estimates hold:

$$|v(z)| \leq C\|v\|_{Y_\alpha} \log(1 + |z|), \ \ in \ \mathbb{R}^2; \tag{102}$$
$$\|\nabla v\|_{L^p} \leq C_p\|v\|_{Y_\alpha}, \ for \ every \ p > 2; \tag{103}$$

where $C > 0$ and $C_p > 0$ are suitable constants depending on α and (α, p) respectively.

See [ChI1] for the proof.

Working with the spaces X_α and Y_α is particularly advantageous for the linear operator
$$L = \Delta + \rho : Y_\alpha \to X_\alpha \tag{104}$$

as we can characterize explicitly $\ker L \subset Y_\alpha$ and $\text{Im } L \subset X_\alpha$. To this purpose consider the family of functions:

$$U_{\mu,a}(z) = u^0_{\varepsilon=0,a}(\mu) + \log \mu^2, \ \mu > 0, \ \text{and} \ a \in \mathbb{C};$$

which satisfy

$$-\Delta U = \lambda e^U - 4\pi N \delta_{z=0} \tag{105}$$

Letting $u^0 = U_{\mu=1,a=0} = \log \rho$ and using polar coordinates we see that the following functions:

$$
\begin{aligned}
\varphi_0 &= \frac{1}{2(N+1)} \frac{\partial}{\partial \mu} U_{\mu,a}|_{\mu=1,a=0} = \frac{1-r^{2(N+1)}}{1+r^{2(N+1)}}, \\
\varphi_+ &= -\frac{1}{4} \frac{\partial}{\partial \alpha} U_{\mu,a}|_{\mu=1,a=0} = \frac{r^{N+1} \cos((N+1)\theta)}{1+r^{2(N+1)}}, \ (\alpha = Re \ a); \\
\varphi_- &= -\frac{1}{4} \frac{\partial}{\partial \beta} U_{\mu,a}|_{\mu=1,a=0} = \frac{r^{N+1} \sin((N+1)\theta)}{1+r^{2(N+1)}}, \ (\beta = Im \ a);
\end{aligned}
\tag{106}
$$

belong to $\ker L$ in Y_α, $\forall \alpha > 0$. More interestingly, the following holds:

Proposition 30 *For $\alpha \in (0,1)$ the operator L in (104) satisfies:*

(a) $\ker L = span\ \{\varphi_0, \varphi_+, \varphi_-\} \subset Y_\alpha$
(b) $Im\ L = \{f \in X_\alpha : \int_{\mathbb{R}^2} f\varphi_\pm = 0\}$

To derive Proposition 30 we start by describing the behaviour of the operator L over radial functions. Hence denote by $L^r : Y^r_\alpha \to X^r_\alpha$ the operator

$$L^r \varphi = \frac{d^2}{dr^2}\varphi + \frac{1}{r}\frac{d}{dr}\varphi + \rho\varphi, \ \varphi \in Y^r_\alpha \tag{107}$$

where Y^r_α and X^r_α denote the subspaces of radial functions in Y_α and X_α respectively.

Lemma 31 *Let $\alpha \in (0,1)$ and $n \in \mathbb{Z}^+$, then*

(a) $\varphi \in Y^r_\alpha$ satisfies $L^r\varphi = 0$ if and only if $\varphi \in span\ \{\varphi_0\}$
(b) $L^r : Y^r_\alpha \to X^r_\alpha$ is onto.

More precisely, for $f \in X_\alpha$ let,

$$
w(r) = \left(\varphi_0(r) \log r + \frac{2}{N+1}\frac{1}{(1+r^{2(N+1)})} \right) \int_0^r \varphi_0(t) f(t) t\, dt +
$$

$$
-\varphi_0(r) \int_0^r \left(\varphi_0(t) \log t + \frac{2}{N+1}\frac{1}{(1+t^{2(N+1)})} f(t) t \right) dt \tag{108}
$$

then $w \in Y^r_\alpha$, and satisfies: $L^r w = f$.

Lemma 31 follows by a variation of parameter type argument. We omit the proof and refer to [ChI1] for details, see also [T8].

Observe that, since $w(r)$ and $\dot{w}(r)$ extend with continuity at $r = 0$, and we find: $w(0) = 0 = \dot{w}(0)$. Furthermore, setting:

$$c_f = \int_0^{+\infty} \varphi_0(t) f(t) t\, dt \tag{109}$$

(well defined) we have

Corollary 32 *The function w in (108) admits the following asymptotic behavior:*

$$w(r) = -c_f \log r + O(1), \ as \ r \to +\infty, \tag{110}$$

$$w'(r) = -\frac{c_f}{r} + O(1), \ as \ r \to +\infty. \tag{111}$$

In particular, by taking $f(r) = \frac{1}{\lambda}\rho^2$ in (108) we see that:

$$w_0(r) = \frac{1}{\lambda(1 + r^{2(N+1)})}\left[\left((1 - r^{2(N+1)})\log r + \frac{2}{N+1}\right)\int_0^r \varphi_0(t)t\rho^2(t)dt + \right.$$

$$\left. -(1 - r^{2(N+1)})\int_0^r \left(\varphi_0(t)\log t + \frac{2}{(N+1)(1 + t^{2(N+1)})}\right)\rho^2(t)tdt\right], \quad (112)$$

defines a solution for problem (97) in Y_α^r, such that,

$$w_0(r) = -\frac{c_0}{\lambda}\log r + O(1) \ as \ r \to +\infty, \tag{113}$$

$$w_0'(r) = -\frac{1}{\lambda}\frac{c_0}{r} + O(1), \ as \ r \to +\infty, \tag{114}$$

with

$$c_0 = \int_0^{+\infty} \varphi_0(t)\rho^2(t)tdt = (8(N+1)^2)^2 \int_0^{+\infty} \frac{(1 - t^{2(N+1)})}{(1 + t^{2(N+1)})^5}t^{4N+1}dt =$$

$$32(N+1)^4 \int_0^{+\infty} \frac{1 - s^{N+1}}{(1 + s^{N+1})^5}s^{2N} ds =$$

$$32(N+1)^4 \left(\int_0^{+\infty} \frac{s^{2N}}{(1 + s^{N+1})^4} - 2\int_0^{+\infty} \frac{s^N}{(1 + s^{N+1})^5}s^{2N+1}\right)$$

$$32(N+1)^4 \left(\int_0^{+\infty} \frac{s^{2N}}{(1 + s^{N+1})^4} + \frac{1}{2(N+1)}\int_0^{+\infty} \frac{d}{ds}\left(\frac{1}{(1 + s^{N+1})^4}\right)s^{2N+1}ds\right) =$$

$$32(N+1)^4 \left(\int_0^{+\infty} \frac{s^{2N}}{(1 + s^{N+1})^4}ds - \frac{2N+1}{2(N+1)}\int_0^{+\infty} \frac{s^{2N}}{(1 + s^{N+1})^4}\right) =$$

$$16(N+1)^3 \int_0^{+\infty} \frac{s^{2N}}{(1 + s^{N+1})^4}ds.$$

From now on we shall use such a solution w_0 into the definition of the operator P given in (100).

Proof of Proposition 30: We start to establish (a) and so let $v \in Y_\alpha$ such that $Lv = 0$. We can use standard elliptic regularity theory to see that $v \in C^2(\mathbb{R}^2)$. We write v according to its Fourier decomposition:

$$v(z) = \sum_{k \in \mathbb{Z}} v_k(r) e^{ik\theta}, \ z = re^{i\theta} \tag{115}$$

with complex valued functions $v_k = v_k(r)$ such that $v_{-k} = \overline{v}_k$ and whose real and immaginary part satisfy:

$$L^r \varphi - \frac{k^2}{r^2} \varphi = 0. \tag{116}$$

For $k = 0$ the real valued function $v_0(r) \in Y_\alpha^r$, and by Lemma 31 we see that $v_0(r) \in \operatorname{span} \{\varphi_0\}$. For $k \in \mathbb{N}$, to determine a fundamental set of solutions to (116) we use the following solutions of the (singular) Liouville equations:

$$\psi_{a,k}(z) = \log \frac{8|(N+1)z^N + (k+N+1)az^{N+k}|^2}{(1 + |z^{N+1} + az^{N+k+1}|^2)^2},$$

for $a \in \mathbb{C}$ and $k \in \mathbb{N}$. Notice that $\psi_{a=0,k} = \log \rho$ and, according to Liouville formula [Lio], $\psi_{a,k}$ satisfies:

$$-\Delta \psi_{a,k} = e^{\psi_{a,k}} - 4\pi N \delta_{z=0} - 4\pi \sum_{j=1}^{k} \delta_{z_j^a}, \ \text{in } \mathbb{R}^2$$

where $z_j^a, j = 1, \ldots, k$, defines the k-distinct non-zero roots of the polynomial: $(N+1)z^N + (k+N+1)az^{N+k}$. Notice in particular that, $|z_j^a| \to +\infty$, as $a \to 0$, $\forall j = 1, \ldots, k$. Therefore, for each test function $\varphi \in C_0^\infty(\mathbb{R}^2)$, and $|a|$ sufficiently small, we have:

$$-\Delta \psi_{a,k} \varphi = e^{\psi_{a,k}} \varphi - 4\pi N \varphi(0).$$

By differentiating this expression with respect to $\alpha = Re(a)$ and $\beta = Im(a)$, we obtain

$$\varphi_{1,k} = \frac{\partial \psi_{a,k}}{\partial \alpha} \Big|_{a=0} = \frac{2}{N+1} \varphi_k(r) \cos k\theta;$$

$$\varphi_{2,k} = \frac{\partial \psi}{\partial \beta} \Big|_{a=0} = \frac{2}{N+1} \varphi_k(r) \sin k\theta;$$

where

$$\varphi_k(r) = \frac{k+N+1+(k-N-1)r^{2(N+1)}}{1+r^{2(N+1)}} r^k. \tag{117}$$

They satisfy:

$$L\varphi_{1,k} = 0 = L\varphi_{2,k}, \ \forall k \in \mathbb{N},$$

so that $\varphi_k(r)$ verifies (116). By replacing k with $-k$ in (117) we still obtain a solution $\tilde{\varphi}_k$ for (116) and we check that $\tilde{\varphi}_k(r) = \varphi_k\left(\frac{1}{r}\right)$. Thus, $\varphi_k(r)$ and $\tilde{\varphi}_k(r)$ define a fundamental set of solutions for (116). But the real and immaginary part of v_k cannot include the component $\tilde{\varphi}_k(r)$, since it admits a $\frac{1}{r^k}$ singularity at the origin, $\forall k \in \mathbb{N}$. Furthermore, for $\alpha \in (0,1)$ and $k \neq N+1$ the function $\varphi_k \notin Y_\alpha^r$, (this is due to its r^k behavior as $r \to +\infty$) and so we conclude that $v_k = 0$, $\forall k \neq N+1$. While, for $k = N+1$, $Re(v_{k=N+1})$ and $Im(v_{k=N+1})$ belong to span$\{\varphi_{N+1}\} =$ span $\{\frac{r^{N+1}}{1+r^{2(N+1)}}\}$. Thus, we conclude that $v \in$ span $\{\varphi_0, \varphi_+, \varphi_-\}$ as claimed.

To establish (b), we show the following:

<u>Claim</u> Im L is closed in X_α.

Let $\varphi_n \in (\ker L)^\perp \subset Y_\alpha$ be such that $L\varphi_n = f_n \in X_\alpha$ and $f_n \to f$ in X_α. We claim that: $\int_{\mathbb{R}^2} \frac{\varphi_n^2}{(1+|z|^{\alpha+2})} \leq C$, $\forall n \in \mathbb{N}$ for suitable $C > 0$. Indeed, by contradiction we assume that (along a subsequence)

$$c_n = \left(\int_{\mathbb{R}^2} \frac{\varphi_n^2}{(1+|z|^{\alpha+2})}\right)^{\frac{1}{2}} \to +\infty, \text{ and set } \Phi_n = \frac{\varphi_n}{c_n}.$$

Then

$$\int_{\mathbb{R}^2} \frac{\Phi_n^2}{(1+|z|^{\alpha+2})} = 1, \ \Phi_n \in (\ker L)^\perp \subset Y_\alpha, \text{ and } L\Phi_n \to 0 \text{ in } X_\alpha. \quad (118)$$

Hence, Φ_n is uniformly bounded in Y_α and (along a subsequence) we have:

$$\Phi_n \to \varphi \text{ weakly in } Y_\alpha,$$

with $\varphi \in Y_\alpha$. We can assume also that the convergence above holds in $L^2_{\text{loc}}(\mathbb{R}^2)$.

Since

$$|\Phi_n(z) - \varphi(z)| \leq c(\|\Phi_n\|_{Y_\alpha} + \|\varphi\|_{Y_\alpha}) \log(1+|z|) \leq C \log(1+|z|)$$

we see that,

$$\int_{\mathbb{R}^2} \frac{(\Phi_n - \varphi)^2}{(1+|z|^{1+\alpha})^2} \leq \|\Phi_n - \varphi\|_{L^2(B_R)}^2 + C \int_{|z| \geq R} \frac{\log(1+|z|)}{(1+|z|^{1+\alpha})^2} =$$

$$= \|\Phi_n - \varphi\|_{L^2(B_R)}^2 + o(1)$$

as $R \to +\infty$. Thus, $\int_{\mathbb{R}^2} \frac{(\Phi_n-\varphi)^2}{(1+|z|^{1+\alpha})^2} \to 0$ as $n \to \infty$, and we conclude that

$$\int_{\mathbb{R}^2} \frac{|\varphi|^2}{(1+|z|^{1+\alpha})^2} = 1. \quad (119)$$

Similarly we see that $\int_{\mathbb{R}^2} \rho \Phi_n \to \int_{\mathbb{R}^2} \rho \varphi$ as $n \to \infty$. Hence $L\varphi = 0$ and $\varphi \in (\ker L)^\perp$. So we conclude that $\varphi = 0$, in contradiction to (119). In conclusion, the sequence $\varphi_n \in (\ker L)^\perp : L\varphi_n = f_n$ is uniformly bounded in Y_α. So we can argue exactly as above to find $\varphi \in Y_\alpha : \varphi_n \to \varphi$ weakly in Y_α with $L\varphi = f$ and the Claim is proved.

Consequently, we may decompose:

$$X_\alpha = \mathrm{Im}L \oplus (\mathrm{Im}L)^\perp$$

according to the scalar product in X_α. Thus, for $\xi \in (\mathrm{Im}L)^\perp$ we have:

$$0 = (Lu, \xi)_{X_\alpha} = (Lu, (1 + |z|)^{2+\alpha}\xi), \ \forall u \in Y_\alpha.$$

The density of C_0^∞ in Y_α implies that the function

$$\psi = (1 + |z|)^{2+\alpha}\xi \in Y_\alpha \text{ and } L\psi = 0.$$

Therefore by part (a), we may write: $\psi = a_0\varphi_0 + a_+\varphi_+ + a_-\varphi_-$ for suitable constants a_0, a_+ and a_-. Furthermore, by Lemma 31 we know that the radial function $f(r) = \varphi_0(r)\frac{1}{(1+r^2)^2} \in \mathrm{Im}L$ and satisfies $(f, \varphi_\pm)_{L^2} = 0$. Consequently

$$0 = (f, \xi)_{X_\alpha} = (f, \psi)_{L^2} = a_0 \left(2\pi \int_0^{+\infty} \left(\frac{\varphi_0(r)}{1+r^2} \right)^2 r dr \right),$$

that is $a_0 = 0$, and also part (b) of our claim follows.   ∎

At this point we can complete our perturbation analysis, as the operator:

$$P : Y_\alpha \times \mathbb{C} \times \mathbb{R} \to X_\alpha$$

in (100) is well defined, smooth and extends with continuity at $\varepsilon = 0$ with $P(0,0,0) = 0$. Moreover, the linearized operator:

$$A : \frac{\partial P}{\partial(u_1, a)}(0,0,0) : Y_\alpha \times \mathbb{C} \to X_\alpha \tag{120}$$

takes the form:

$$A(\varphi, b) = L\varphi + M(b) \tag{121}$$

with $\varphi \in Y_\alpha$, $b = b_1 + ib_2 \in \mathbb{C}$ and

$$M(b) = -4 \left(\rho w_0 - \frac{2}{\lambda}\rho^2 \right) \varphi_+ b_1 - 4 \left(\rho w_0 - \frac{2}{\lambda}\rho^2 \right) \varphi_- b_2. \tag{122}$$

Observe that:

Lemma 33

$$\int_{\mathbb{R}^2} \left(\rho w_0 - \frac{2}{\lambda} \rho^2 \right) \varphi_\pm^2 = \pi \int_0^{+\infty} \left(\rho(r) w_0(r) - \frac{2}{\lambda} \rho^2(r) \right) \frac{r^{2(N+1)}}{(1+r^{2(N+1)})^2} r \, dr < 0.$$

Proof. By easy calculation, we find:

$$L^r \left(\frac{1}{16(1+r^{2(N+1)})^2} \right) = \frac{(N+1)^2 r^{4N+2}}{(1+r^{2(N+1)})^4}.$$

Furthermore we can take advantage of the decay estimates (113) and (114) to use integration by part and obtain:

$$\int_{\mathbb{R}^2} \left(\rho w_0 - \frac{2}{\lambda} \rho^2 \right) \varphi_\pm^2 =$$

$$= \pi \int_0^{+\infty} \left(\frac{8(N+1)^2 r^{4N+2}}{(1+r^{2(N+1)})^4} w_0(r) - \frac{2}{\lambda}\rho^2 \frac{r^{2(N+1)}}{(1+r^{2(N+1)})^2} \right) r \, dr =$$

$$= \pi \int_0^{+\infty} \left(\frac{1}{2} L^r \left(\frac{1}{(1+r^{2(N+1)})^2} \right) w_0(r) - \frac{2}{\lambda}\rho^2 \frac{r^{2(N+1)}}{(1+r^{2(N+1)})^2} \right) r \, dr =$$

$$= \int_0^{+\infty} \left(\frac{1}{2} L^r w_0 \frac{1}{(1+r^{2(N+1)})^2} - \frac{2}{\lambda}\rho^2 \frac{r^{2(N+1)}}{(1+r^{2(N+1)})^2} \right) r \, dr =$$

$$= \frac{1}{\lambda} \int_0^{+\infty} \rho^2(r) \left(\frac{1}{2(1+r^{2(N+1)})^2} - \frac{2r^{2(N+1)}}{(1+r^{2(N+1)})^2} \right) r \, dr,$$

where, to derive the last identity we have used the fact that w_0 satisfies (97). The sign of the integral above can be determined by means of change of variable $t = r^2$,

$$\int_0^{+\infty} \rho^2(r) \frac{1 - 4r^{2(N+1)}}{2(1+r^{2(N+1)})^2} r \, dr = 16(N+1)^4 \int_0^{+\infty} \frac{1-4t^{N+1}}{(1+t^{N+1})^6} t^{2N} \, dt =$$

$$= 16(N+1)^4 \left(\int_0^{+\infty} \frac{t^{2N}}{(1+t^{N+1})^5} - 5 \int_0^{+\infty} \frac{t^{3N+1}}{(1+t^{N+1})^6} \, dt \right) =$$

$$= 16(N+1)^4 \left(\int_0^{+\infty} \frac{t^{2N}}{(1+t^{N+1})^5} + \frac{1}{N+1} \int_0^{+\infty} t^{2N+1} \frac{d}{dt} \left(\frac{1}{(1+t^{N+1})^5} \right) \, dt \right) =$$

$$16(N+1)^4 \left(\int_0^{+\infty} \frac{t^{2N}}{(1+t^{N+1})^5} - \frac{2N+1}{N+1} \int_0^{+\infty} \frac{t^{2N}}{(1+t^{N+1})^5} \, dt \right) =$$

$$-16(N+1)^3 N \int_0^{+\infty} \frac{t^{2N}}{(1+t^{N+1})^5} \, dt < 0,$$

and the desired conclusion follows. ∎

We are now ready to show the following,

Proposition 34 *For $\alpha \in (0,1)$, the operator $A : Y_\alpha \to X_\alpha$ in (120) and (121) is* <u>onto</u> *and* $\ker A = \ker L = \text{span}\ \{\varphi_0, \varphi_+, \varphi_-\}$.

Proof. Let $f \in X_\alpha$, we need to find $\varphi \in Y_\alpha$ and $b = b_1 + ib_2 \in \mathbb{C}$ such that:

$$A(\varphi, b) = L\varphi + M(b) = f, \text{ with } M(b) \text{ in (122).} \tag{123}$$

To this purpose, multiply (123) by φ_+ and integrate over $\mathbb{R}^2$ to find:

$$\int_{\mathbb{R}^2} f\varphi_+ =$$

$$= \int_{\mathbb{R}^2} L\varphi\varphi_+ - 4b_1 \int_{\mathbb{R}^2} \left(\rho w_0 - \frac{2}{\lambda}\rho^2\right) \varphi_+^2 - 4b_2 \int_{\mathbb{R}^2} \left(\rho w_0 - \frac{2}{\lambda}\rho^2\right) \varphi_+\varphi_- =$$

$$= -4\pi b_1 \int_0^{+\infty} \left(\rho(r)w_0(r) - \frac{2}{\lambda}\rho^2(r)\right) \frac{r^{2(N+1)}}{(1 + r^{2(N+1)})^2}dr.$$

By Lemma 33, we may solve for b_1 as follows,

$$b_1 = -\frac{1}{4\pi} \int_{\mathbb{R}^2} f\varphi_+ \left(\int_0^{+\infty} \left(\rho(r)w_0(r) - \frac{2}{\lambda}\rho^2(r)\right) \frac{r^{2(N+1)}}{(1 + r^{2(N+1)})^2}dr\right)^{-1}. \tag{124}$$

Analogously we derive:

$$b_2 = -\frac{1}{4\pi} \int_{\mathbb{R}^2} f\varphi_- \left(\int_0^{+\infty} \left(\rho(r)w_0(r) - \frac{2}{\lambda}\rho^2(r)\right) \frac{r^{2(N+1)}}{(1 + r^{2(N+1)})^2}dr\right)^{-1}. \tag{125}$$

Set $g = f - M(b)$, where $M(b)$ is given in (122) with b_1 and b_2 specified in (124) and (125) respectively. Since $\int_{\mathbb{R}^2} g\varphi_\pm = 0$, by Proposition 30 (b), we find $\varphi \in X_\alpha : L\varphi = g$. Thus we have checked that Im $A = X_\alpha$. If we take $f = 0$ in the argument above, we find that $b_1 = 0 = b_2$, and so $A\varphi = 0$ if and only if $L\varphi = 0$, that is $\ker A = \ker L = \text{span}\ \{\varphi_0, \varphi_+, \varphi_-\}$. ∎

Setting,

$$U_\alpha = (\text{span}\{\varphi_0, \varphi_+, \varphi_-\})^\perp,$$

we obtain to the following existence result for (92):

Proposition 35 *For every* $\lambda > 0$ *and* $\alpha \in (0,1)$, *there exists* $\varepsilon_0 > 0$ *sufficiently small and smooth functions:*

$$a_\varepsilon : (-\varepsilon_0, \varepsilon_0) \to \mathbb{C}, \ u_{1,\varepsilon} : (-\varepsilon_0, \varepsilon_0) \to U_\alpha$$

with $a_{\varepsilon=0} = 0$ *and* $u_{1,\varepsilon=0} = 0$ *such that*

$$u_\varepsilon(z) = u_{\varepsilon, a_\varepsilon}^0(\varepsilon z) + \log \varepsilon^2 + \varepsilon^2 w_0(\varepsilon|z|) + \varepsilon^2 u_{1,\varepsilon}(\varepsilon z) \tag{126}$$

defines a solution for (92), with w_0 in (112) satisfying (113)–(114). Furthermore, as $\varepsilon \to 0$ the following estimates hold:

$$|u_{1,\varepsilon}(z)| \leq o(1)\log(1+|z|), \ |\nabla u_{1,\varepsilon}(z)| = \frac{o(1)}{1+|z|}, \ \forall z \in \mathbb{R}^2;$$

$$\lambda \int_{\mathbb{R}^2} e^{u_\varepsilon}(1 - e^{u_\varepsilon}) = 8\pi(N+1) + o(1) \qquad (127)$$

Proof. By a straightforward application of the Implicit Function Theorem (cf. [Nir]) we obtain u_ε in (126). To check the validity of (127) we use (102) to estimate:

$$|u_{1,\varepsilon}(z)| \leq C_1 \|u_{1,\varepsilon}\|_{Y_\alpha} \log(1+|z|), \ \forall z \in \mathbb{R}^2$$

with $\|u_{1,\varepsilon}\|_{Y_\alpha} \to 0$, as $\varepsilon \to 0$. Furthermore, we see that,

$$-\Delta u_{1,\varepsilon} = f_{1,\varepsilon}$$

with $f_{1,\varepsilon} \to 0$ in X_α and $(1+|z|^2)|f_{1,\varepsilon}| \leq c_\lambda \varepsilon$ in $\mathbb{R}^2$, with a suitable constant c_λ independent of ε. Since $u_{1,\varepsilon} \in Y_\alpha$, we can write

$$\nabla u_{1,\varepsilon}(z) = \frac{1}{2\pi} \int_{\mathbb{R}^2} \frac{y-z}{|y-z|^2} f_{1,\varepsilon}(y)dy,$$

and for $|z| \geq 2$, derive the following estimate:

$$|z||\nabla u_{1,\varepsilon}(z)| \leq \int_{\{|y-z| \leq \frac{|z|}{2}\}} \frac{|z||f_{1,\varepsilon}(y)|}{|y-z|} + \int_{\{|y-z| \geq \frac{|z|}{2}\}\}} \frac{|z||f_{1,\varepsilon}(y)|}{|y-z|}$$

$$\leq 2\int_{\{|y-z| \leq \frac{|z|}{2}\}} \frac{|y||f_{1,\varepsilon}(y)|}{|y-z|} + 2\int_{\{|y-z| \leq \frac{|z|}{2}\}} |f_{1,\varepsilon}(y)|$$

$$\pi|z|\left(\max_{\{|y| \geq \frac{|z|}{2}\}} |y||f_{1,\varepsilon}(y)|\right) + 2\|f_{1,\varepsilon}\|_{L^1(\mathbb{R}^2)}$$

$$\leq c_\lambda \varepsilon + 2\|f_{1,\varepsilon}\|_{L^1(\mathbb{R}^2)} \to 0, \text{ as } \varepsilon \to 0.$$

and the first estimate in (127) follows. Finally, notice that

$$\lambda \int_{\mathbb{R}^2} e^{u_\varepsilon(z)}(1 - e^{u_\varepsilon(z)}) = \lambda \int_{\mathbb{R}^2} e^{u_\varepsilon(\frac{z}{\varepsilon})}(1 - e^{u_\varepsilon(\frac{z}{\varepsilon})})\frac{1}{\varepsilon^2} =$$

$$\lambda \int_{\mathbb{R}^2} e^{u_{\varepsilon,a_\varepsilon}(z)+\varepsilon^2(w_0+u_{1,\varepsilon})} - \lambda\varepsilon^2 \int_{\mathbb{R}^2} e^{2u_{\varepsilon,a_\varepsilon}^0(z)+2\varepsilon^2(w_0+u_{1,\varepsilon})}$$

Therefore, by the estimates established above, we can pass to the limit into the integral sign to conclude that,

$$\lim_{\varepsilon \to 0} \lambda \int_{\mathbb{R}^2} e^{u_\varepsilon(z)}(1-e^{u_\varepsilon(z)}) = \lambda \int_{\mathbb{R}^2} e^{u^0_{\varepsilon=0,a=0}} = 16\pi(N+1) \int_0^{+\infty} \frac{r^{2N+1}}{(1+r^{2(N+1)})^2} dr$$

$$= 8\pi(N+1) \int_0^{+\infty} \frac{dt}{(1+t)^2} = 8\pi(N+1).$$

This completes the proof.   ∎

We have established the following existence result concerning non-topo-logical Chern-Simons vortices:

Theorem 36 *For $k > 0$, $N \in \mathbb{N}$ and a given set of (vortex) points $Z = \{z_1, \ldots, z_N\}$ (repeated according to their multiplicity), there exist $\varepsilon_0 > 0$ such that, for every $\varepsilon \in (0, \varepsilon_0)$ we have a vortex configuration $(\mathcal{A}^\varepsilon, \varphi^\varepsilon)_\pm$ solution to the selfdual equations (33) (with the $\pm$ sign choosen accordingly) such that*

i) $|\varphi^\varepsilon_\pm| < 1$ *in* $\mathbb{R}^2$; $\varphi^\varepsilon_\pm$ *vanishes exactly in the set* Z, *and if* $n_j \in \mathbb{N}$ *is the multeplicity of* z_j *then,*

$$\varphi^\varepsilon_+(z) \text{ and } \overline{\varphi}^\varepsilon_-(z) = O((z-z_j)^{n_j}), \text{ as } z \to z_j, \, j = 1, \ldots, N. \quad (128)$$

ii) There exist constants $C_\varepsilon > 0, R_\varepsilon > 0$ *and* $\beta_\varepsilon \to 0^+$ *as* $\varepsilon \to 0$ *such that*

$$|F^\varepsilon_{12}| + |z|^2 |\nabla|\varphi^\varepsilon_\pm||^2 \leq C_\varepsilon |z|^{-2(N+2\beta_\varepsilon)}, \, \forall |z| \geq R_\varepsilon. \quad (129)$$

In particular $\varphi^\varepsilon_\pm(z) \to 0$ *as* $|z| \to +\infty$.
iii)

$$Magnetic \; flux\text{: } \Phi^\varepsilon_\pm = \int_{\mathbb{R}^2} (F^\varepsilon_{12})_\pm = \pm 4\pi(N+1) + o(1); \quad (130)$$

$$Electric \; charge\text{: } Q^\varepsilon_\pm = \int_{\mathbb{R}^2} (J^0_\varepsilon)_\pm = \pm 4\pi k(N+1) + o(1); \quad (131)$$

$$Total \; energy\text{: } E_\varepsilon = \int_{\mathbb{R}^2} \mathcal{E}_\pm = 4\pi(N+1) + o(1); \quad (132)$$

$\varepsilon \to 0$.

The perturbative approach presented above has been successfully applied to obtain non-topological type solutions in many other context, as one can see for instance in [Ch1], [Ch2], [Ch3], [Ch4], [ChCh1], [ChI2], [ChI3], [ChT1] and [ChT2].
On the other hand an alternative construction of non-topological vortices is carried out in [CFL], and provides a new class of Chern-Simons vortices with the property that they "concentrate" around the vortex points (as $\kappa \to 0$). This is a much desirable property from the point of view of the physical applications, that as we know, it is always satisfied by topological vortices, see (89).

We conclude by observing that a parallel analysis (more or less complete) has been developed for the study of periodic vortices or vortices defined over compact surfaces,in the framework of selfdual abelian models that include and generalize those discussed above. In this respect, see for example: [CY], [ChiR], [DJLPW], [DJLW1], [DJLW2], [DJLW3], [Ha1], [Ha2], [MNR], [NT1], [NT3], [Ol], [Ri1], [Ri2], [Ri3], [RT1], [SY1], [T1], [T2], [T3] and [WY].

Furthermore, such analysis is relevant in connection to other problems from physics and geometry where equations with similar features emerge naturally, see e.g. [Au], [Ban], [CLMP1], [CLMP2], [CK1], [CK2], [CK3], [ChCL], [ChY1], [ChY2], [ChY3], [H], [K], [KW1] [KW2], [Ki1], [Ki2], [Ni], [Su2] and [Wo].

On the contrary, much less has been accomplished in the context of non-abelian selfdual theories, where the study of non-abelian vortex configurations give rise to systems of elliptic equations, which pose some new and delicate analytical difficulties. In this direction, we mention the (partial) contributions of [BT2] and [SY2] towards the understanding of W-condensates for the (non-abelian) Electroweak theory (cf. [La]), as motivated by the work of Ambjorn-Olesen (cf. [AO1], [AO2] and [AO3]); see also [SY3].

While, non-abelian selfdual Chern-Simons vortices are analyzed in [JoLW], [JoW1], [JoW2] [LN], [ChOS], [NT1], [Y3] and [Y4].

But still many unresolved issues remain (see [T8]), as for instance the understanding of the asymptotic behavior of non-abelian Chern-Simons vortices. Indeed, as vortex configurations are likely to develop a "concentration" behavior (for limiting values of the parameters), it is necessary to provide an accurate blow-up analysis for systems, in the same spirit of what it is now available for single Liouville-type equations, see e.g. [BP], [BCLT], [BT1], [BLS], [BM], [CLS], [ChL1], [ChL2], [ChL3], [ChL4], [ChLW], [CD], [CL1], [CL2], [CL3], [Che], [Chn], [CW], [DeKM], [Dj], [Dr], [Es], [EGP], [L1], [L2], [LS], [Li1], [Li2], [LiL1], [LiL2], [LiW], [Lu], [LZ], [MW], [MNR], [MaN], [NS], [OS1], [OS2], [PT], [Sh], [ST], [Su1], [WW1] and [WW2].

This direction of investigation is still quite open, aside from the contributions in [JoLW], [JoW1], [JoW2], [MN], [SW1], [SW2] and [W], which however do not take into account the presence of Dirac measures, that are the cause of even more degenerate behaviours.

References

[Ab]   A.A. Abrikosov, On the magnetic properties of superconductors of second group, *Sov. Phys. JETP* **5** (1957), 1174–1182.

[Ad]   D.R. Adams, A sharp inequality of J. Moser for higher order derivatives, *Ann. of Math.* **128** (1988), 385–398.

[AH]   I. Aitchinson, A. Hey, *Gauge theories in Particle physics*, IoP Publisher vol. 1 (2002), vol. 2 (2003).

[AO1]   J. Ambjorn, P. Olesen, On Electroweak magnetism, *Nucl. Phys. B* **218** (1989), 67–71.

[AO2]   J. Ambjorn, P. Olesen, A condensate solution of the classical electroweak theory which interpolates between the broken and symmetric phase, *Nucl. Phys. B* **330** (1990), 193–204.

[AO3]   J. Ambjorn, P. Olesen A magnetic condensate solution of the classical Electroweak theory, *Phys. Lett. B* **218** (1989), 67–71.

[ADHM]   M.F. Atiyah, V.G. Drinfeld, N.J. Hitchin, Yu.I Mannin, Constraction of Instantons, *Phys. Lett. A* **65** (1978), 185–187.

[AtH]   M.F. Atiyah and N.J. Hitchin, *The Geometry and Dynamics of Magnetic Monopoles*, Princeton Univ. Press, Princeton (1988).

[AHS1]   M.F. Atiyah, N.J. Hitchin, I.M. Singer, Deformation of Instantons, *Proc. Natl. Acad. Sci.* USA 74, (1997), 2662–2663.

[AHS2]   M.F. Atiyah, N.J. Hitchin, I.M. Singer, Selfduality in four dimensional Riemannian geometry, *Proc. Roy Soc. A* **362** (1978), 425–461.

[Au]   T. Aubin, *Some Nonlinear Problems in Riemannian Geometry*, Springer-Verlag, Berlin Heidelberg New York, (1998).

[BP]   S. Baraket, F. Pacard, Construction of singular limit for a semilinear elliptic equation in dimension 2, *Calc. Var. P.D.E*, **6** (1998), 1–38.

[Ban]   C. Bandle, *Isoperimetric Inequalities and Applications*, Pitman A. Publishing **7**, (1980).

[BCLT]   D. Bartolucci, C.C. Chen, C.S. Lin, G. Tarantello, Profile of blow-up solutions to mean field equations with singular data, *Comm. P.D.E.* **29**, n. 7-8 (2004), 1241–1265.

[BT1]   D. Bartolucci, G. Tarantello, The Liouville equation with singular data: a concentration-compactness principle via a local representation formula, *J. Diff. Eq.* **185** (2002), 161–180.

[BT2]   D. Bartolucci, G. Tarantello, Liouville type equations with singular data and their applications to periodic multivortices for the electroweak theory, *Comm. Math. Phys.* **229** (2002), 3–47.

[Be]   W. Beckner, Sharp Sobolev inequalityies on the sphere and the Moser-Trudinger inequality *Ann. of Math.* **138** (1993), 213–242.

[BPST]   A.A. Belavin, A.M. Polyakov, A.S. Schwartz, Yu.S. Tyupkin, Pseudoparticle solutions of the Yang Mills Equations, *Phys. Lett. B* (1975), 85–87.

[BeR]   J. Berger, J. Rubistein, On the zero set of the wave function in superconductivity, *Comm. Math. Phys.* **202**, n. 3 (1999), 621–628.

[BBH]   F. Bethuel, H. Brezis, F. Helein, *Ginzburg Landau Vortices*, Birkhauser, 1994.

[Bo]   E.B. Bogomolnyi, The stability of classical solutions, *Sov. J. Nucl. Phys.* **24** (1976), 449–454.

[Bra1]   S. Bradlow, Vortices in Holomorphic line bundles over closed Kahler Manifolds, *Comm. Math. Phys.* **135** (1990), 1–17.

[Bra2]   S. Bradlow, Special metrics and solvability for holomorphic bundles with global sections, *J. Diff. Geom.* **33** (1991), 169–214.

[BLS]   H. Brezis, Y.Y. Li, I. Shafrir, A sup + inf inequality for some nonlinear elliptic equations involving exponential nonlinearities, *J. Funct. Anal.* **115** (1993), 344–358.

[BM]   H. Brezis, F. Merle, Uniform estimates and blow-up behavior for solutions of $-\Delta u = V(x)e^u$ in two dimensions, *Comm. P.D.E.* **16** (1991), 1223–1253.

[CLS]   X. Cabré, M. Lucia, M. Sanchon, On the minimizers of a Moser-Trudinger type inequality, *Comm. P.D.E.* **30**, n. 7-9 (2005), 1315–1330.

[CY]   L. Caffarelli, Y. Yang, Vortex condensation in the Chern-Simons-Higgs model, *Comm. Math. Phys.*, **168** (1995), 154–182.

[CLMP1]   E. Caglioti, P.L. Lions, C. Marchioro, M. Pulvirenti, A special class of stationary flows for two-dimensional Euler equations, a statistical mechanics description, part I, *Comm. Math. Phys.* **143** (1992), 501–525.

[CLMP2]   E. Caglioti, P.L. Lions, C. Marchioro C., M. Pulvirenti, A special class of
          stationary flows for two-dimensional Euler equations, a statistical mechanics
          description, part II, *Comm. Math. Phys.* **174** (1995), 229–260.

[CaL]   D. Cangemi, C. Lee, Selfdual Chern-Simons solitons and $(2 + 1)$-dimensional
          Einstein gravity, *Phys. Review D* **46**, n. 10 (1992), 4768–4771.

[Ch1]   D. Chae, Existence of multistrings solutions of the selfgravitating massive
          W-boson, *Lett. Math. Phys.* **73**, n. 2 (2005), 123–134.

[Ch2]   D. Chae, On the elliptic system arising from a self-gravitating Born-Infeld
          Abelian Higgs theory, *Nonlinearity* **18**, n. 4 (2005), 1823–1833.

[Ch3]   D. Chae, Existence of the semilocal Chern-Simons vortices, *J. Math. Phys.*
          46, n. 4 (2005), 042303, 10 pp.

[Ch4]   D. Chae, On the multi-string solutions of the self-dual static Einstein-Maxwell-
          Higgs system, *Calc. Var. PDE* **20**, n. 1 (2004), 47–63.

[ChCh1]   D. Chae, K. Choe, Existence of selfgravitating Chern-Simons vortices,
          J. Math. Phys. **44**, n. 12 (2003), 5616–5636.

[ChI1]   D. Chae, O. Imanuvilov, The existence of non-topological multivortex solu-
          tions in the relativistic selfdual Chern-Simons theory, *Comm. Math. Phys.*
          (2000), 119–142.

[ChI2]   D. Chae, O. Imanuvilov, Non-topological solutions in the generalized self-dual
          Chern-Simons-Higgs theory, *Calc. Var. PDE* **16**, n. 1 (2003), 47–61.

[ChI3]   D. Chae, O. Imanuvilov, Non-topological multivortex solutions to the self-
          dual Maxwell-Chern-Simons-Higgs systems, *J. Funct. Anal.* **196**, n. 1 (2002),
          87–118.

[ChK1]   D. Chae, N. Kim, Topological multivortex solutions of the selfdual Maxwell-
          Chern-Simons-Higgs system, *Jour. Diff. Eq.* **134** (1997), 154–182.

[ChK2]   D. Chae, N. Kim, Vortex condensates in the relativistic selfdual Maxwell-
          Chern-Simons Higgs system, Preprint.

[ChNa]   D. Chae, H.S. Nam, On the condensate multivortex solutions of the self-dual
          Maxwell-Chern-Simons $CP(1)$ model, *Ann. Henri Poincaré* **2**, n. 5 (2001),
          887–906.

[ChOS]   D. Chae, H. Ohtsuka, T. Suzuki, Some existence results for solutions to SU(3)
          Toda system. *Calc. Var. PDE* **24**, n. 4 (2005), 403-429.

[ChT1]   D. Chae, G. Tarantello, On Planar Electroweak vortices *Ann. IHP Analyse
          Non Lineaire*, AN21 (2004), 187–207.

[ChT2]   D. Chae, G. Tarantello, Selfgravitating Electroweak Strings, *J. Diff. Eq.* **213**
          (2005), 146–170.

[CFL]   H. Chan, C.C. Fu, C.S. Lin, Non-topological multi-vortex solutions to the
          selfdual Chern-Simons-Higgs equation, *Comm. Math. Phys.* **231** (2002),
          189–221.

[ChNe]   M. Chai, Chain-N.P. Nelipa, *Introduction to Gauge Field Theory*, Springer,
          Berlin New York, (1984).

[ChCL]   A. Chang, C.C. Chen, C.S. Lin, Extremal functions for a mean field equation
          in two dimension, *Lectures on partial differential equations*, New Stud. Adv.
          Math., 2, Int. Press, Somerville, MA, (2003), 61–93.

[ChY1]   A. Chang, P. Yang, Conformal deformation of metric on S^2, *J. Diff. Geom.*
          27 (1988), 259–296.

[ChY2]   A. Chang, P. Yang, Prescribing Gaussian curvature on S^2, *Acta Math.*, **159**
          (1987), 215–259.

[ChY3]   A. Chang, P. Yang, The inequality of Moser-Trudinger and Applications to
          Conformal Geometry, *Comm. Pure Appl. Math.* **56** (2003), 1135–1150.

[CK1]   S. Chanillo, M. Kiessling, Rotational symmetry of solutions of some nonlinear
          problems in statistical mechanics and in geometry, *Comm. Math. Phys.* **160**
          (1994), 217–238.

[CK2]   S. Chanillo, M. Kiessling, Conformally invariant systems of nonlinear PDE of
          Liouville type *Geom. Funct. Analysis* **5** (1995) 924–947.

[CK3]   S. Chanillo, M. Kiessling, Surfaces with prescribed scalar curvature *Duke Math. J.* **105** (2002) 309–353.

[ChL1]   C.C. Chen, C.S. Lin, Sharp Estimates for Solutions of Multi Bubbles in Compact Riemann Surfaces, *Comm. Pure Appl. Math.* **55** (2002) 728–771.

[ChL2]   C.C. Chen, C.S. Lin, Topological Degree for a Mean Field Equation on Riemann Surfaces, *Comm. Pure Appl. Math.* **56** (2003), 1667–1727.

[ChL3]   C.C. Chen, C.S. Lin, On the symmetry of blow up solutions to a mean field equation, *Ann. IHP Analyse Nonlineaire* **18** (2001), 271–296.

[ChL4]   C.C. Chen, C.S. Lin, A sharp sup + inf estimate for a nonlinear equation in the plane *Comm. Anal. Geom.* **6** (1998), 1–19.

[ChLW]   C.C. Chen, C.S. Lin, G. Wang, Concentration phenomena of two-vortex solutions in a Chern-Simons model. *Ann. Sc. Norm. Super. Pisa Cl. Sci.* **5** 3 n. 2 (2004), 367–397.

[CD]   W. Chen, W. Ding, Scalar curvature on S^2, *Trans. AMS*(1987), 365–382.

[CL1]   W. Chen, C. Li, Classification of solutions of some nonlinear elliptic equations, *Duke Math. J.* **63** (1991), 615–623.

[CL2]   W. Chen, C. Li, Qualitative properties of solutions of some nonlinear elliptic equations in R^2, *Duke Math. J.* **71** (1993), 427–439.

[CL3]   W. Chen, C. Li, Prescribing Gaussian curvature on surfaces with conical singularities, *J. Geom. Anal.* **1** (1991), 359–372.

[Che]   X.A. Chen, A Trudinger inequality on surfaces with conical singularities, *Proc. Amer. Math. Soc.* **108** (1990), 821–832.

[Chn]   X.X. Chen, Remarks on the existence of branch bubbles on the blow up analysis of equation $-\Delta u = e^u$ in dimension two, *Comm. Anal. Geom.* **7** (1999), 295–302.

[CHMcLY] X. Chen, S. Hastings, J. McLeod, Y. Yang, A nonlinear elliptic equation arising from gauge field theory and cosmology, *Proc Roy. Soc. Lond. A* (1994), 453–478.

[ChiR]   F. Chiacchio, T. Ricciardi Multiplicity for a selfdual $CP(1)$ Maxwell-Chern-Simons model, *NoDEA* **13** (2007), 563–584.

[CSW]   M. Chipot, I. Shafrir, G. Wolansky, On the solutions of Liouville systems, *Jour. Diff. Eq.* **140** (1997), 59–105, Erratum *Jour. Diff. Eq.* **178** (2002), 630.

[CM]   Y.M. Cho, D. Maison, Monopole configurations in Weinberg-Salam model, *Phys. Rev. Lett. B* **391** (1997), 360–365.

[Cho1]   K. Choe, Uniqueness in Chern-Simons theory, *J. Math. Phys.* **46** (2005), n. 1, 012305.

[Cho2]   K. Choe, Asymptotic behavior of condensate solutions in the Chern-Simons-Higgs theory, *J. Math. Phys.* to appear.

[ChoK]   K. Choe, N. Kim, Blow-up analysis of the selfdual Chern-Simons -Higgs vortex equation, *Ann. IHP Analyse Non Lineaire*, to appear.

[ChoN]   K. Choe, H.S. Nam, Existence and uniqueness of Topological Multivortex solutions of the Selfdual Chern-Simons CP(1) Model, *Nonlinear Anal.* **66** (2007), 2794–2813.

[CW]   K.S. Chou, T.Y.H. Wan, Asymptotic radial symmetry for solutions of $\Delta u + \exp u = 0$ in a punctured disc, *Pacific J. of Math.* **163** (1994), 269–276.

[DET]   J. Dalbeault, M.J. Esteban, G. Tarantello, The role of Onofri type inequalities in the symmetry properties of extremals for Caffarelli-Kohn-Nirenberg inequalities in two space dimensions, *Ann. Sc. Nor. Pisa vol VII*, (2008), 313–341.

[DeKM]   M. Del Pino, M. Kowalczyk, M. Musso Singular limits in Liouville-type equations, *Calc. Var. PDE*, **24** (2005), 47–81.

[DJLPW] W. Ding, J. Jost, J. Li, X. Peng, G. Wang, Self duality equations for Ginzburg-Landau and Seiberg-Witten type functionals with 6th order potential, *Comm. Math. Phys.* **217** (2001), 383–407.

[DJLW1] W. Ding, J. Jost, J. Li, G. Wang, The differential equation $\Delta u = 8\pi - 8\pi e^u$ on a compact Riemann surface, *Asian. J. Math.* **1** (1997), 230–248.

[DJLW2] W. Ding, J. Jost, J. Li, G. Wang, An analysis of the two-vortex case in the Chern-Simons-Higgs model, *Calc. Var. and P.D.E.* **7** (1998), 87–97.

[DJLW3] W. Ding, J. Jost, J. Li, G. Wang, Existence results for mean field equations, *Ann. IHP Analyse Non Linéaire* **16** (1999), 653–666.

[Dj] Z. Djadli, Existence results for the mean field problem in Riemann surfaces of all genus, *Comm. Contemp. Math.*, Vol. 10 (2008), 205–220.

[Dr] O. Druet, Multibumps analysis in Dimension 2: Quantification of blow-up levels, *Duke Math. J.* **132** (2006), 217–269.

[DK] S. Donaldson, P. Kronheimer, *The geometry of four-manifolds*, Oxford Univ. Press (1990).

[DGP] Q. Du, M.D. Gunzburger, J.S. Peterson, Analysis and approximation of the Ginzburg-Landau model of superconductivity. *SIAM Review*, **34**, 1, (1992), 54–81.

[D1] G. Dunne, *Self-Dual Chern-Simons Theories*, Lect. Notes in Phys., M **36** New Series, Springer, (1995).

[D2] G. Dunne, Mass degeneracies in self-dual models, *Phys. Lett. B* **345** (1995), 452–457.

[D3] G. Dunne, Aspects of Chern Simons theory, *Les Houches·Lectures on section LXIX: "Topological Aspects of low dimensional systems*, Eds A. Comtet, T.Jolicoeur, S. Ouvry and F. David, EDP Sciences Springer (1998), 55–175.

[Es] P. Esposito, Blow up solutions for a Liouville equation with singular data, *SIAM J. Math. Anal.* **36**, n. 4 (2005), 1310–1345.

[EGP] P. Esposito, M. Grossi, A. Pistoia, On the existence of blowing-up solutions for a mean field equation. *Ann. Inst. H. Poincaré Anal. Non Linéaire* **22**, n. 2 (2005), 227–257.

[Fel] B. Felsager, *Geometry, Particle and Fields*, Springer, Berlin and New York (1998).

[Fo] L. Fontana, Sharp borderline Sobolev inequalities on compact Riemannian manifolds, *Comment. Math. Helv.* **68** (1993), 415–454.

[Fro] J. Frohlich, The fractional Quantum Hall effect, Chern-Simons theory and integral lettice, *Proc. Internat. Congr. Math.* Birkhauser, Basil (1995), 75–105.

[FM1] J. Frohlich, P. Marchetti, Quantum field theory of anyons, *Lett. Math. Phys.* **16** (1988), 347–358.

[FM2] J. Frohlich, P. Marchetti, Quantum field theory of vortices and anyons, *Comm. Math. Phys.* **121** (1989), 177–223.

[Ga1] O. Garcia-Prada, A direct existence proof for the vortex equation over a compact Riemannian surface, *Bull. London Math. Soc.* **26** (1994), 88–96.

[Ga2] O. Garcia-Prada, Invariant connections and vortices, *Comm. Math. Phys.* **156** (1993), 527–546.

[Ga3] O. Garcia-Prada, Dimensional reduction of stable bundles vortices and stablepairs, *Intern. J. Math.* **5** (1994), 1–52.

[GL] V. Ginzburg and L. Landau, On the theory of Supercondictivity, *Zh. Eksper. Theor. Fiz.* **20** (1950) 1064–1082. Translated in "Collected papers of L. Landau" ed D. Ter Haar Pergamon, New York (1965), 546–568.

[GT] D. Gilbarg, N.S. Trudinger, *Elliptic partial differential equations of second order*, Springer Verlag, (1983).

[GS] M. Gockeller, T. Schucker, *Differential Geometry, Gauge Theory and Gravity*, Cambridge Univ. Press (1990).

[GO] P. Goddard, D.I. Olive, Magnetic monopoles in gauge field theories, *Rep. Prog. Phys.* **41** (1978) 1360–1473.

[Ha1] J. Han, Asymptotics for the vortex condensate solutions in Chern-Simons-Higgs theory, *Asymp. Anal.* **28** (2001) 31–48.

[Ha2]   J. Han, Asymptotic limit for condensate solutions in the abelian Chern-Simons-Higgs model, *Proc. AMS* **131** n. 6 (2003) 1839–1845.

[Ha3]   J. Han, Existence of topological multivortex solutions in the selfdual gauge theories, *Proc. Royal soc. Edinburgh*, **130A** (2000), 1293–1309.

[HaK]   J. Han, N. Kim, Non-selfdual Chern-Simons and Maxwell-Chern-Simons vortices in bounded domains, *J. Funct. Anal.* **221** (2005), 167–204.

[H]   Z.C. Han, Prescribing Gaussian curvature on S^2, *Duke Math. J.* **61** (1990), 679–703.

[HJS]   M.C. Hong, J. Jost, M. Struwe, Asymtotic limits of a Ginzburg-Landau type functional, *Geometric Analysis and the Calculus of Variations for S. Hildebrandt* (J. Jost ed.), International Press Boston (1996), 99–123.

[Hi]   N.J. Hitchin, The selfduality equations on a Riemann surface, *Proc. London Math. Soc.* **55** (1987), 59–126.

[HKP]   J. Hong, Y. Kim, P.Y. Pac, Multi-vortex solutions of the Abelian Chern-Simons theory, *Phys. Rev. Lett.* **64** (1990), 2230–2233.

[JW]   R. Jackiw, E.J. Weinberg, Self-dual Chern-Simons vortices, *Phys. Rev. Lett.* **64** (1990), 2234–2237.

[JT]   A. Jaffe, C. Taubes, *Vortices and monopoles*, Birkhauser, Boston, (1980).

[JS1]   R.L. Jerrard, H.M. Soner, Dynamics of Ginzburg-Landau vortices. *Arch. Rational Mech. Anal.* **142**, n. 2, (1998), 99–125.

[JS2]   R.L. Jerrard, H.M. Soner, The Jacobian and the Ginzburg-Landau energy. *Calc. Var. PDE* **14**, n. 2 (2002), 151–191.

[JoLW]   J. Jost, C.S. Lin, G. Wang, Analytic aspects of the Toda system: II. Bubbling Behavior and existence of solutions, *Comm. Pure Appl. Math.* **59** (2006), 526–558.

[Jo]   J. Jost, *Riemannian Geometry and Geometric Analysis*, Spriger-Verlag Berlin, Heidelberg (1998) (second Edition).

[JoW1]   J. Jost, G. Wang, Analytic aspects of the Toda system: I. A Moser-Trudinger inequality, *Comm. Pure Appl. Math.* **54** (2001), 1289–1319.

[JoW2]   J. Jost, G. Wang, Classification of solutions of a Toda system in R^2, *Int. Math. Res. Not.* **6** (2002), 277–290.

[K]   J. Kazdan, Prescribing the curvature of a Riemannian manifold, *CBMS Lectures AMS* **57**, (1984).

[KW1]   J. Kazdan, F. Warner, Existence and conformal deformations of metric with prescribed Gaussian and scalar curvature, *Ann. of Math.* **101** (1975), 317–331.

[KW2]   J. Kazdan, F. Warner, Curvature functions for compact 2-manifolds, *Ann. of Math.* **99** (1974), 14–47.

[Ki1]   M.K.H. Kiessling, Statistical mechanics of classical particles with logarithmic interaction, *Comm. Pure Appl. Math.* **46** (1991), 27–56.

[Ki2]   M.K.H. Kiessling, Statistical mechanics approach to some problems in conformal geometry, *Phys. A* **79** (2000), 353–368.

[KiKi]   S. Kim, Y. Kim, Selfdual Chern-Simons vortices on Riemann surfaces, *J. Math. Phys.* **43** (2002), 2355–2362.

[KS1]   M. Kurzke, D. Sprin, Gamma limit of the non-selfdual Chern-Simons-Higgs energy, preprint (2005).

[KS2]   M. Kurzke, D. Sprin, Scaling limit of the Chern-Simons-Higgs energy, preprint (2005).

[La]   C.H. Lai (ed.), *Selected Papers on Gauge Theory of Weak and Electromagnetic Interactions*, World Scientific Singapore. 1981.

[LLM]   C. Lee, K. Lee, H. Min, Self-dual Maxwell-Chern-Simons solitons, *Phys. Lett. B* **252** (1990), 79–83.

[L1]   Y.Y. Li, On Nirenberg's problem and related topics, *Top. Meth. Nonlin. Anal.* **3**, n. 2 (1994), 21–233.

[L2]   Y.Y. Li, Harnack type inequality: the method of moving planes, *Comm. Math. Phys.* **200** (1999), 421–444.

[LS]   Y.Y. Li, I. Shafrir, Blow up analysis for solutions of $-\Delta u = V(x)e^u$ in dimension two, *Ind. Univ. Math. Jour.* **43**, n. 4 (1994), 1255–1270.

[Li1]   C.S. Lin, Uniqueness of solutions to the mean field equations for the spherical Osanger vortex, *Arch. Rat. Mech. Anal.* **153** (2000), 153–176.

[Li2]   C.S. Lin, Topological degree for the mean field equation on S^2, *Duke Math J.* **104** (2000), 501–536.

[LiL1]   C.S. Lin, M. Lucia, Uniqueness of solutions for a mean field equations on torus, *J. Diff. Eq.* **229** (2006), 172–185.

[LiL2]   C.S. Lin, M. Lucia, One-dimensional symmetry of periodic minimizers for a mean field equations, *Ann. Sc. Norm. Pisa* **5**, n. 6 (2007), 269–290.

[LiW]   C.S. Lin, C.L. Wang, Elliptic functions, Green functions an the mean field equation on tori, preprint (2006).

[Lin1]   F.H. Lin, Some dynamical properties of Ginzburg-Landau vortices. *Comm. Pure Appl. Math.* **49**, n. 4 (1996), 323–359.

[Lin2]   F.H. Lin, Vortex dynamics for the nonlinear wave equation, *Comm. Pure Appl. Math.* **52**, n. 6, (1999), 737–761.

[LR1]   F.H. Lin, T. Riviere, Quantization property for moving line vortices, *Comm. Pure App. Math.* **54**, (2001), 826–850.

[LR2]   F.H. Lin, T. Riviere, A Quantization property for static Ginzburg-Landau Vortices, *Comm. Pure App. Math.* **54**, (2001), 206–228.

[Lio]   J. Liouville, Sur l'equation aux derivées partielles $\frac{\partial^2 \log \lambda}{\partial u \partial v} \pm \frac{\lambda}{2a^2} = 0$, *J. Math. Pure Appl.* **18** (1853), 71–72.

[Lu]   M. Lucia, A blowing-up branch of solutions for a mean field equation, *Calc. Var. and P.D.E.* **26** (2006), 313–333.

[LN]   M. Lucia, M. Nolasco, $SU(N)$ Chern-Simons vortex theory and Toda systems, *Jour. Diff. Eq.* **184**, n. 2 (2002), 443–474.

[LZ]   M. Lucia, L. Zhang, A priori estimates and uniqueness for some mean field equations, *J.D.E.* **217** (2005), 154–178.

[MW]   L. Ma, J. Wei, Convergence for a Liouville equation, *Comment. Math. Helv.* **76** (2001), 506–514.

[MaN]   M. Macri', M. Nolasco, Uniqueness of topological solutions for a class of selfdual vortex theories, *Proc. Royal Soc. Edin.* **137** (2007), 847–866.

[MNR]   M. Macri', M. Nolasco, T. Ricciardi, Asymptotics for selfdual vortices on the torus and on the plane: a technique, *SIAM Math. Anal.* **37**, n. 1 (2005), 1–16.

[MN]   A. Malchiodi, C.B. Ndiaye, Some existence results for the Toda system on closed surfaces, preprint (2005).

[Mo]   J. Moser, A sharp form of an inequality by N. Trudinger, *Indiana Univ. J.* **20** (1971), 1077–1092.

[NS]   K. Nagashi, T. Suzuki, Asymptotic analysis for a two dimensional elliptic eigenvalue problem with exponentially dominated nonlinearity, *Asymp. Anal.* **3** (1990), 173–188.

[NO]   H. Nielsen, P. Olesen, Vortex-Line models for dual strings, *Nucl. Phys.* B **61** (1973), 45–61.

[Ni]   W.M. Ni, On the elliptic equation $\Delta u + Ke^u = 0$ and conformal metrics with prescribed Gaussian curvature, *Invent. Math.* **66** (1982), 343–352.

[Nir]   L. Nirenberg, *Topics in nonlinear analysis*, Courant Lecture Notes AMS (2001).

[NT1]   M. Nolasco, G. Tarantello, Vortex condensates for the $SU(3)$ Chern-Simons theory, *Comm. Math. Phys.* **213** (2000), 599–639.

[NT2]   M. Nolasco, G. Tarantello, On a sharp Sobolev type inequality on two dimensional compact manifolds, *Arch. Rat. Mech. Anal.* **145** (1998), 161–195.

[NT3]   M. Nolasco, G. Tarantello, Double vortex condensates in the Chern-Simons-Higgs theory, *Calc. Var. P.D.E.* **9** (1999), 31–94.

[Ob]   M. Obata, The conjectures on conformal transformations of Riemannian manifolds, *J. Diff. Geom.* **6** (1971), 247–258.

[OS1]   H. Ohtsuka, T. Suzuki, Palais-Smale sequences relative to the Trudinger Moser inequality, *Calc. Var. P.D.E.* **17** (2003), 235–255.

[OS2]   H. Ohtsuka, T. Suzuki, Blow-up analysis for Liouville-type equations in selfdual-gauge field theories, *Comm. Contemp. Math.* **7** (2005), 117–205.

[Ol]   P. Olesen, Soliton condensation in some selfdual Chern-Simons theories, *Phys. Lett. B* **265** (1991), 361–365.

[On]   E. Onofri, On the positivity of the effective action in a theory of random surfaces, *Comm. Math. Phys.* **86** (1982), 321–326.

[PR]   F. Pacard, T. Riviere, Linear and nonlinear aspects of vortices. The Ginzburg-Landau model, *Progress in Nonlinear Differential Equations and their Applications* **39**, Birkhauser Boston, Inc., Boston, MA, (2000).

[Park]   R.D. Parks, *Superconductivity*, vol. 1 and 2, Marcel Dekker publ. (1969).

[PiR]   L.M. Pismen, J. Rubinstein, Motion of vortex lines in the Ginzburg-Landau model. *Phys. D.* **47**, n. 3 (1991), 353–360.

[Po]   S. Pokorski et al. *Gauge Field Theories*, Cambridge Monographs on Mathematical Physics, Cambridge Univ. Press (2000).

[PT]   J. Prajapat, G. Tarantello, On a class of elliptic problems in R^2: Symmetry and Uniqueness results, *Proc. Roy. Soc. Edinburgh* **131A** (2001), 967–985.

[PS]   M.K. Prasad, C.M. Sommerfield, Exact classical solutions for the 't Hooft monopole and the Julia-Zee dyon, *Phys. Rev. Lett.* **35** (1975), 760–762.

[Q]   C.Quigg, *Gauge theory of Strong, Weak and Electroweak interactions*, Westview Press (1997).

[Ra]   R. Rajaraman, *Solitons and Instantons*, North Holland publ. (1982).

[Ri1]   T. Ricciardi, Asymptotics for Maxwell-Chern-Simons Multivortices, *NonLinear Analysis TMA* **50** (2002), 193–1106.

[Ri2]   T. Ricciardi, Multiplicity for a nonlinear fourth order elliptic equation in Maxwell-Chern-Simons vortex theory, *Diff. Int. Eq.* **17**, n. 3-4 (2004), 369–390.

[Ri3]   T. Ricciardi, On a nonlinear elliptic system from Maxwell-Chern-Simons vortex theory, *Asympt. Anal.* **2** (2003), 113–126.

[RT1]   T. Ricciardi, G. Tarantello, Self-dual vortices in the Maxwell-Chern-Simons-Higgs theory, *Comm. Pure Appl. Math.* **53** (2000), 811–851.

[RT2]   T. Ricciardi, G. Tarantello, On a periodic boundary value problem with exponential nonlinearity, *Diff. Int. Eqs.* **11**, n. 5 (1998), 745–753.

[SS]   E. Sandier, S. Serfaty, *Vortices in the magnetic Ginzburg-Landau model*, Progress in Nonlinear Differential Equations and their Applications, vol. 70 Birkhauser, Boston (2007).

[Sch]   J.R. Schrieffer, *The Theory of Superconductivity*, Benjamin publ. (1964).

[Sh]   I. Shafrir, Une inegalité de type sup + inf pour l'equation $-\Delta u = V(x)e^u$, *C.R. Acad. Sci. Paris* **315** (1992), 159–164.

[SW1]   I. Shafrir, G. Wolansky, Moser Trudinger and logarithmic HLS inequalities for systems, *J. Europ. Math. Sc.* **7**, n. 4 (2005), 413–448.

[SW2]   I. Shafrir, G. Wolansky, The logarithmic HLS inequalities for systems on compact manifolds, *J. Funct. Anal.* **227** (2005), 200–226.

[SY1]   J. Spruck, Y. Yang, The existence of non-topological solutions in the self-dual Chern-Simons theory, *Comm. Math. Phys.* **149** (1992), 361–376.

[SY2]   J. Spruck, Y. Yang, On Multivortices in the Electroweak Theory I: Existence of Periodic Solutions, *Comm. Math. Phys.* **144** (1992), 1–16.

[SY3]   J. Spruck, Y. Yang, On Multivortices in the Electroweak Theory II:Existence of Bogomol'nyi solutions in R^2, *Comm. Math. Phys.* **144** (1992), 215–234.

[St]   M. Struwe, *Variational Methods, Application to Partial Differential Equations and Hamiltonian systems*, 3rd edition Springer 34, (2000).

[ST]   M. Struwe, G. Tarantello, On the multivortex solutions in the Chern-Simons gauge theory, *Boll. U.M.I. Sez. B Artic. Ric. Mat.* **1** (1998), 109–121.

[Su1]   T. Suzuki, Global analysis for a two dimensional elliptic eigenvalue problem
          with exponential nonlinearity, *Ann. I.H.P. Analyse Non Linéaire* **9** (1992),
          367–398.

[Su2]   T. Suzuki, Two dimensional Emden-Fowler equations with exponential nonlin-
          earities, *Nonlinear Diffusion Equations and their equilibrium state* **3** (1992),
          493–512, Birkauser, Boston.

['tH1]   G. 't Hooft, Computation of the quantum effects due to a four dimensional
          pseudoparticle, *Phys. Rev. D* **14** (1976), 3432–3450.

['tH2]   G. 't Hooft, A property of electric and magnetic flux in non abelian gauge
          theories, *Nucl. Phys. D* **153** (1979), 141–160.

[T1]   G. Tarantello, Multiple condensates solutions for the Chern-Simons-Higgs
          theory, *J. Math. Phys.* **37**, n. 8 (1996), 3769–3796.

[T2]   G. Tarantello, On Chern Simons vortex theory, Nonlinear PDE's and physical
          modeling: Superfluidity, Superconductivity and Reactive flows, H. Berestycki
          ed. Kluver Academic publ., (2002), 507–526.

[T3]   G. Tarantello, Selfdual Maxwell-Chern-Simons vortices, *Milan J. Math.* **72**
          (2004), 29–80.

[T4]   G. Tarantello, *Analitycal aspects of Liouville-type equations with singular
          sources*, Handbook of Differential Equations. Stationary partial differential
          equations, vol 1. Elsevier Sciences, M.Chipot, P.Quittner Eds.

[T5]   G. Tarantello, A quantization property for blow up solutions of singular
          Liouville-type equations, *Journal Func. Anal.* **219** (2005), 368–399.

[T6]   G. Tarantello, An Harnack inequality for Liouville-type equations with
          singular sources, *Indiana Univ. Math J.* **54**, n.2 (2005), 599–615.

[T7]   G. Tarantello, Uniqueness of Selfdual periodic Chern-Simons vortices of
          topological-type, *Calc. Var. PDE*, **29** (2007), 191–217.

[T8]   G. Tarantello, *Selfdual gauge field vortices: an analytical approach*, Progress
          in Nonlinear Differential Equations and their Applications, vol. 72 Birkhauser,
          Boston (2008).

[Ta1]   C. Taubes, Arbitrary N-vortex solutions for the first order Ginzburg-Landau
          equations, *Comm. Math. Phys.* **72** (1980), 277–292.

[Ta2]   C.H. Taubes, On the equivalence of the first and second order equations for
          gauges theories, *Comm. Math. Phys.* **75** (1980), 207–227.

[Ta3]   C.H. Taubes, The existence of a non-minimal solution to the SU(2) Yang-
          Mills-Higgs equations in R^3, Part I and II, *Comm. Math. Phys.* **86** (1982),
          257–320.

[Tra]   A. Trautmann, *Differential geometry for physicists* Stony Brook lectures.
          Monographs and Textbooks in Physical Science, 2. Bibliopolis, Naples, 1984.

[Tr]   N.D. Trudinger, On imbedding into Orlicz spaces and some applications, *J.
          Math. Mech.* **17** (1967), 473–483.

[Va]   P. Valtancoli, Classical and Chern-Simons Vortices on curved spaces, *Int. J.
          Mod. Phys. A*, **7**, n. 18 (1990), 4335–4352.

[W]   G. Wang, Moser-Trudinger inequality and Liouville systems, *C.R. Acad. Sci.
          Paris* **328** (1999), 895–900.

[WW1]   G. Wang, J. Wei, On a conjecture of Wolanski, *Nonlinear Analysis TMA* **48**
          (2002), 927–937.

[WW2]   G. Wang, J. Wei, Steady state solutions of a reaction-diffusion system
          modelling chemotaxis, *Math. Nachr.* 233–234 (2002), 221–236.

[Wa]   R. Wang, The existence of Chern-Simons vortices, *Comm. Math. Phys.* **137**
          (1991), 587–597.

[WY]   S. Wang, Y. Yang, Abrikosov's vortices in the critical coupling, *SIAM J. Math.
          Anal.* **23** (1992), 1125–1140.

[Wit]   E. Witten, Some exact Multipseudoparticle solutions of classical Yang-Mills
          theory, *Phys. Rev. Lett.* **38** (1997), 121–124.

[Wo]   G. Wolanski, On the evolution of self-interacting clusters and applications to semilinear equations with exponential nonlinearity, *J. Anal. Math.* **59** (1992), 251–272.

[YM]   C.N. Yang, R. Mills, Conservation of isotopic spin and isotopic invariance, *Phys. Rev. Lett.* **96** (1954), 191–195.

[Y1]   Y. Yang, *Solitons in Field Theory and Nonlinear Analysis*, Springer Monographs in Mathematics, Springer-Verlag New York, (2001).

[Y2]   Y. Yang, Topological solitons in the Weinberg-Salam theory, *Physica* D **101** (1997), 55–94.

[Y3]   Y. Yang, The relativistic non-abelian Chern-Simons equations, *Comm. Math. Phys.* **186** (1997), 199–218.

[Y4]   Y. Yang, On a system of nonlinear elliptic equations arising in theoretical physics, *J. Func. Anal.* (2000), 1–36.

The k-Hessian Equation

Xu-Jia Wang

Abstract The k-Hessian is the k-trace, or the kth elementary symmetric polynomial of eigenvalues of the Hessian matrix. When $k \geq 2$, the k-Hessian equation is a fully nonlinear partial differential equations. It is elliptic when restricted to k-admissible functions. In this paper we establish the existence and regularity of k-admissible solutions to the Dirichlet problem of the k-Hessian equation. By a gradient flow method we prove a Sobolev type inequality for k-admissible functions vanishing on the boundary, and study the corresponding variational problems. We also extend the definition of k-admissibility to non-smooth functions and prove a weak continuity of the k-Hessian operator. The weak continuity enables us to deduce a Wolff potential estimate. As an application we prove the Hölder continuity of weak solutions to the k-Hessian equation. These results are mainly from the papers [CNS2, W2, CW1, TW2, Ld] in the references of the paper.

Key Words: Hessian equation, a priori estimates, Sobolev inequality, variational problem, Hessian measure, potential estimate.

AMS subject classification: 35J60, 35J20, 35A15, 28A33.

1 Introduction

Let Ω be a bounded, smooth domain in the Euclidean space $\mathbb{R}^n$. In this note we study the *k-Hessian equation*

$$S_k[u] = f \quad \text{in} \quad \Omega, \tag{1.1}$$

X.-J. Wang
Mathematical Sciences Institute, Australian National University,
Canberra, ACT 0200, Australia
e-mail: wang@maths.anu.edu.au
This work was supported by the Australian Research Council.

S.-Y.A. Chang et al. (eds.), *Geometric Analysis and PDEs*,
Lecture Notes in Mathematics 1977, DOI: 10.1007/978-3-642-01674-5_5,
© Springer-Verlag Berlin Heidelberg 2009

where $1 \le k \le n$, $S_k[u] = \sigma_k(\lambda)$, $\lambda = (\lambda_1, \cdots, \lambda_n)$ are the eigenvalues of the Hessian matrix $(D^2 u)$, and

$$\sigma_k(\lambda) = \sum_{i_1 < \cdots < i_k} \lambda_{i_1} \cdots \lambda_{i_k} \tag{1.2}$$

is the *k-th elementary symmetric polynomial*. The k-Hessian equation includes the Poisson equation ($k = 1$)

$$-\Delta u = f, \tag{1.3}$$

and the Monge-Ampère equation ($k = n$)

$$\det D^2 u = f, \tag{1.4}$$

as special examples.

We say a second order partial differential equation

$$F(D^2 u, Du, u, x) = 0 \tag{1.5}$$

is *fully nonlinear* if $F(r, p, z, x)$ is nonlinear in r. The k-Hessian equation is fully nonlinear when $k \ge 2$. We say F is *elliptic* (or *degenerate elliptic*) with respect to a solution u if the matrix $\{F_{ij}\}$ is positive definite (or positive semi-definite) at $(r, p, z, x) = (D^2 u(x), Du(x), u(x), x)$, where $F_{ij} = \{\frac{\partial F}{\partial r_{ij}}\}$. We say F is *uniformly elliptic* if there exist positive constants Λ and λ such that

$$\lambda I \le \{F^{ij}\} \le \Lambda I, \tag{1.6}$$

where I is the unit matrix. We also say F is elliptic if $-F$ is.

The Monge-Ampère equation (1.4) is elliptic if and only if the function u is uniformly convex or concave. For the k-Hessian equation, it is elliptic when u is k-admissible [CNS2], namely the eigenvalues $\lambda(D^2 u)$ lie in the convex cone Γ_k, which will be introduced in Section 2 below. Fully nonlinear equations of mixed type are very difficult. In this note we restrict ourself to k-admissible solutions to the k-Hessian equation.

There are many other important fully nonlinear equations, see §11 below for examples. But the k-Hessian equation (1.1) is variational, and when restricted to k-admissible solutions, it enjoys many nice properties which are similar to those of the Poisson equation. In this paper we discuss the regularity, variational properties, and local behaviors of solutions to the k-Hessian equation.

We divide this note into a number of sections.

In §2 we introduce the notion of k-admissible functions, and show that the k-Hessian equation is elliptic at k-admissible functions. We also collect some inequalities related to the polynomial σ_k.

In §3 we establish the global a priori estimates and prove the existence of solutions to the Dirichlet problem.

In §4 we establish the interior gradient and second derivative estimates. From the interior gradient estimate we also deduce a Harnack inequality.

In §5 we use gradient flow to prove Sobolev type inequalities for k-admissible functions which vanish on the boundary. That is

$$\|u\|_{L^p(\Omega)} \le C\left[\int_\Omega (-u)S_k[u]\right]^{1/(k+1)}, \tag{1.7}$$

where C depends only on n, k, Ω; $p = \frac{n(k+1)}{n-2k}$ if $k < \frac{n}{2}$, $p < \infty$ if $k = \frac{n}{2}$; and $p = \infty$ if $k > \frac{n}{2}$. Moreover, the corresponding embedding of k-admissible functions into L^p space is compact when p is below the critical exponent. As an application we give an L^∞ estimate for solutions to the k-Hessian equation (1.1) when $f \in L^p(\Omega)$ with $p > \frac{n}{2k}$ if $k \le \frac{n}{2}$, or $p = 1$ if $k > \frac{n}{2}$.

In §6 we use the Sobolev type inequality (1.7) to study variational problems of the k-Hessian equation. We prove the existence of a min-max solution to the Hessian equation in the sub-critical and critical growth cases.

In §7 we present some local integral estimates. In particular we show that a k-admissible function belongs to $W_{loc}^{1,p}(\Omega)$ for any $p < \frac{nk}{n-k}$.

In §8 we extend the notion of k-admissible functions to nonsmooth functions; and prove that for any k-admissible function u, we can assign a measure $\mu_k[u]$ to u such that if a sequence of k-admissible functions $\{u_j\}$ converges to u almost everywhere, then $\mu_k[u_j]$ converges to $\mu_k[u]$ weakly as measures. As an application we prove the existence of weak solutions to the k-Hessian equation.

This weak continuity has many other applications as well, in particular it enables us to establish various potential theoretical results for k-admissible functions. In §9 we prove a Wolff potential estimate, and deduce a necessary and sufficient condition for a weak solution to be Hölder continuous.

In §10, we include some a priori estimates for the parabolic Hessian equations used in previous sections.

In the last Section 11, we give more examples of fully nonlinear elliptic equations.

Main references for this note are [CNS2, W2, CW1, TW2, Ld]. There are many other works on the k-Hessian equations. The materials in §2 and §3 are mostly taken from [CNS2], but for the key double normal derivative estimate we adapt the approach from [T]. See also [I] for the k-Hessian equation for some k. The interior derivative estimates in §4 are from [CW1], but for the Monge-Ampère equation they were first established by Pogorelov [P]. The Sobolev type inequalities in §5 were proved by K.S. Chou for convex functions, and in [W2] for general k-admissible functions by a gradient flow method. The existence of min-max solutions in §6 was first obtained by K.S. Chou [Ch1] for the Monge-Ampère equation and later in [CW1] for $2 \le k \le \frac{n}{2}$. See also [W1] for the Monge-Ampère equation by a degree theory method, which also applies to the case $\frac{n}{2} < k < n$ by the embedding in Theorem 5.1. The local integral estimates in §7 and weak continuity in §8 can be found in [TW2]. The Wolff potential estimate and Hölder continuity of k-admissible solutions in §9 were proved in [Ld].

The result in §6.4 on the variational problem in the critical growth case was not published before, it was included in the preprint [CW2]. The proof of the weak continuity in §8, which uses ideas from [TW1,TW5], is different from that in [TW2]. As the reader will see below, most results in the note are generalization of the counterparts for the Poisson equation. But the study of fully nonlinear equations requires new techniques and is usually more complicated, in particular for estimates near the boundary. These results and techniques can also be used in other problems. See e.g., [FZ, KT, STW].

2 Admissible Functions

2.1 Admissible Functions

We say a function $u \in C^2(\Omega) \cap C^0(\overline{\Omega})$ is k-admissible if

$$\lambda(D^2 u) \in \overline{\Gamma}_k, \tag{2.1}$$

where Γ_k is an open symmetric convex cone in $\mathbb{R}^n$, with vertex at the origin, given by

$$\Gamma_k = \{(\lambda_1, \cdots, \lambda_n) \in \mathbb{R}^n \mid \sigma_j(\lambda) > 0 \ \forall \ j = 1, \cdots, k\}. \tag{2.2}$$

Clearly $\sigma_k(\lambda) = 0$ for $\lambda \in \partial \Gamma_k$,

$$\Gamma_n \subset \cdots \subset \Gamma_k \subset \cdots \subset \Gamma_1,$$

Γ_n is the positive cone,

$$\Gamma_n = \{(\lambda_1, \cdots, \lambda_n) \in \mathbb{R}^n \mid \lambda_1 > 0, \cdots, \lambda_n > 0\},$$

and Γ_1 is the half space $\{\lambda \in \mathbb{R}^n \mid \Sigma \lambda_i > 0\}$. A function is 1-admissible if and only if it is sub-harmonic, and an n-admissible function must be convex. For any $2 \le k \le n$, a k-admissible function is sub-harmonic, and the set of all k-admissible functions is a convex cone in $C^2(\Omega)$.

The cone Γ_k may also be equivalently defined as the component $\{\lambda \in \mathbb{R}^N \mid \sigma_k(\lambda) > 0\}$ containing the vector $(1, \cdots, 1)$, and characterized as

$$\Gamma_k = \{\lambda \in \mathbb{R}^n \mid 0 < \sigma_k(\lambda) \le \sigma_k(\lambda + \eta) \ \text{ for all } \eta_i \ge 0, \ \in \mathbb{R}\}. \tag{2.3}$$

We note that the k-Hessian operator S_k is also elliptic or degenerate elliptic if $\lambda(D^2 u) \in -\overline{\Gamma}_k$. But by making the change $u \to -u$ it suffices to consider functions with eigenvalues $\lambda \in \Gamma_k$. In this note we consider functions with eigenvalues in Γ_k only.

2.2 Admissible Solution is Elliptic

We show that if u is k-admissible, the matrix

$$\{S_k^{ij}(A)\} = \{\frac{\partial}{\partial a_{ij}}\sigma_k(\lambda(A))\} \geq 0 \qquad (2.4)$$

is positive semi-definite at $A = D^2 u$ and so the k-Hessian operator is (degenerate) elliptic. To prove (2.4), note that the k-Hessian operator can also be written in the form

$$S_k[u] = [D^2 u]_k, \qquad (2.5)$$

where for a matrix $A = (a_{ij})$, $[A]_k$ denotes the sum of the kth principal minors. Therefore

$$S_k^{nn}[u] = [D^2 u]'_{k-1}, \qquad (2.6)$$

where $[D^2 u]' = \{u_{x_i x_j}\}_{1 \leq i,j \leq n-1}$. Denote

$$\bar{D}^2 u = \begin{pmatrix} [D^2 u]', 0 \\ 0, \qquad u_{nn} \end{pmatrix}.$$

One easily verifies that

$$[\bar{D}^2 u]_m \geq [D^2 u]_m \quad \forall \ 1 \leq m \leq k.$$

Hence by (2.2), $\lambda(\bar{D}^2 u) \in \overline{\Gamma}_k$. By (2.3) it follows that

$$S_k^{nn}[u] = [D^2 u]'_{k-1} = \frac{\partial}{\partial \lambda_n}\sigma_k(\lambda) \geq 0 \quad (\lambda = \lambda(\bar{D}^2 u)). \qquad (2.7)$$

Note that (2.7) also holds after a rotation of coordinates, so the k-Hessian equation is (degenerate) elliptic if u is k-admissible.

When u is k-admissible, $S_k[u]$ is nonnegative. Therefore in our investigation of the k-Hessian equation, we always assume that f is nonnegative. If f is positive and $u \in C^2(\Omega)$, $S_k[u]$ is elliptic. Note that we allow that the eigenvalues $\lambda(D^2 u)$ lie on the boundary of Γ_k, and in such case the k-Hessian equation may become degenerate elliptic.

2.3 Concavity

When u is k-admissible,

$$S_k^{1/k}[u] = \left[\sigma_k(\lambda(D^2 u))\right]^{1/k},$$

is concave when regarded as a function of $r = D^2 u$. In other words,

$$\sum a_{ij} a_{st}\, \partial^2_{u_{ij} u_{st}} S_k^{1/k}[u] \leq 0 \tag{2.8}$$

for any symmetric matrix $\{a_{ij}\}$. This property follows from the concavity of $\sigma_k^{1/k}(\lambda)$ in Γ_k (see (xii) in §2.5 below). Indeed, when u_{ij} is diagonal, one can verify (2.8) directly by the expression (2.5). When u_{ij} is not diagonal, by a rotation of coordinates $y_\alpha = c_{\alpha i} x_i$ such that $u_{\alpha\beta}$ is diagonal, one has

$$\sum a_{ij} a_{st}\, \partial^2_{u_{ij} u_{st}} S_k^{1/k}[u] = \sum a_{\alpha\beta}^* a_{\gamma\delta}^*\, \partial^2_{u_{\alpha\beta} u_{\gamma\delta}} S_k^{1/k}[u] \leq 0,$$

where $a_{\alpha\beta}^* = a_{ij} c_{\alpha i} c_{\beta j}$, subscripts i, j, s, t mean derivatives in x and subscripts $\alpha, \beta, \gamma, \delta$ mean derivatives in y. The concavity is needed in establishing the regularity of fully nonlinear elliptic equations.

2.4 A Geometric Assumption on the Boundary

In order that there exists a smooth k-admissible function which vanishes on $\varphi\Omega$, the boundary $\varphi\Omega$ must satisfy a geometric condition, that is

$$\sigma_{k-1}(\kappa) \geq c_0 > 0 \quad \text{on} \quad \varphi\Omega \tag{2.9}$$

for some positive constant c_0, where $\kappa = (\kappa_1, \cdots, \kappa_{n-1})$ denote the principal curvatures of $\varphi\Omega$ with respect to its inner normal. Indeed, let $u \in C^2(\overline{\Omega})$ be a k-admissible function which vanishes on $\varphi\Omega$. For any fixed point $x_0 \in \varphi\Omega$, by a translation and rotation of coordinates, we may assume that x_0 is the origin and locally $\varphi\Omega$ is given by $x_n = \rho(x')$ such that $e_n = (0, \cdots, 0, 1)$ is the inner normal of $\varphi\Omega$ at x_0, where $x' = (x_1, \cdots, x_{n-1})$. Differentiating the boundary condition $u(x', \rho(x')) = 0$, we get

$$u_{ij}(0) + u_n \rho_{ij}(0) = 0. \tag{2.10}$$

By our choice of coordinates, the principal curvatures of $\varphi\Omega$ at x_0 are the eigenvalues of $\{\rho_{ij}(0)\}_{1 \leq i,j \leq n-1}$. When u is k-admissible, it is subharmonic and so $u_n(x_0) < 0$. We obtain

$$S_k^{nn}[u] = |u_n|^{k-1} \sigma_{k-1}(\kappa). \tag{2.11}$$

Hence (2.9) follows from (2.4) provided $\lambda(D^2 u) \in \Gamma_k$.

In this note we call a domain whose boundary satisfies (2.9) $(k-1)$-*convex*. When $k = n$, it is equivalent to the usual convexity. In the following we always assume that Ω is $(k-1)$-convex.

If Ω is $(k-1)$-convex, then for any smooth function φ on $\varphi\Omega$, there is a function $\underline{u}$, which is k-admissible in a neighborhood of $\varphi\Omega$ and satisfies $\underline{u} = \varphi$ on $\varphi\Omega$. Indeed, if $\varphi = 0$, let $\underline{u}(x) = -d_x + t d_x^2$, where $x \in \Omega$ and d_x is distance from x to $\varphi\Omega$. Then $\underline{u}$ is k-admissible near $\varphi\Omega$ provided t is sufficiently large. We refer the reader to [GT] for the computation of the second derivatives of the distance function. For a general boundary value φ, extend φ to Ω such that it is harmonic in Ω. Then $\varphi + \sigma\underline{u}$ is k-admissible near $\varphi\Omega$ for large σ, and $S_k[\varphi + \sigma\underline{u}]$ can be as large as we want provided σ is sufficiently large.

Note that the function $\underline{u}$ is defined only in a neighborhood of $\varphi\Omega$. But it suffices for the a priori estimates in §3. By the existence of solutions to the Dirichlet problem (Theorem 3.4), there is a k-admissible function u defined in the whole domain Ω such that $u = \varphi$ on $\varphi\Omega$.

2.5 Some Algebraic Inequalities

We collect some inequalities related to the polynomial $\sigma_k(\lambda)$, which are needed in our investigation of the k-Hessian equation.

Denote $\sigma_0 = 1$ and $\sigma_k = 0$ for $k > n$. Assume $\lambda \in \Gamma_k$. Arrange $\lambda = (\lambda_1, \cdots, \lambda_n)$ in descending order, namely $\lambda_1 \geq \cdots \geq \lambda_n$. Denote $\sigma_{k;i} = \sigma_k(\lambda)_{|\lambda_i=0}$, so that $\frac{\partial}{\partial\lambda_i}\sigma_k(\lambda) = \sigma_{k-1,i}(\lambda)$. The following ones are easy to verify

$$(i) \quad \sigma_k(\lambda) = \sigma_{k;i}(\lambda) + \lambda_i\sigma_{k-1;i}(\lambda),$$

$$(ii) \quad \sum_{i=1}^{n} \sigma_{k;i}(\lambda) = (n-k)\sigma_k(\lambda),$$

$$(iii) \quad \sigma_{k-1,n}(\lambda) \geq \cdots \geq \sigma_{k-1,1}(\lambda) > 0,$$

$$(iv) \quad \lambda_k \geq 0 \quad \text{and} \quad \sigma_k(\lambda) \leq C_{n,k}\lambda_1\cdots\lambda_k.$$

We also have

$$(v) \quad \sigma_k(\lambda)\sigma_{k-2}(\lambda) \leq C_{n,k}[\sigma_{k-1}(\lambda)]^2,$$

$$(vi) \quad \sigma_k(\lambda) \leq C_{n,k}[\sigma_l(\lambda)]^{k/l}, \quad 1 \leq l < k.$$

Furthermore we have

$$(vii) \quad \lambda_1\sigma_{k-1,1}(\lambda) \geq C_{n,k}\sigma_k(\lambda).$$

$$(viii) \quad \sigma_{k-1;k}(\lambda) \geq C_{n,k}\sum_{i=1}^{n}\sigma_{k-1;i}(\lambda),$$

$$(ix) \quad \sigma_{k-1;k}(\lambda) \geq C_{n,k}\sigma_{k-1}(\lambda),$$

$$(x) \quad \prod_{i=1}^{n}\sigma_{k;i}(\lambda) \geq C_{n,k}[\sigma_k(\lambda)]^{n(k-1)/k}.$$

In the above the constant $C_{n,k}$ may change from line to line. There are more inequalities useful in the study of the k-Hessian equation. For example, we have

$$(xi) \ \sum \mu_i \sigma_{k-1,i} \geq k[\sigma_k(\mu)]^{1/k}[\sigma_k(\lambda)]^{1-1/k} \ \ \forall \, \lambda, \mu \in \Gamma_k,$$

$$(xii) \ \{\frac{\partial^2}{\partial\lambda_i\partial\lambda_j}\sigma_k(\lambda)\} \leq 0 \ \ \forall \, \lambda \in \Gamma_k.$$

the last inequality means that $\sigma_k^{1/k}(\lambda)$ is concave in Γ_k. We refer the reader to [CNS2, LT, Lg] for these and more inequalities related to σ_k.

3 The Dirichlet Problem

In this section we study the existence and regularity of solutions to the Dirichlet problem of the k-Hessian equation,

$$S_k[u] = f(x) \ \ \text{in} \ \Omega, \tag{3.1}$$
$$u = \varphi \ \ \text{on} \ \varphi\Omega,$$

where Ω is a bounded, $(k-1)$-convex domain in $\mathbb{R}^n$ with $C^{3,1}$ boundary, $\varphi \in C^{3,1}(\varphi\Omega)$, $f \geq 0$, $f \in C^{1,1}(\overline{\Omega})$.

3.1 A priori Estimates

First we establish the global estimate for the second derivatives.

Theorem 3.1 (CNS2, T) *Let $u \in C^{3,1}(\overline{\Omega})$ be a k-admissible solution to the Dirichlet problem (3.1). Assume that Ω is $(k-1)$-convex, $\varphi\Omega \in C^{3,1}$, $\varphi \in C^{3,1}(\varphi\Omega)$, $f \geq f_0 > 0$, and $f^{1/k} \in C^{1,1}(\overline{\Omega})$. Then we have the a priori estimate*

$$\|u\|_{C^{1,1}(\overline{\Omega})} \leq C, \tag{3.2}$$

where C depends only on n, k, Ω, f_0, $\|\varphi\|_{C^{3,1}(\varphi\Omega)}$ and $\|f\|_{C^{1,1}(\overline{\Omega})}$.

Proof. First consider the L^∞ estimate. Let $w = \frac{1}{2}a|x|^2 - b$, where the constants a, b are chosen large such that $S_k[w] > f$ in Ω and $w \leq \varphi$ on $\varphi\Omega$. Then $w - u$ satisfies the elliptic equation $\sum a_{ij}(w-u)_{ij} > 0$ in Ω and $w - u \leq 0$ on $\varphi\Omega$, where $a_{ij} = \int_0^1 S_k^{ij}[u+t(w-u)]dt$. It follows that $w \leq u$ in Ω. Extend φ to Ω such that it is harmonic. By the comparison principle we have $w \leq u \leq \varphi$ in Ω.

Next consider the gradient estimate. Denote $F[u] = S_k^{1/k}[u]$, $\hat{f} = f^{1/k}$. Differentiating the equation

$$F[u] = \hat{f} \tag{3.3}$$

in direction x_l, one obtains

$$L[u_l] = \hat{f}_l,$$

where $L = F_{ij}\partial_{ij}$ is the linearized equation of F, $F_{ij} = F_{u_{ij}}$. So $|L[u_l]| \le C$. Let $w = \frac{1}{2}a|x|^2$. By (ii) and (vi) above, $L[w] \ge c_1 a > 0$ for some positive constant $c_1 > 0$ depends only on n, k. Hence $L[w \pm u_l] \ge 0$, provided a is chosen suitably large. It follows that $w \pm u_l$ attains its maximum on the boundary $\varphi\Omega$. Hence

$$\sup_{x \in \Omega} |Du(x)| \le C(1 + \sup_{x \in \varphi\Omega} |Du(x)|). \tag{3.4}$$

Next let $\hat{w} = \varphi + \sigma\underline{u}$ be the function in §2.4. Denote $\mathcal{N} = \{x \in \Omega \mid \hat{w}(x) > w(x)\}$. Then when σ is sufficiently large, $\mathcal{N}$ is a neighborhood of $\varphi\Omega$, and $S_k[\hat{w}] > f$ in $\mathcal{N}$. Therefore by the comparison principle, $\hat{w} \le u \le \varphi$ in $\mathcal{N}$. Hence by the boundary condition $\hat{w} = u = \varphi$ on $\varphi\Omega$, we infer that $\partial_\gamma \varphi \le \partial_\gamma u \le \partial_\gamma \hat{w}$, where γ is the unit outer normal to $\varphi\Omega$. Hence Du is bounded on $\varphi\Omega$.

Finally consider the second derivative estimate. Since u is sub-harmonic, it suffices to prove that $u_{\xi\xi} \le C$ for any unit vector ξ. Differentiating equation (3.3) twice in direction ξ, we obtain, by the concavity of F,

$$L[u_{\xi\xi}] \ge \hat{f}_{\xi\xi}.$$

Hence $L[Cw + u_{\xi\xi}] \ge 0$ for a suitably large constant C and so

$$\sup_\Omega u_{\xi\xi} \le C + \sup_{\varphi\Omega} u_{\xi\xi}. \tag{3.5}$$

Therefore we reduce the estimate to the boundary.

For any given boundary point $x_0 \in \varphi\Omega$, by a translation and a rotation of the coordinates we assume that x_0 is the origin and locally $\varphi\Omega$ is given by

$$x_n = \rho(x') \tag{3.6}$$

such that $D\rho(0) = 0$, where $x' = (x_1, \cdots, x_{n-1})$. Differentiating the boundary condition $u = \varphi$ on $\varphi\Omega$ twice, we have, for $1 \le i, j \le n-1$,

$$u_{ij}(0) + u_n(0)\rho_{ij}(0) = \varphi_{ij}(0) + \varphi_n(0)\rho_{ij}(0). \tag{3.7}$$

Hence

$$|D_{ij}u(0)| \le C \quad i, j \le n-1. \tag{3.8}$$

Next we establish

$$|u_{in}(0)| \le C \quad i < n. \tag{3.9}$$

By a rotation of the $x_1, \cdots, x_{n-1}$ axes, we assume that $x_1, \cdots, x_{n-1}$ are the principal directions of $\varphi\Omega$ at the origin. Let $T = \partial_i + \kappa_i(0)(x_i\partial_n - x_n\partial_i)$, where κ_i is the principal curvature of $\varphi\Omega$ in direction x_i, $1 \le i \le n-1$. One can verify that

$$|T(u - \varphi)| \le C|x'|^2|\partial_\gamma(u - \varphi)| \le C|x'|^2| \quad \text{on } \varphi\Omega.$$

Next observing that S_k is invariant under rotation of coordinates and $(x_i\partial_n - x_n\partial_i)$ is an infinitesimal generator of a rotation, we have $TF[u] = L[T(u)]$. Hence

$$|L(T(u - \varphi))| \le C(1 + \Sigma_i F_{ii}).$$

Let

$$w = \rho(x') - x_n - \delta|x'|^2 + Kx_n^2, \tag{3.10}$$

where $K > 1$ large and $\delta > 0$ small are constants. By the assumption that Ω is $(k-1)$-convex, the function w is k-admissible in $B_\varepsilon(0) \cap \Omega$ for small $\varepsilon > 0$. By the concavity of F,

$$L[w] \ge F[u + w] - F[u] \ge F[w] - F[u]$$
$$\ge c_1 K^{1/k} - C \ge \frac{1}{2}c_1 K^{1/k}$$

for some constants c_1 depending on n, k, and δ, provided K is sufficiently large. Choose a K' large such that $L[K'w \pm T(u - \varphi)] \ge 0$. It follows that the maximum of $K'w \pm T(u - \varphi)$ in $B_\varepsilon \cap \Omega$ is attained on the boundary $\partial(B_\varepsilon \cap \Omega)$. But on the boundary $\partial(B_\varepsilon \cap \Omega)$, it is easy to see that

$$w \le -\frac{1}{2}\delta|x'|^2 \quad \text{on } \varphi\Omega \cap B_\varepsilon(0),$$
$$w < 0 \quad \text{on } \Omega \cap \partial B_\varepsilon(0).$$

Hence $K'w \pm T(u - \varphi) \le 0$ provided K' is chosen large enough. Hence $K'w \pm T(u - \varphi)$ attains its maximum 0 at the origin and we obtain

$$|\partial_n(T(u - \varphi))| \le K'|\partial_n w| \le C,$$

from which (3.9) follows.

Finally we consider the double normal derivative estimate

$$u_{nn}(0) \le C. \tag{3.11}$$

If $\varphi = 0$, by (3.7) we have $u_{ij}(0) = (-u_n)\rho_{ij}$. By the geometric assumption (2.9), we have

$$S_k^{nn}[u] = \sigma_{k-1}[\lambda(D^2u)'] = |u_n|^{k-1}\sigma_{k-1}(\kappa) > 0,$$

where $(D^2u)' = (u_{ij})_{1 \le i,j \le n-1}$. Note that

$$S_k[u] = u_{nn}\sigma_{k-1}\{\lambda[(D^2u)']\} + R = f, \qquad (3.12)$$

where R is the rest terms which do not involve u_{nn}, and so is bounded by (3.8) and (3.9). Hence $u_{nn}(0)$ must be bounded.

For general boundary function φ, we adapt the approach from [T]. By (3.12) it suffices to prove $\sigma_{k-1}\{\lambda[(D^2u)']\} > 0$ on $\varphi\Omega$. For any boundary point $x \in \varphi\Omega$, let $\xi^{(1)}, \cdots, \xi^{(n-1)}$ be an orthogonal vector field on $\varphi\Omega$. Denote $\nabla_i = \xi_m^{(i)}D_m u$,

$$\nabla_{ij}u = \xi_m^{(i)}\xi_l^{(j)}D_{ml}u, \quad \mathcal{C}_{ij} = \xi_m^{(i)}\xi_l^{(j)}D_m\gamma_l,$$

and $\nabla^2 u = \{\nabla_{ij}u\}$, $\mathcal{C} = \{\mathcal{C}_{ij}\}$, where γ is the unit inner normal of $\varphi\Omega$ at x. Then we have

$$\lambda[(D^2u)'] = \lambda[\nabla^2 u](x).$$

Similar to (3.7) we have

$$\nabla^2 u = D_\gamma(u - \varphi)\mathcal{C} + \nabla^2\varphi. \qquad (3.13)$$

For any $(n-1) \times (n-1)$-matrix r with eigenvalues $(\lambda_1, \cdots, \lambda_{n-1})$, denote

$$G(r) = [\sigma_{k-1}(\lambda)]^{1/(k-1)}.$$

and $G^{ij} = \frac{\partial G}{\partial r_{ij}}$. Assume that $\inf_{x \in \varphi\Omega} G(\nabla^2 u)$ is attained at x_0. Then by (3.13) and the concavity of G,

$$G_0^{ij}[D_\gamma(u - \varphi)\mathcal{C}_{ij}(x) + \nabla_{ij}\varphi(x)] \ge G_0^{ij}[D_\gamma(u - \varphi)\mathcal{C}_{ij}(x_0) + \nabla_{ij}\varphi(x_0)]$$

for any $x \in \varphi\Omega$, where $G_0^{ij} = G^{ij}(\nabla^2 u(x_0))$. We can also write the above formula in the form

$$G_0^{ij}\mathcal{C}_{ij}(x_0)[D_\gamma(u - \varphi)(x) - D_\gamma(u - \varphi)(x_0)]$$
$$\ge G_0^{ij}\{[D_\gamma(u - \varphi)(x) - D_\gamma(u - \varphi)(x_0)][\mathcal{C}_{ij}(x_0) - \mathcal{C}_{ij}(x)]$$
$$+ D_\gamma(u - \varphi)(x_0)][\mathcal{C}_{ij}(x_0) - \mathcal{C}_{ij}(x)] - [\nabla_{ij}\varphi(x) - \nabla_{ij}\varphi(x_0)]\}$$

Assume that near x_0, $\varphi\Omega$ is given by (3.6) with

$$\rho(x') = \frac{1}{2}\sum_{i=1}^{n-1}\kappa_i x_i^2 + O(|x'|^3).$$

Then we have $\mathcal{C}_{ij}(x_0) = \partial_i\gamma_j = \kappa_i\delta_{ij}$. Recall that Ω is $(k-1)$-convex. The eigenvalues of $\{\mathcal{C}_{ij} - c_1\delta_{ij}\}$ (as a vector in $\mathbb{R}^{n-1}$) lies in Γ_{k-1}, provided c_1 is sufficiently small. Hence $G_0^{ij}(\mathcal{C}_{ij} - c_1\delta_{ij}) \ge 0$ at x_0, and so

$$G_0^{ij}\mathcal{C}_{ij}(x_0) \ge c_1\sum G_0^{ii} \ge \delta_0 > 0.$$

Therefore we obtain

$$D_n(u - \varphi)(x) - D_n(u - \varphi)(x_0) \leq \ell(x') + C|x'|^2,$$

where ℓ is a linear function of x' with $\ell(0) = 0$. Denote

$$v(x) = D_n(u - \varphi)(x) - D_n(u - \varphi)(x_0) - \ell(x').$$

We have

$$v(x) \leq C|x'|^2 \quad \forall \, x \in \varphi\Omega. \tag{3.14}$$

Differentiating equation (3.3) we have

$$|L(v)| \leq C(1 + \sum F^{ii}), \tag{3.15}$$

where $L = \sum F^{ij} \partial_{ij}$ is the linearized operator of F.

Let w be the function given in (3.10). Then by (3.15) we can choose K' sufficiently large such that $L(K'w) \geq \pm L(v)$ in $B_\varepsilon \cap \Omega$. By (3.14), we can also choose K' large such that $K'w + v \leq 0$ on $\partial(B_\varepsilon \cap \Omega)$. By the comparison principle it follows that $K'w + v \leq 0$ in $B_\varepsilon \cap \Omega$. Hence $K'w + v$ attains its maximum at x_0. We obtain $\partial_n(K'w + v) \leq 0$ at x_0, namely $u_{nn}(x_0) \leq C$.

To complete the proof, one observes that in (3.12),

$$R = -\sum u_{1i}^2 \frac{\partial^2}{\partial u_{11} \partial u_{ii}} S_k[u] \leq 0.$$

Hence

$$\sigma_{k-1}\{\lambda[(D^2 u)']\}(x_0) \geq \frac{f}{u_{nn}}(x_0) \geq \frac{f_0}{u_{nn}(x_0)}. \tag{3.16}$$

Recall that $\sigma_{k-1}\{\lambda[(D^2 u)']\}$ attains its minimum at x_0. Hence by (3.12) we obtain $u_{\gamma\gamma}(x) < C$ at any boundary point $x \in \varphi\Omega$.

By the a priori estimate (3.2), equation (3.1) becomes uniformly elliptic if f is strictly positive. The uniform ellipticity follows from inequality (iii) in §2.5. To get the higher order derivative estimates, we employ the regularity theory of fully nonlinear, uniformly elliptic equations.

3.2 Regularity for Fully Nonlinear, Uniformly Elliptic Equation

We say a fully nonlinear elliptic operator F is *concave* if F, as a function of $r = D^2 u$, is a concave function. From §2.3, the k-Hessian equation is concave when u is k-admissible and the equation is written in the form (3.3).

The regularity theory of fully nonlinear elliptic equations was established by Evans and Krylov independently. Their proof is based on Krylov-Safonov's Hölder estimates for linear, uniformly elliptic equation of non-divergent form.

Theorem 3.2 *Consider the fully nonlinear, uniformly elliptic equation*

$$F(D^2u) = f(x) \quad in \ \Omega. \tag{3.17}$$
$$u = \varphi \ on \ \varphi\Omega,$$

Suppose F is concave, $F \in C^{1,1}$, $f \in C^{1,1}(\Omega)$, and $u \in W^{4,n}(\Omega)$ is a solution of (3.17). Then there exists $\alpha \in (0,1)$ depending only on n, λ, Λ (the constants in (1.6)) such that for any $\Omega' \subset\subset \Omega$,

$$\|u\|_{C^{2,\alpha}(\Omega')} \le C, \tag{3.18}$$

where C depends only on $n, \lambda, \Lambda, \alpha, \Omega, \ dist(\Omega', \varphi\Omega), \|f\|_{C^{1,1}(\Omega)}$, and $\sup_\Omega |u|$.
If furthermore $\varphi \in C^{3,1}(\overline{\Omega})$, $\varphi\Omega \in C^{3,1}$, and $f \in C^{1,1}(\overline{\Omega})$, then

$$\|u\|_{C^{2,\alpha}(\overline{\Omega})} \le C, \tag{3.19}$$

where C depends only on $n, \lambda, \Lambda, \alpha, \varphi\Omega, \|f\|_{C^{1,1}(\Omega)}, \|\varphi\|_{C^{3,1}(\Omega)}$ and $\sup_\Omega |u|$.

From (3.19) one also obtains $C^{3,\alpha}$ estimates by differentiating the equation (3.17) and apply the Schauder theory for linear, uniformly elliptic equations. Theorem 3.2 also extends to more general equations of the form (1.1) provided F satisfies certain structural conditions. We refer the readers to [E, K1, GT] for details.

As a corollary of Theorem 3.2, we obtain the higher order derivative estimate for the k-Hessian equation.

Theorem 3.3 *Let $u \in C^{3,1}(\overline{\Omega})$ be a k-admissible solution of (3.1). Assume that Ω is $(k-1)$-convex, $f \in C^{1,1}(\overline{\Omega})$, and $f \ge f_0 > 0$ in Ω. Then we have*

$$\|u\|_{C^{3,\alpha}(\overline{\Omega})} \le C, \tag{3.20}$$

where $\alpha \in (0,1)$, C depends only on $n, k, \alpha, f_0, \Omega, \|\varphi\|_{C^{3,1}(\varphi\Omega)}$, and $\|f\|_{C^{1,1}(\overline{\Omega})}$.

3.3 Existence of Smooth Solutions

By Theorem 3.3 and the continuity method, we obtain the existence of smooth solutions to the Dirichlet problem (3.1).

Theorem 3.4 *Assume that Ω is $(k-1)$-convex, $\varphi\Omega \in C^{3,1}$, $f \in C^{1,1}(\overline{\Omega})$, and $f \ge f_0 > 0$. Then there is a unique k-admissible solution $u \in C^{3,\alpha}(\overline{\Omega})$ to the Dirichlet problem (3.1).*

Proof. We apply the continuity method to the Dirichlet problem

$$S_k[u_t] = f_t \quad \text{in} \quad \Omega,$$
$$u_t = \varphi_t \quad \text{on} \quad \varphi\Omega,$$

where $t \in [0,1]$, $f_t = C_n^k(1-t) + tf$, $\varphi_t = \frac{1-t}{2}|x|^2 + t\varphi$. Then when $t = 0$, $u_0 = \frac{1}{2}|x|^2$ is the solution to the above Dirichlet problem at $t = 0$. To apply the continuity method, we consider solution $u = v + \varphi_t$ so that $v \in C^{3+\alpha}(\overline{\Omega})$ with $v = 0$ on $\varphi\Omega$. Note that the uniqueness of k-admissible solutions follows from the comparison principle.

3.4 Remarks

(i) In the proof of Theorem 3.1, the assumption $f \geq f_0$ was used only once in (3.16). Therefore this assumption can be relaxed to $f \geq 0$ for the zero boundary value problem. By approximation and Theorems 3.1 and 3.4, it follows that there is a k-admissible solution $u \in C^{1,1}(\overline{\Omega})$ to the k-Hessian equation (2.1) which vanishes on $\varphi\Omega$, provided Ω is $(k-1)$-convex and $f^{1/k} \in C^{1,1}(\overline{\Omega})$, $f \geq 0$.

The above results are also true for a general boundary function $\varphi \in C^{3,1}(\varphi\Omega)$. Indeed Krylov [K2] established the a priori estimate (3.2), not only for solutions to the k-Hessian equation, but also for solutions to the Dirichlet problem (3.17) for general functions $\varphi \in C^{3,1}(\overline{\Omega})$, provided $f \geq 0$ and $f \in C^{1,1}(\varphi\Omega)$. The main difficulty is again the estimation on the boundary. For the k-Hessian equation, Krylov's proof was simplified in [ITW].

We also note that the geometric assumption (2.6) can be replaced by the existence of a subsolution $\underline{u}$ to (3.1) with $\underline{u} = \varphi$ [G].

(ii) The estimate (3.2) also extends to the Hessian quotient equation [T]

$$S_{k,l}[u] = \frac{S_k[u]}{S_l[u]} = f, \tag{3.21}$$

where $0 \leq l < k \leq n$ and we define $S_l[u] = 1$ when $l = 0$.

(iii) For the second boundary value problem of the k-Hessian equation, and some other boundary value problems, we refer the reader to [J,S,U3]

(iv) Much more can be said about the regularity of the Monge-Ampère equation. The interior regularity was established by Calabi and Pogorelov [GT, P]. The global regularity for the Dirichlet problem was obtained independently by Caffarelli, Nirenberg and Spruck [CNS1], and by Krylov [K1], assuming all data are smooth enough. Caffarelli [Ca] established the interior $C^{2,\alpha}$ and $W^{2,p}$ estimates for strictly convex solutions, assuming that $f \in C^\alpha$ and $f \in C^0$, respectively. The continuity of f is also necessary for the $W^{2,p}$ estimate [W3].

The boundary $C^{2,\alpha}$ estimate for the Dirichlet problem was established in [TW6], assuming that $f > 0, \in C^\alpha(\overline{\Omega})$, the boundary $\varphi\Omega$ is uniformly convex and C^3 smooth, and boundary function $\varphi \in C^3$. If either $\varphi\Omega$ or φ is only $C^{2,1}$, the solution may not belong to $W^{2,p}(\Omega)$ for large p, even f is a positive constant.

4 Interior a Priori Estimates

In this section we establish interior gradient and second derivative estimates for the k-Hessian equation

$$S_k[u] = f(x, u). \tag{4.1}$$

These estimates were previously proved in [CW1]. From the interior gradient estimate, we also deduce a Harnack inequality. Estimates in this section will be repeatedly used in subsequent sections.

4.1 Interior Gradient Estimate

Theorem 4.1 *Let $u \in C^3(B_r(0))$ be a k-admissible solution of (4.1). Suppose that $f \geq 0$ and f is Lipschitz continuous. Then*

$$|Du(0)| \leq C_1 + C_2 \frac{M}{r}, \tag{4.2}$$

where $M = 4\sup|u|$, C_2 is a constant depending only on n, k; C_1 depends on n, k, M, r and $\|f\|_{C^{0,1}}$. Moreover, if f is a constant, then $C_1 = 0$.

Proof. Introduce an auxiliary function

$$G(x, \xi) = u_\xi(x)\varphi(u)\rho(x),$$

where $\rho(x) = (1 - \frac{|x|^2}{r^2})^+$, $\varphi(u) = 1/(M - u)^{1/2}$, and $M = 4\sup|u|$. Suppose G attains its maximum at $x = x_0$ and $\xi = e_1$, the unit vector in the x_1 axis. Then at x_0, $G_i = 0$ and $\{G_{ij}\} \leq 0$. That is

$$u_{1i} = -\frac{u_1}{\varphi\rho}(u_i\varphi'\rho + \varphi\rho_i), \tag{4.3}$$

$$\tag{4.4}$$

$$
\begin{aligned}
0 \geq S_k^{ij} G_{ij} &= \varphi\rho\partial_1 f + ku_1 f\varphi'\rho + u_1\varphi''\rho S_k^{ij}u_iu_j + u_1\varphi S_k^{ij}\rho_{ij} \\
&\quad + u_1\varphi' S_k^{ij}(u_i\rho_j + u_j\rho_i) + 2S_k^{ij}u_{1i}(u_j\varphi'\rho + \varphi\rho_j) \\
&= \varphi\rho\partial_1 f + ku_1 f\varphi'\rho + u_1\rho(\varphi'' - \frac{2\varphi'^2}{\varphi})S_k^{ij}u_iu_j + u_1\varphi S_k^{ij}\rho_{ij} \\
&\quad - u_1\varphi' S_k^{ij}(u_i\rho_j + u_j\rho_i) - \frac{2u_1\varphi}{\rho}S_k^{ij}\rho_i\rho_j,
\end{aligned}
$$

where we used the relations $S_k^{ij} u_{ij} = kf$ and $S_k^{ij} u_{ij1} = \partial_1 f$, which follows by differentiating equation (4.1).

By our choice of φ, $\varphi'' - \frac{2\varphi'^2}{\varphi} \geq \frac{1}{16} M^{-5/2}$. Denote $\mathcal{S} = \Sigma_i S_k^{ii}$. Note that the term $ku_1 f\varphi'\rho$ is nonnegative. From (4.4) we obtain

$$0 \geq -16M^{5/2}\varphi\rho|\partial_1 f| + \rho S_k^{11} u_1^3 - C\mathcal{S}(\frac{M^2}{\rho r^2} u_1 + \frac{M}{r} u_1^2), \qquad (4.5)$$

where C is independent of r, M. To prove (4.2), we assume that $|Du(0)| > CM/r$, otherwise we are through. Then by $G(x_0) \geq G(0)$, we have $u_1\rho(x_0) > CM/r$. Hence by (4.3) we have

$$u_{11} \leq -\frac{\varphi'}{2\varphi} u_1^2 \quad \text{at } x_0. \qquad (4.6)$$

Hence by (ix) above, $S_k^{11} \geq C\mathcal{S}$.

To control $\partial_1 f$ by $\mathcal{S}$, by a rotation of the coordinates, we assume that $D^2 u$ is diagonal in the new coordinates y, and $u_{y_1 y_1} \geq \cdots \geq u_{y_n y_n}$. Then at the point x_0 where G reaches its maximum,

$$u_{y_n y_n} \leq u_{x_1 x_1} \leq -\frac{\varphi'}{2\varphi} u_{x_1}^2 \leq -\frac{1}{4M} u_{x_1}^2$$

by (4.6). From equation (4.1),

$$f = u_{y_n y_n}\sigma_{k-1;n}(\lambda) + \sigma_{k;n}(\lambda), \quad \lambda = \lambda(D^2 u).$$

By §2 (vi), we obtain

$$0 \leq u_{y_n y_n}\sigma_{k-1;n}(\lambda) + C[\sigma_{k-1;n}(\lambda)]^{k/(k-1)}.$$

Hence

$$\sigma_{k-1;n}(\lambda) \geq C|u_{y_n y_n}|^{k-1} \geq Cu_{x_1}^{2k-2}.$$

We obtain

$$\mathcal{S} \geq Cu_1^{2k-2} \geq C\frac{u_{x_1}^{2k-2}}{M^{k-1}} \quad \text{at } x_0.$$

Recall that in (4.6), we assumed that $u_{x_1} \geq CM/r$. Hence $\mathcal{S} \geq CM^{k-1}/r^{2k-2}$ and $\mathcal{S}^{-1}|\partial_1 f|$ is bounded. Multiplying (4.5) by $\rho^2/\mathcal{S}$, we obtain (4.2).

4.2 Harnack Inequality

From the interior gradient estimate, we obtain a Harnack inequality for the k-Hessian equation. First we prove a lemma, which also follows from the interpolation inequality (2.12) in [TW2].

Lemma 4.1 *Suppose $u \in C^1(B_R(0))$ is a function which satisfies for any $B_r(x) \subset B_R(0)$,*

$$|Du(x)| \leq \frac{C_1}{r} \sup_{B_r(x)} |u|. \tag{4.7}$$

Then

$$|u(0)| \leq \frac{C_2}{|B_R|} \int_{B_R} |u|, \tag{4.8}$$

where C_2 depends only on C_1 and n.

Proof. There is no loss of generality in assuming that $R = 1$, $\int_{B_1} |u| = 1$, and u is a C^1 function defined in $B_{1+\varepsilon}(0)$ for some small $\varepsilon > 0$. Let K be the largest constant such that $|u(x)| \geq K(1 - |x|)^{-n}$ for some $x \in B_1(0)$, namely $K = \sup(1 - |x|)^n |u(x)|$. Choose $y \in B_1(0)$ such that $|u(y)| = K(1 - |y|)^{-n}$ and $|y| = \sup\{|x| \in B_1(0) \mid |u(x)| = K(1 - |x|)^{-n}\}$. Then we have $|u| \leq 2^n |u(y)| = Kr^{-n}$ in $B_r(y)$, where $r = \frac{1}{2}(1 - |y|)$. Therefore by applying the interior gradient estimate to u in $B_r(y)$, we get $|Du(x)| \leq CKr^{-n-1}$. Hence $|u(x)| > \frac{1}{2}Kr^{-n}$ whenever $|x - y| \leq r/2C$. It follows that $\int_{B_r(y)} |u| \geq K/C$. But by assumption, $\int_{B_R} |u| \leq 1$, we obtain an upper bound for K and Lemma 4.1 follows.

Theorem 4.2 *Let u be a non-positive, k-admissible solution to*

$$S_k[u] = c \quad in \quad B_R(0), \tag{4.9}$$

where $c \geq 0$ is a constant. Then we have

$$\sup_{B_{R/2}(0)} (-u) \leq C \inf_{B_{R/2}(0)} (-u), \tag{4.10}$$

where C depends only on n, k.

Proof. By Lemma 4.1,

$$\sup_{B_{R/2}} (-u) \leq C \int_{3R/4} (-u).$$

Since u is subharmonic, we have [GT]

$$\int_{3R/4} (-u) \leq C \inf_{B_{R/2}} (-u).$$

From the above two inequalities we obtain (4.10).

The interior gradient estimate also implies the following Liouville Theorem.

Corollary 4.1 *Let $u \in C^3(\mathbb{R}^n)$ be an entire solution to $S_k[u] = 0$. If $u(x) = o(|x|)$ for large x, then $u \equiv$ constant.*

4.3 Interior Second Derivative Estimate

Theorem 4.3 *Let $u \in C^4(\Omega)$ be a k-admissible solution of (4.1). Suppose $f \in C^{1,1}(\overline{\Omega} \times \mathbb{R})$ and $f \geq f_0 > 0$. Suppose there is a k-admissible function w such that*

$$w > u \quad \text{in } \Omega, \quad \text{and} \quad w = u \quad \text{on } \varphi\Omega. \tag{4.11}$$

Then

$$(w - u)^4(x)|D^2 u(x)| \leq C, \tag{4.12}$$

where C depends only on n, k, f_0, $\sup_\Omega(|Dw| + |Du|)$, and $\|f\|_{C^{1,1}(\overline{\Omega})}$.

Proof. Writing equation (4.1) in the form

$$F[u] = \hat{f},$$

where $\hat{f} = f^{1/k}(x, u)$, and differentiating twice, we get

$$F_{ii} u_{ii\gamma\gamma} + (F_{ij})_{rs} u_{ij\gamma} u_{rs\gamma} = \hat{f}_{\gamma\gamma}.$$

Suppose $(D^2 u)$ is diagonal. Then

$$(F_{ij})_{rs} = \begin{cases} \mu' \sigma_{k-2;ir}(\lambda) + \mu'' \sigma_{k-1;i} \sigma_{k \angle 1;r} & \text{if } i = j, r = s, \\ -\mu' \sigma_{k-2;ij}(\lambda) & \text{if } i \neq j, r = j, \text{and } s = i, \\ 0 & \text{otherwise,} \end{cases}$$

where $\mu(t) = t^{1/k}$. Hence

$$F_{ii} u_{ii\gamma\gamma} = \hat{f}_{\gamma\gamma} + \sum_{i,j=1}^n \mu' \sigma_{k-2;ij} u_{ij\gamma}^2 - \sum_{i,j=1}^n [\mu'' \sigma_{k-1;i} \sigma_{k-1;j} + \mu' \sigma_{k-2;ij}] u_{ii\gamma} u_{jj\gamma}. \tag{4.13}$$

Let

$$G(x) = \rho^\beta(x) \varphi(\tfrac{1}{2}|Du|^2) u_{\xi\xi},$$

where $\rho = w - u$, $\beta = 4$, $\varphi(t) = (1 - \frac{t}{M})^{-1/8}$, and $M = 2\sup_{x \in \Omega} |Du|^2$. Suppose G attains its maximum at x_0 and in the direction $\xi = (1, 0, \cdots, 0)$. By a rotation of axes we assume that $D^2 u$ is diagonal at x_0 with $u_{11} \geq \cdots \geq u_{nn}$. Then at x_0,

$$0 = (\log G)_i = \beta \frac{\rho_i}{\rho} + \frac{\varphi_i}{\varphi} + \frac{u_{11i}}{u_{11}}, \tag{4.14}$$

$$0 \geq F_{ii}(\log G)_{ii} = \beta F_{ii}\left[\frac{\rho_{ii}}{\rho} - \frac{\rho_i^2}{\rho^2}\right] + F_{ii}\left[\frac{\varphi_{ii}}{\varphi} - \frac{\varphi_i^2}{\varphi^2}\right] + F_{ii}\left[\frac{u_{11ii}}{u_{11}} - \frac{u_{11i}^2}{u_{11}^2}\right]. \quad (4.15)$$

Case 1: $u_{kk} \geq \varepsilon u_{11}$ for some $\varepsilon > 0$.

By (4.14) we have

$$\frac{u_{11i}}{u_{11}} = -\left(\frac{\varphi_i}{\varphi} + \beta\frac{\rho_i}{\rho}\right). \quad (4.16)$$

Hence by (4.15),

$$0 \geq \beta F_{ii}\left[\frac{\rho_{ii}}{\rho} - (1 + 2\beta)\frac{\rho_i^2}{\rho^2}\right] + F_{ii}\left[\frac{\varphi_{ii}}{\varphi} - 3\frac{\varphi_i^2}{\varphi^2}\right] + F_{ii}\frac{u_{11ii}}{u_{11}}. \quad (4.17)$$

By the concavity of F,

$$F_{ii}u_{11ii} \geq \hat{f}_{11} \geq -C(1 + u_{11}).$$

We have

$$F_{ii}\left[\frac{\varphi_{ii}}{\varphi} - 3\frac{\varphi_i^2}{\varphi^2}\right] = \left(\frac{\varphi''}{\varphi} - 3\frac{\varphi'^2}{\varphi^2}\right)F_{ii}u_i^2 u_{ii}^2 + \frac{\varphi'}{\varphi}u_\gamma F_{ii}u_{ii\gamma} + \frac{\varphi'}{\varphi}F_{ii}u_{ii}^2$$

$$\geq \frac{\varphi'}{\varphi}F_{ii}u_{ii}^2 + \frac{\varphi'}{\varphi}u_\gamma \hat{f}_\gamma,$$

where by §2 (ix)

$$\sum F_{ii}u_{ii}^2 > F_{jj}u_{jj}^2 \geq \theta \mathcal{F}u_{11}^2,$$

$\mathcal{F} = \sum_{i=1}^n F_{ii}$, $\theta = \theta(n, k, \varepsilon)$. Hence

$$F_{ii}\left[\frac{\varphi_{ii}}{\varphi} - 3\frac{\varphi_i^2}{\varphi^2}\right] \geq \theta \mathcal{F}u_{11}^2 - C.$$

Since $\rho = w - u$ and w is k-admissible, we have

$$F_{ii}\rho_{ii} \geq -F_{ii}u_{ii} = -\mu' S_k^{ii}u_{ii} = -k\mu'f.$$

Inserting the above estimates to (4.17) we obtain

$$0 \geq \sum F_{ii}(\log G)_{ii} \geq \theta \mathcal{F}u_{11}^2 - C\mathcal{F}\frac{\rho_i^2}{\rho^2} - \frac{k\beta\mu'f}{\rho} - C. \quad (4.18)$$

Note that $u_{kk} \geq \varepsilon u_{11}$, we have

$$\mathcal{F} \geq F_{nn} \geq \theta \mu' u_{11} \cdots u_{k-1,k-1} \geq \theta_1 u_{11}^{k-1}.$$

Multiplying (4.18) by $\rho^{2\beta}\varphi^2$, we obtain $G(x_0) \leq C$.

Case 2: $u_{kk} \leq \varepsilon u_{11}$.

Since $(u_{11}, \cdots, u_{nn}) \in \overline{\Gamma}_k$, we have $u_{kk} \geq 0$ and so $|u_{kk}| \leq \varepsilon u_{11}$. By the arrangement $u_{11} \geq \cdots \geq u_{nn}$, we have $u_{jj} \leq \varepsilon u_{11}$ for all $j = k, \cdots, n$. Noting that

$$\frac{\partial^{k-1}}{\partial \lambda_1 \cdots \partial \lambda_{k-1}} \sigma_k[\lambda] = \sum_{j=k}^{n} \lambda_j \geq 0$$

we obtain

$$|u_{jj}| \leq C\varepsilon u_{11} \quad \text{for} \quad j = k, \cdots, n.$$

By (4.14),

$$\frac{\rho_i}{\rho} = -\frac{1}{\beta}\left(\frac{\varphi_i}{\varphi} + \frac{u_{11i}}{u_{11}}\right) \quad i = 2, \cdots, n. \tag{4.19}$$

Applying (4.16) for $i = 1$ and (4.19) for $i = 2, \cdots, n$ to (4.15), we obtain

$$0 \geq \left\{ \sum_{i=1}^{n} \left[\beta F_{ii} \frac{\rho_{ii}}{\rho} + F_{ii}\left(\frac{\varphi_{ii}}{\varphi} - 3\frac{\varphi_i^2}{\varphi^2}\right) \right] - \beta(1 + 2\beta)F_{11} \frac{\rho_1^2}{\rho^2} \right\}$$
$$+ \left\{ \sum_{i=1}^{n} F_{ii} \frac{u_{11ii}}{u_{11}} - \left(1 + \frac{2}{\beta}\right) \sum_{i=2}^{n} F_{ii} \frac{u_{11i}^2}{u_{11}^2} \right\} =: I_1 + I_2 \tag{4.20}$$

As in case I we have

$$I_1 \geq \theta F_{ii} u_{ii}^2 - F_{11} \frac{C}{\rho^2} - \frac{k\beta\mu' f}{\rho} - C$$
$$\geq \frac{1}{2}\theta F_{11} u_{11}^2 - \frac{k\beta\mu' f}{\rho} - C$$

provided $\rho^2 u_{11}^2$ is suitably large. By (vii) in §2 we obtain

$$I_1 \geq \theta_1 \mu' f u_{11} - \frac{k\beta\mu' f}{\rho} - C.$$

We claim

$$I_2 \geq \hat{f}_{11}/u_{11}. \tag{4.21}$$

If (4.21) is true then (4.20) reduces to

$$0 \geq \theta_1 \mu' f u_{11} + \frac{\hat{f}_{11}}{u_{11}} - \frac{k\beta\mu' f}{\rho} - C. \tag{4.22}$$

Multiplying the above inequality by $\rho^\beta \varphi$ we obtain $G(x_0) \leq C$.

To verify (4.21) we first note that by the concavity of F,

$$-\sum_{i,j=1}^{n} [\mu'' \sigma_{k-1;i} \sigma_{k-1;j} + \mu' \sigma_{k-2;ij}] u_{ii1} u_{jj1} = -\sum \frac{\partial^2}{\partial \lambda_i \partial \lambda_j} \mu(S_k(\lambda)) u_{ii1} u_{jj1} \geq 0.$$

Hence by (4.13),

$$
u_{11}I_2 \geq \hat{f}_{11} + \sum_{i,j=1}^{n} \mu' \sigma_{k-2;ij} u_{ij1}^2 - \left(1 + \frac{2}{\beta}\right) \sum_{i=2}^{n} F_{ii} \frac{u_{11i}^2}{u_{11}}
$$

$$
\geq \hat{f}_{11} + \sum_{i=2}^{n} \mu' \left(2\sigma_{k-2;1i} - \left(1 + \frac{2}{\beta}\right) \frac{\sigma_{k-1;i}}{u_{11}} \right) u_{11i}^2.
$$

Since $\beta = 4$, we need only

$$
\sigma_{k-2;1i} - \frac{3}{4} \frac{\sigma_{k-1;i}}{u_{11}} \geq 0. \tag{4.23}
$$

But (4.23) follows from the following lemma.

Lemma 4.2 *Suppose $\lambda \in \Gamma_k$ and $\lambda_1 \geq \cdots \geq \lambda_n$. Then for any $\delta \in (0,1)$, there exists $\varepsilon > 0$ such that if*

$$
S_k(\lambda) \leq \varepsilon \lambda_1^k \quad or \quad |\lambda_i| \leq \varepsilon \lambda_1 \ for \ i = k+1, \cdots, n
$$

we have

$$
\lambda_1 S_{k-1;1} \geq (1-\delta) S_k. \tag{4.24}
$$

Proof. To prove (4.24) we first consider the case $S_k(\lambda) \leq \varepsilon \lambda_1^k$. We may suppose $S_k(\lambda) = 1$. If (4.24) is not true,

$$
S_{k-1;1} < (1-\delta)\lambda_1^{-1} \leq \varepsilon^{1/k}.
$$

Hence

$$
S_{k;1} \leq C S_{k-1;1}^{k/(k-1)} \leq C \varepsilon^{1/(k-1)}.
$$

Noting that

$$
S_k = S_{k-1;1}\lambda_1 + S_{k;1},
$$

we obtain (4.24).

Next we consider the case $|\lambda_i| \leq \varepsilon \lambda_1$ for $i = k+1, \cdots, n$. Observing that if $\lambda_k << \lambda_1$, we have $S_k(\lambda) << \lambda_1^k$ and so (4.24) holds. Hence we may suppose $|\lambda_i| << \lambda_k$ for $i = k+1, \cdots, n$. In this case both sides in (4.24) $= \lambda_1 \cdots \lambda_k(1 + o(1))$ with $o(1) \to 0$ as $\varepsilon \to 0$. Hence (4.24) holds.

In Section 6 we will investigate the existence of nonzero solutions to equation (4.1) with zero Dirichlet boundary condition. Assume that $f \in C^{1,1}(\overline{\Omega} \times \mathbb{R})$, $f(x,u) > 0$ when $u < 0$. Then by choosing $w = -\delta$ for small constant δ in (4.12), we obtain a local second derivative estimate. Therefore by the regularity theory for fully nonlinear, uniformly elliptic equations, one also obtains local $C^{3,\alpha}$ estimate for the solution u. That is

Theorem 4.4 *Let $u \in C^4(\Omega) \cap C^0(\overline{\Omega})$ be a k-admissible solution of (4.1). Suppose $u = 0$ on $\varphi\Omega$, $f \in C^{1,1}$ and $f > 0$ when $u < 0$. Then u satisfies a priori estimates in $C_{loc}^3(\Omega) \cap C^{0,1}(\overline{\Omega})$, namely for any $\Omega' \subset\subset \Omega$,*

$$\|u\|_{C^3(\Omega')} + \|u\|_{C^{0,1}(\overline{\Omega})} \le C, \tag{4.25}$$

where C depends only on n, k, f, $\sup|u|$, and $dist(\Omega', \varphi\Omega)$. If $f^{1/k} \in C^{1,1}(\overline{\Omega})$, $\varphi\Omega \in C^{3,1}$ and Ω is uniformly $(k-1)$-convex, then

$$\|u\|_{C^{1,1}(\overline{\Omega})} \le C.$$

Remark. Theorem 4.3 was established in [P] for the Monge-Ampère equation, and in [CW1] for the k-Hessian equations. The condition (4.11) in Theorem 4.3 is necessary when $k \ge 3$ [P, U1], but may be superfluous when $k = 2$ [WY]. Instead of (4.11), Urbas [U2] established the interior second derivative estimate under the assumption $D^2 u \in L^p(\Omega)$, $p > \frac{1}{2}k(n-1)$.

5 Sobolev Type Inequalities

The k-Hessian operator can also be written in the form

$$S_k[u] = [D^2 u]_k, \tag{5.1}$$

see (2.5). Hence by direct computation, one has [R]

$$\sum_i \partial_i S_k^{ij}[u] = 0 \quad \forall j. \tag{5.2}$$

It follows that the k-Hessian operator is of divergence form

$$\begin{aligned}
S_k[u] &= \frac{1}{k} \sum u_{ij} S_k^{ij}[u] \\
&= \frac{1}{k} \sum \partial_{x_i}(u_{x_j} S_k^{ij}[u]),
\end{aligned} \tag{5.3}$$

Denote by $\Phi^k(\Omega)$ the set of all k-admissible functions in Ω, and by $\Phi_0^k(\Omega)$ the set of all k-admissible functions vanishing on $\varphi\Omega$. Let

$$\begin{aligned}
I_k(u) &= \int_\Omega (-u) S_k[u] dx \\
&= \frac{1}{k} \int_\Omega u_i u_j S_k^{ij}[D^2 u].
\end{aligned} \tag{5.4}$$

By (5.2), we can compute the first variation of I_k,

$$\langle \delta I_k(u), h \rangle = (k+1) \int_\Omega (-h) S_k[u] \tag{5.5}$$

for any smooth h with compact support. Hence the Hessian equation (3.1) is variational, namely it is the Euler equation of the functional

$$J(u) = \frac{1}{k(k+1)} \int_\Omega u_i u_j S_k^{ij}[u] + \int_\Omega fu. \tag{5.6}$$

The second variation is also easy to compute. Indeed by (5.2) we have

$$\frac{d^2}{dt^2} I_k(u + t\varphi) = (k+1) \int_\Omega \varphi_i \varphi_j S_k^{ij}[u] \tag{5.7}$$

for any $u \in C^2(\Omega), \varphi \in C_0^\infty(\Omega)$, or any $u, \varphi \in C^2(\overline{\Omega})$, both vanishing on $\varphi\Omega$. In particular if u is k-admissible, then $\frac{d^2}{dt^2} I_k(u + t\varphi) \geq 0$.

Denote

$$\|u\|_{\Phi_0^k} = [I_k(u)]^{1/(k+1)}, \qquad u \in \Phi_0^k. \tag{5.8}$$

One can easily verify that $\| \cdot \|_{\Phi_0^k}$ is a norm in Φ_0^k [W2]. In this section we prove Sobolev type inequalities for the functional I_k.

5.1 Sobolev Type Inequalities

The following Theorem 5.1 was proved in [W2]. The proof below is also from there. For convex functions, the theorem was first established in [Ch2].

Theorem 5.1 *Let $u \in \Phi_0^k(\Omega)$.*
(i) If $1 \leq k < \frac{n}{2}$, we have

$$\|u\|_{L^{p+1}(\Omega)} \leq C\|u\|_{\Phi_0^k} \quad \forall \ p+1 \in [1, k^*], \tag{5.9}$$

where C depends only on n, k, p, and $|\Omega|$,

$$k^* = \frac{n(k+1)}{n - 2k}.$$

When $p+1 = k^*$, the best constant C is attained when $\Omega = \mathbb{R}^n$ by the function

$$u(x) = [1 + |x|^2]^{(2k-n)/2k}. \tag{5.10}$$

(ii) If $k = \frac{n}{2}$,

$$\|u\|_{L^p(\Omega)} \leq C\|u\|_{\Phi_0^k} \tag{5.11}$$

for any $p < \infty$, where C depends only on n, p, and $diam(\Omega)$.
(iii) If $\frac{n}{2} < k \leq n$,

$$\|u\|_{L^\infty(\Omega)} \leq C\|u\|_{\Phi_0^k}, \tag{5.12}$$

where C depends on n, k, and $diam(\Omega)$.

Remark. Our proof of Theorem 5.1 reduces the above inequalities to rotationally symmetric functions. When $k = \frac{n}{2}$, we have accordingly the embedding of $\Phi_0^k(\Omega)$ in the Orlicz space associated with the function $e^{|t|^{(n+2)/n}}$.

Proof. *Step 1.* When u is radial and $\Omega = B_1(0)$,

$$\|u\|_{\Phi_0^k(\Omega)} = C\left(\int_0^1 r^{n-k}|u'|^{k+1}\right)^{1/(k+1)}. \tag{5.13}$$

One can verify Theorem 5.1 for k-admissible, radial functions vanishing on $\partial B_1(0)$. For details see [W2].

Step 2. We prove that Theorem 5.1 holds for general k-admissible functions when $\Omega = B_1(0)$. Indeed, let

$$T_p = \inf\{\|u\|_{\Phi_0^k}^{k+1}/\|u\|_{L^{p+1}(\Omega)}^{k+1} \mid u \in \Phi_0^k(\Omega)\},$$

$$T_{p,r} = \inf\{\|u\|_{\Phi_0^k}^{k+1}/\|u\|_{L^{p+1}(\Omega)}^{k+1} \mid u \in \Phi_0^k(\Omega) \text{ is radial}\}.$$

Suppose to the contrary that $T_p < T_{p,r}$. Choose a constant $\lambda \in (T_p, T_{p,r})$ and consider the functional

$$J(u) = J(u,\Omega) = \int_\Omega \frac{-u}{k+1} S_k[u] - \frac{\lambda}{k+1}\left[(p+1)\int_\Omega F(u)\right]^{\frac{k+1}{p+1}}, \tag{5.14}$$

where

$$F(u) = \int_0^{|u|} f(t)dt,$$

and f is a smooth, positive function such that

$$f(t) = \begin{cases} \delta^p & |t| < \frac{1}{2}\delta \\ |t|^p & \delta < |t| < M, \\ \varepsilon t^{-2} & |t| > M + \varepsilon, \end{cases}$$

where $M > 0$ is any fixed constant, $\delta, \varepsilon > 0$ are small constants. We also assume that f is monotone increasing when $\frac{1}{2}\delta < |t| < \delta$, and $\varepsilon M^{-2} \le f(t) \le |t|^p$ when $M < |t| < M + \varepsilon$. The introduction of ε, δ is such that f is positive and uniformly bounded, so the global a priori estimates for parabolic Hessian equations (Theorem 10.1) applies. Obviously F is also uniformly bounded and J is bounded from below. The Euler equation of the functional is

$$S_k[u] = \lambda\beta(u)f(u), \tag{5.15}$$

where

$$\beta(u) = \left[(p+1)\int_\Omega F(u)\right]^{\frac{k-p}{p+1}}.$$

Note that for a given u, $\beta(u)$ is a constant. By our choice of the constant λ, we have

$$\inf\{J(u) \mid u \in \Phi_0^k(\Omega)\} < -1 \quad (\text{if } M >> 1), \tag{5.16}$$
$$\inf\{J(u) \mid u \in \Phi_0^k(\Omega), u \text{ is radial}\} \to 0 \quad \text{as } \delta \to 0.$$

Consider the parabolic Hessian equation

$$\log S_k[u] - u_t = \log\{\lambda\beta(u)f(u)\} \quad (x,t) \in \Omega \times [0,\infty), \tag{5.17}$$

subject to the boundary condition

$$u(\cdot,t) = 0 \quad \text{on } \varphi\Omega \ \forall \, t \geq 0.$$

We say a function $u(x,t)$ is k-admissible with respect to the parabolic equation (5.17) if for any $t \in [0,\infty)$, $u(\cdot,t)$ is k-admissible. Equation (5.17) is a descent gradient flow of the functional J. Indeed, let $u(x,t)$ be a k-admissible solution. We have

$$\frac{d}{dt}J(u(\cdot,t)) = -\int_\Omega (S_k[u] - \psi) \log \frac{S_k[u]}{\psi} \leq 0,$$

and equality holds if and only if u is a solution to the elliptic equation (5.15), where $\psi(u) = \lambda\beta(u)f(u)$.

Let $u_0 \in \Phi_0^k(\Omega)$ be such that

$$J(u_0) \leq \inf_{\Phi_0^k(\Omega)} J(u) + \varepsilon' < -1.$$

By a slight modification (see Remark 5.1 below), we may assume that the compatibility condition $S_k[u_0] = \lambda\beta(u_0)f(u_0)$ holds on $\varphi\Omega \times \{t = 0\}$. In the parabolic equation (5.17), $\beta(u)$ is a function of t. By (5.16) and since $F(u)$ is uniformly bounded, we have

$$C_1 \leq \beta(u) \leq C_2,$$

for some positive constants C_1, C_2 independent of time t. Note that C_1, C_2 may depend on M but are independent of the small constants ε and δ. Therefore by Theorem 10.1, there is a global smooth k-admissible solution u to the parabolic Hessian equations (5.17).

By the global a priori estimates and since (5.17) is a descent gradient flow, $u(\cdot,t)$ sub-converges to a solution u_1 of the elliptic equation (5.15). By the Aleksandrov's moving plane method, see also [D] (p.327) for the Monge-Ampère equation, we infer that u_1 is a radial function. Therefore we have

$$\inf\{J(u) \mid u \in \Phi_0^k(\Omega), u \text{ is radial}\} \leq -1.$$

We reach a contradiction when ε, δ are small.

Step 3. Denote

$$T_p(\Omega) = \inf\{\|u\|_{\Phi_0^k}^{k+1}/\|u\|_{L^{p+1}(\Omega)}^{k+1} \mid u \in \Phi_0^k(\Omega)\}.$$

We claim that for any $(k-1)$-convex domains Ω_1, Ω_2 with $\Omega_1 \subset \Omega_2$,

$$T_p(\Omega_1) \geq T_p(\Omega_2). \tag{5.18}$$

Suppose to the contrary that $T_p(\Omega_1) < T_p(\Omega_2)$. Let $\lambda \in (T_p(\Omega_1), T_p(\Omega_2))$ be a constant. Let $J(u, \Omega)$ be the functional given in (5.14). Then we have

$$\inf\{J(u, \Omega_1) \mid u \in \Phi_0^k(\Omega_1)\} < -1 \quad \text{(when } M \gg 1\text{)}, \tag{5.19}$$
$$\inf\{J(u, \Omega_2) \mid u \in \Phi_0^k(\Omega_2)\} \to 0 \quad \text{as } \delta \to 0.$$

Let $u_1 \in \Phi_0^k(\Omega_1)$ be the solution to (5.15) obtained in Step 2 which satisfies

$$J(u_1, \Omega_1) \leq -1.$$

Let

$$w(x) = -M - \varepsilon - \frac{1}{2}\varepsilon^{1/2k}(R^2 - |x|^2),$$

where R is chosen large such that $\Omega_1 \subset B_R(0)$. Recall that $f(t) = \varepsilon t^{-2}$ when $|t| > M + \varepsilon$, and $C_1 \leq \beta(u_1) \leq C_2$, where C_1, C_2 are independent of ε. By equation (5.15) we have $S_k[u_1] \leq C\varepsilon$. Hence $S_k[w] \geq C\varepsilon^{1/2} \geq S_k[u_1]$ when $u_1 < -M - \varepsilon$. Applying the comparison principle to u_1 and w in $\{u_1 < -M - \varepsilon\}$, we obtain a lower bound for u_1,

$$u_1 \geq -M - R^2\varepsilon^{1/2k}. \tag{5.20}$$

Hence when ε is sufficiently small, $F(u_1) = |u_1|^{p+1} + o(1)$ if ε, δ is small, though $f(u_1)$ may violate strongly. In particular we have

$$\beta(u_1) = (1 + o(1))\left[\int_{\Omega_1} |u_1|^{p+1}\right]^{\frac{k-p}{p+1}} \tag{5.21}$$

with $o(1) \to 0$ as $\varepsilon, \delta \to 0$.

Extending u_1 to Ω_2 such that $u_1 = 0$ in $\Omega_2 - \Omega_1$ (so u_1 is not k-admissible in Ω_2). Let $\psi(x) = S_k[u_1]$ in Ω_1 and $\psi(x) = 0$ in $\Omega_2 - \Omega_1$. Denote

$$E(\varphi) = \int_{\Omega_2} (-\varphi)\psi - \lambda\left[\int_{\Omega_2} |\varphi|^{p+1}\right]^{\frac{k+1}{p+1}}.$$

Then, since $u_1 = 0$ outside Ω_1,

$$E(u_1) = \int_{\Omega_1} (-u_1) S_k[u_1] - \lambda \left[\int_{\Omega_1} |u_1|^{p+1} \right]^{\frac{k+1}{p+1}}$$

$$\leq \int_{\Omega_1} (-u_1) S_k[u_1] - \lambda \left[(p+1) \int_{\Omega_1} F(u_1) \right]^{\frac{k+1}{p+1}}$$

$$= J(u_1, \Omega_1) \leq -1,$$

where we have used, by the construction of f, the fact that $F(u) \leq \frac{1}{p+1} |u|^{p+1}$. Let $u_2 = u_{2,m} \in \Phi_0^k(\Omega_2)$ be the solution of

$$S_k[u] = f_m \quad \text{in} \quad \Omega_2,$$

where f_m be a sequence of smooth, positive functions which converges monotone decreasingly to ψ. By the maximum principle we have $\|u_2\|_{L^\infty(\Omega_2)} \leq C$ for some $C > 0$ independent of m. By the comparison principle we have $u_2 < u_1 \leq 0$ in Ω_1.

By our choice of λ and by approximation and uniform boundedness of u_2, we have

$$E(u_2) = \int_{\Omega_2} (-u_2) \psi - \lambda \left[\int_{\Omega_2} |u_2|^{p+1} \right]^{\frac{k+1}{p+1}}$$

$$\geq \int_{\Omega_2} (-u_2) S_k[u_2] - \lambda \left[\int_{\Omega_2} |u_2|^{p+1} \right]^{\frac{k+1}{p+1}} - \frac{1}{8}$$

$$\geq -\frac{1}{8}$$

provided m is sufficiently large.

Denote $\rho(t) = E[u_1 + t(u_2 - u_1)]$. Then $\rho(0) = E(u_1) \leq -1$ and $\rho(1) = E(u_2) \geq -\frac{1}{8}$. We compute

$$\rho'(0) = \int_{\Omega_2} (u_1 - u_2) S_k[u_1] - (k+1)\lambda \left[\int_{\Omega_2} |\varphi|^{p+1} \right]^{\frac{k-p}{p+1}} \int_{\Omega_2} |u_1|^p (u_1 - u_2).$$

Since u_1 is a solution of (5.15), by (5.21) we have

$$\int_{\Omega_2} (u_1 - u_2) S_k[u_1] = \lambda \beta(u_1) \int_{\Omega_1} |u_1|^p (u_1 - u_2)$$

$$= \lambda (1 + o(1)) \left[\int_{\Omega_1} |u_1|^{p+1} \right]^{\frac{k-p}{p+1}} \int_{\Omega_2} |u_1|^p (u_1 - u_2)$$

$$< (k+1)\lambda \left[\int_{\Omega_1} |u_1|^{p+1} \right]^{\frac{k-p}{p+1}} \int_{\Omega_2} |u_1|^p (u_1 - u_2)$$

We obtain $\rho'(0) < 0$. Note that the functional E is linear in the first integral and convex in the second integral, we have $\rho''(t) \leq 0$ for all $t \in (0,1)$. Therefore we must have $\rho(1) < \rho(0)$. We reach a contradiction. Hence (5.18) holds.

Finally we remark that when $k < \frac{n}{2}$ and $p + 1 = k^*$, the best constant in (5.9) is achieved by the function in (5.10). This assertion follows from Step 2 by solving an ode. By the Hölder inequality, one also sees that when $k < \frac{n}{2}$ and $p + 1 < k^*$, the constant in (5.9) depends on the volume $|\Omega|$ but not the diameter of Ω. When $k > \frac{n}{2}$, The above proof implies the embedding $\Phi_0^k(\Omega) \hookrightarrow L^\infty(\Omega)$. Indeed, in Step 2 we have shown that the best constant T_p is achieved by radial functions, and so the assertion follows from Step 1.

Remark 5.1. For any initial function $u_0 \in \Phi_0^k(\Omega)$ satisfying $J(u_0) < 0$, we can modify u_0 slightly near $\varphi\Omega$ such that it satisfies the compatibility condition $S_k[u_0] = \lambda\beta(u_0)f(u_0)$ on $\varphi\Omega \times \{t = 0\}$. Indeed, it suffices to replace u_0 by the solution $\hat{u}_0 \in \Phi_0^k(\Omega)$ of $S_k[u] = g$, where $g(x) = (1 + a)S_k[u_0]$ when $\mathrm{dist}(x, \varphi\Omega) > \delta_1$ and $g(x) = \lambda\beta(u_0)f(u_0)$ when $\mathrm{dist}(x, \varphi\Omega) < \frac{1}{2}\delta_1$. We choose δ_1 a sufficiently small constant and a also small such that $\beta(\hat{u}_0) = \beta(u)$.

5.2 Compactness

In this section we prove the embedding $\Phi_0^k(\Omega) \hookrightarrow L^p(\Omega)$ is compact when $k < \frac{n}{2}$ and $p < k^*$. First we quote a theorem from [TW4]

Theorem 5.2 *Suppose Ω is $(k-1)$-convex. Then*

$$\|u\|_{\Phi_0^l(\Omega)} \leq C\|u\|_{\Phi_0^k(\Omega)} \tag{5.22}$$

for any $1 \leq l < k \leq n$, and any $u \in \Phi_0^k(\Omega)$. The best constant C is achieved by the solution $u \in \Phi_0^k(\Omega)$ to the Hessian quotient equation

$$\frac{S_k[u]}{S_l[u]} = 1 \quad in \ \ \Omega. \tag{5.23}$$

Proof. The proof is based on the global existence of smooth solutions to initial boundary problem of the parabolic equation [TW4]

$$u_t - \log\frac{S_k[u]}{S_l[u]} = 0 \ \ in \ \Omega \times [0,\infty), \tag{5.24}$$

subject to the boundary condition $u = 0$ on $\varphi\Omega \times (0,\infty)$. As above, a solution to (5.24) is k-admissible if for any t, $u(\cdot,t) \in \Phi^k(\Omega)$. The a priori estimation for the parabolic equation (5.24) is very similar to that for the elliptic equation (3.1).

By constructing appropriate super- and sub-barriers, we also infer that for any initial function $u(\cdot, 0)$ satisfying the compatibility condition on $S_k[u] = S_l[u]$ on $\varphi\Omega$, the solution $u(\cdot, t)$ converges to the solution u^* of (5.23). Note that $u(\cdot, t) \in \Phi_0^k(\Omega)$ implies the boundary condition $u = 0$ on $\varphi\Omega \times [0, \infty)$.

With the above results for the parabolic equation (5.24), Theorem 5.2 follows immediately. Indeed, let

$$J(u) = \frac{1}{k+1} \int_\Omega (-u) S_k[u] - \frac{1}{l+1} \int_\Omega (-u) S_l[u].$$

For any $u_0 \in \Phi_0^k(\Omega)$, modify u slightly near $\varphi\Omega$ such that $S_k[u_0] = S_l[u_0]$ on $\varphi\Omega$. Let $u(\cdot, t) \in \Phi_0^k(\Omega)$ be the solution to the parabolic equation (5.24). Then

$$\frac{d}{dt} J(u(\cdot, t)) = -\int_\Omega \{S_k[u] - S_l[u]\} \log \frac{S_k[u]}{S_l[u]} \leq 0.$$

It follows that $J(u^*) \leq J(u_0)$ for any $u_0 \in \Phi_0^k(\Omega)$. Replacing u_0 by $u_0 \|u^*\|_{\Phi_0^k(\Omega)} / \|u_0\|_{\Phi_0^k(\Omega)}$, we obtain Theorem 5.2.

Theorem 5.3 *The embedding $\Phi_0^k(\Omega) \hookrightarrow L^p(\Omega)$ is compact when $k < \frac{n}{2}$ and $p < k^*$.*

By the Hölder inequality, Theorem 5.3 follows from Theorem 5.2 and the compactness of the embedding $W^{1,2}(\Omega) \hookrightarrow L^p(\Omega)$ for $p < \frac{2n}{n-2}$.

Next we show that when $k > \frac{n}{2}$, a k-admissible function is Hölder continuous.

Theorem 5.4 *Suppose $u \in \Phi^k(\Omega) \cap L^\infty(\Omega)$ and $k > \frac{n}{2}$. Then $u \in C_{loc}^\alpha(\Omega)$ with $\alpha = 2 - \frac{n}{k}$, and for any $x, y \in \Omega' \subset\subset \Omega$,*

$$|u(x) - u(y)| \leq C|x - y|^\alpha, \tag{5.25}$$

where C depends only on $n, k, \Omega, \operatorname{dist}(\Omega', \varphi\Omega)$, and $\|u\|_{L^\infty(\Omega)}$.

Proof. Let

$$w(x) = |x|^{2 - n/k}.$$

By direct computation, w is k-admissible and $S_k[w] = 0$ when $x \neq 0$. For any interior point $x_0 \in \Omega$, Applying the comparison principle to u and $\hat{u}(x) = u(x_0) + Cw(x - x_0)$, where C is chosen large such that $\hat{u} > u$ on $\varphi\Omega$, we obtain Theorem 5.4.

5.3 An L^∞ Estimate

As an application of Theorem 5.1, we prove an L^∞ estimate for solutions to the k-Hessian equation. See Theorem 2.1 in [CW1]. The proof below is essentially the same as that in [CW1].

Theorem 5.5 *Let* $u \in C^2(\Omega) \cap C^0(\overline{\Omega})$ *be a* k-*admissible solution of*

$$\begin{cases} S_k(D^2 u) = f(x) & in \ \Omega, \\ u = \varphi & on \ \varphi\Omega. \end{cases} \tag{5.26}$$

Suppose $f \geq 0$, $f \in L^p(\Omega)$, *where* $p > n/2k$ *if* $k \leq \frac{n}{2}$, *or* $p = 1$ *if* $k > \frac{n}{2}$. *Then there exists* $C > 0$ *depending only on* n, k, p, Ω *such that*

$$|\inf_{\Omega} u| \leq |\inf_{\Omega} \varphi| + C\|f\|_{L^p(\Omega)}^{1/k}. \tag{5.27}$$

Proof. By replacing the boundary function φ by inf φ and by the comparison principle, we need only to prove (5.27) for $\varphi \equiv 0$. Since the k-Hessian equation is homogeneous, we may assume that $\|f\|_{L^p(\Omega)} = 1$.

First we prove (5.27) for $k = \frac{n}{2}$. Multiplying (5.26) by $-u$ and taking integration, we obtain,

$$\|u\|_{\Phi_0^k(\Omega)}^{k+1} = |\int_{\Omega} f(x)u(x)dx| \leq \|f\|_{L^p}\|u\|_{L^q}$$
$$\leq |\Omega|^{\frac{1}{q}(1-\frac{1}{\beta})}\|u\|_{q\beta} \leq C|\Omega|^{\frac{1}{q}(1-\frac{1}{\beta})}\|u\|_{\Phi_0^k(\Omega)},$$

where $\frac{1}{p} + \frac{1}{q} = 1$ and $\beta > 1$ will be chosen large. Hence

$$\|u\|_{\Phi_0^k} \leq C|\Omega|^{\frac{1}{qk}(1-\frac{1}{\beta})}.$$

By the Sobolev type inequality (5.11),

$$\|u\|_{L^1(\Omega)} \leq |\Omega|^{1-\frac{1}{\beta}}\|u\|_{L^\beta(\Omega)} \leq C|\Omega|^{1-\frac{1}{\beta}}\|u\|_{\Phi_0^k} \leq C|\Omega|^{1+\delta}, \tag{5.28}$$

where $\delta = \frac{1}{qk} - \frac{1}{\beta}(1 + \frac{1}{qk}) > 0$ provided β is sufficiently large. Hence

$$|\{u(x) < -K\}| \leq \frac{C}{K}|\Omega|^{1+\delta}. \tag{5.29}$$

From Sard's theorem, the level set $\{u(x) < t\}$ has smooth boundary for almost all t. Therefore we may assume all the level sets involved in the proof below have smooth boundary.

Denote $u_1 = u + K$, $\Omega_1 = \{u_1 < 0\}$. When K is large enough, we have $|\Omega_1| \leq \frac{1}{2}|\Omega|$. For $j > 1$ we define inductively u_j and Ω_j by $u_j = u_{j-1} + 2^{-\delta j}$ and $\Omega_j = \{u_j < 0\}$. Then similarly to (5.28) we have

$$\|u_j\|_{L^1(\Omega_j)} \leq C|\Omega_j|^{1+\delta}$$

for some C independent of j. Therefore

$$|\Omega_{j+1}| \leq C2^{\delta(j+1)}|\Omega_j|^{1+\delta},$$

where $\Omega_{j+1} = \{u_j(x) < -2^{-\delta(j+1)}\}$.

Assume by induction that $|\Omega_i| \leq \frac{1}{2}|\Omega_{i-1}|$ for all $i = 1, 2, \cdots, j$, then by (5.29),

$$|\Omega_j|^\delta \leq 2^{-\delta(j-1)}|\Omega_1| \leq \frac{2^{-\delta(j-1)}C}{K}|\Omega|^{1+\delta}.$$

When K is large, we obtain $|\Omega_{j+1}| \leq \frac{1}{2}|\Omega_j|$. Therefore the set $\{x \in \Omega \mid u(x) < -K - \sum_{j=1}^\infty 2^{-\delta j}\}$ has measure zero. In other words, we have

$$|\inf u| \leq K + \sum_{j=1}^\infty 2^{-\delta j}.$$

Therefore (5.27) is established for $k = \frac{n}{2}$.

When $k < \frac{n}{2}$, let $w \in \Phi_0^{n/2}$ be the solution to

$$S_{n/2}[w] = f^{n/2k} \quad \text{in} \quad \Omega.$$

By inequality (vi) in §2, $S_k[w] \geq C_{n,k}S_{n/2}^{2k/n}[w] \geq C_{n,k}f$. Hence by the comparison principle we also obtain (5.27).

When $k > \frac{n}{2}$, multiplying (5.26) by $-u$ and taking integration, we have

$$\|u\|_{\Phi_0^k}^{k+1} = |\int_\Omega f(x)u(x)dx| \leq \|f\|_{L^1}\|u\|_{\Phi_0^k}$$

and (5.27) follows from (5.12). This completes the proof.

We will prove in Section 9 that the solution in Theorem 5.5 is Hölder continuous.

Theorem 5.5 was extended by Kuo and Trudinger to more general elliptic equations [KT]. In their paper [KT], Kuo and Trudinger considered the linear elliptic inequality

$$L[u] = \sum a_{ij}(x)u_{x_ix_j} \leq f \quad \text{in} \quad \Omega. \tag{5.30}$$
$$u \leq 0 \quad \text{on} \quad \varphi\Omega$$

Assume that the eigenvalues $\lambda(\mathcal{A}) \in \Gamma_k^*$, where $\mathcal{A} = -\{a_{ij}(x)\}$, Γ_k^* is the dual cone of Γ_k, given by

$$\Gamma_k^* = \{\lambda \in \mathbb{R}^n \mid \lambda \cdot \mu \geq 0 \ \forall \mu \in \Gamma_k\}$$

Denote

$$\rho_k^*(\mathcal{A}) = \inf\{\lambda \cdot \mu \mid \mu \in \Gamma_k, \sigma_k(\mu) \geq 1\}.$$

They proved the following maximum principle

Theorem 5.6 *Let $u \in C^2(\Omega) \cap C^0(\overline{\Omega})$ be a solution of (5.30). Assume that $\lambda(\mathcal{A}) \in \Gamma_k^*$ and $\rho_k^*(\mathcal{A}) > 0$. Then we have the estimate*

$$\sup_{\Omega} u \leq C \| \frac{f}{\rho_k^*(\mathcal{A})} \|_{L^q(\Omega)},$$

where $q = k$ if $k > \frac{n}{2}$, and $q > \frac{n}{2}$ if $k \leq \frac{n}{2}$, where C depends only on n, k, q, and Ω.

Theorem 5.6 extended the well-known Aleksandrov maximum principle.

6 Variational Problems

Consider the Dirichlet problem

$$\begin{cases} S_k(D^2u) = f(x, u) & \text{in } \Omega, \\ u = 0 & \text{on } \varphi\Omega, \end{cases} \tag{6.1}$$

where $f(x, u) \in C^{1,1}(\overline{\Omega} \times \mathbb{R})$ is a nonnegative function in $\overline{\Omega} \times \mathbb{R}$. There has been a huge amount of works on the existence of positive solutions to semilinear elliptic equations, namely equation (6.1) with $k = 1$. In this section we show that there are similar existence results for the k-Hessian equation. Materials in this section are taken from in [CW1, CW2], except the eigenvalue problem in §6.2, which was previously treated in [W2]. We note that the published paper [CW1] is a part of the preprint [CW2]. The preprint [CW2] also contains the existence of solutions in the critical growth case, presented in §6.4 below.

As shown in §5, a solution of (6.1) is a critical point of the Hessian functional

$$J(u) = \frac{-1}{k+1} \int_{\Omega} u S_k[u] - \int_{\Omega} F(x, u), \tag{6.2}$$

where $F(x, u) = \int_u^0 f(x, t)dt$. The functional J is defined on the convex cone $\Phi_0^k(\Omega)$. We don't know the behavior of the functional near the boundary of $\Phi_0^k(\Omega)$, and so we cannot use the variational theory directly. To find a critical point of J, we employ a descent gradient flow of the functional, which was previously used by Chou [Ch1] for the Monge-Ampère equation. That is a parabolic Hessian equation of the form

$$\mu(S_k[u]) - u_t = \mu(f(u)). \tag{6.3}$$

We assume that μ is a smooth function defined on $(0, \infty)$, satisfying $\mu'(t) > 0$, $\mu''(t) < 0$ for all $t > 0$,

$$\mu(t) \to -\infty \quad \text{as } t \to 0,$$
$$\mu(t) \to +\infty \quad \text{as } t \to +\infty,$$

and such that $\mu(S_k[u])$ is concave in D^2u. As we consider solutions in Φ_0^k, the boundary condition for (6.3) is

$$u = 0 \quad \text{on} \quad \varphi\Omega \times [0, \infty).$$

Let $u \in \Phi_0^k(\Omega \times \mathbb{R}^+)$ be a k-admissible solution to (6.3). Then

$$\frac{d}{dt}J(u(\cdot, t)) = -\int_\Omega (S_k(\lambda) - f)u_t \tag{6.4}$$

$$= -\int_\Omega (S_k(\lambda) - f)(\mu(S_k(\lambda)) - \mu(f)) \leq 0.$$

As before, we say a solution u is k-admissible with respect to (6.3) if for any $t \geq 0$, $u(\cdot, t)$ is k-admissible. To simplify the notation, we will denote $u \in \Phi_0^k(\Omega \times \mathbb{R}^+)$ if $u(\cdot, t) \in \Phi_0^k(\Omega)$ for all $t \in \mathbb{R}^+ = [0, \infty)$.

A typical example of μ is $\mu(t) = \log t$ [Ch1]. But for the k-Hessian equation we have to choose a different μ in our treatment below. For the a priori estimates for the parabolic equation (6.3), we always need to assume that f is strictly positive. But in studying the variational problem (6.1), typically f vanishes when $u = 0$. To avoid such situation, we add a small positive constant to f, or modify f slightly near $u = 0$.

To study the variational problem associated with the k-Hessian equation, similar to the Laplace equation, we divide the problem into three cases, namely the sublinear case, the eigenvalue problem, and the superlinear case.

6.1 The Sublinear Growth Case

We say $f(x, u)$ is sublinear with respect to the k-Hessian operator if

$$\lim_{u \to -\infty} |u|^{-k} f(x, u) \to 0 \tag{6.5}$$

uniformly for $x \in \Omega$. Note that the power k is due to that the k-Hessian operator is homogeneous of degree k.

Theorem 6.1 *Let Ω be $(k-1)$-convex with $C^{3,1}$-boundary. Suppose $f(x, u) \in C^{1,1}(\overline{\Omega} \times \mathbb{R})$, $f(x, u) > 0$ when $u < 0$, f satisfies (6.5), and $\inf_{\Phi_0^k(\Omega)} J(u) < 0$. Then there is a nonzero solution $u \in C^{0,1}(\overline{\Omega}) \cap C^{3,\alpha}(\Omega)$ to (6.1), which is the minimizer of the functional J.*

Proof. We sketch the proof, as it was essentially included in the proof of Theorem 5.1.

Replace f by $f + \delta$ for some small $\delta > 0$, so that f is strictly positive. Observe that in the sublinear growth case, by the Sobolev type inequality (Theorem 5.1), we have $J(tu) \to +\infty$ as $t \to \infty$ for any $u \in \Phi_0^k(\Omega)$, $u \neq 0$, and J is bounded from below. As the infimum of J is negative, one can choose an initial function $u_0 \in \Phi_0^k(\Omega)$ such that $J(u_0) < \inf_{u \in \Phi_0^k(\Omega)} J(u) + \delta$. By Remark 5.1, we may assume the compatibility condition $S_k[u_0] = f$ on $\varphi\Omega$ at $t = 0$ is satisfied. In the sublinear growth case, one can construct a sub-barrier $\underline{u}$ to the parabolic equation (6.3) such that $\underline{u} \leq u_0$. By the comparison principle one gets a global uniform estimate, and also derivative estimates up to the third order, for solutions to (6.3). Therefore there is a global smooth solution to (6.3). By (6.4), the solution sub-converges to a nonpositive solution $u = u_\delta$ of (6.1).

We claim that all the solutions u_δ are uniformly bounded for $\delta > 0$ small. Indeed, if $m_\delta = -\inf u_\delta \to \infty$, the function $v_\delta = u_\delta/m_\delta$ satisfies the equation

$$S_k[v] = m_\delta^{-k}[f(x, m_\delta v_\delta) + \delta].$$

By (6.5), the right hand side converges to zero uniformly. Hence by the comparison principle, one infers that $\inf v_\delta \to 0$, which contradicts with the fact that $\inf v_\delta = -1$. Next by the assumption that $\inf J < 0$, it is easily seen that u_δ does not converge to zero.

Sending $\delta \to 0$, by the interior a priori estimates in §4, we conclude that u_δ converges to a solution $u \in C^{0,1}(\overline{\Omega}) \cap C^{3,\alpha}(\Omega)$ of (6.1) which is a minimizer of J.

A particular case in Theorem 6.1 is when $f \in C^{1,1}(\overline{\Omega})$ is a function of x, independent of u [B]. Then for any initial u_0 satisfying the compatibility condition, the solution $u \in \Phi_0^k(\Omega \times \mathbb{R}^+)$ of (6.3) is uniformly bounded and converges to a solution u^* of (6.1). By the convexity of the functional J (see (5.7)) and the uniqueness of solutions to the Dirichlet problem, u^* is a minimizer of the functional J in $\Phi_0^k(\Omega)$.

6.2 The Eigenvalue Problem

Similar to the Laplace operator, the k-Hessian operator admits a positive eigenvalue λ_1 such that

$$S_k[u] = \lambda|u|^k \quad \text{in} \ \ \Omega, \tag{6.6}$$
$$u = 0 \quad \text{on} \ \ \varphi\Omega,$$

has a nonpositive k-admissible solution when $\lambda = \lambda_1$. The following theorem was proved in [W2] for $k < n$ and in [Lp] for $k = n$. Here we provide a proof which uses Theorem 6.1.

Theorem 6.2 *Let Ω be $(k-1)$-convex with $C^{3,1}$-boundary. Then there exists $\lambda_1 > 0$ depending only on n, k, Ω, such that*

  (i) (6.6) has a nonzero k-admissible solution $\varphi_1 \in C^{1,1}(\overline{\Omega}) \cap C^{3,\alpha}(\Omega)$ when $\lambda = \lambda_1$.

  (ii) If $(\lambda^*, \varphi^*) \in [0, \infty) \times (C^{1,1}(\overline{\Omega}) \cap C^{3,\alpha}(\Omega))$ is another solution to (6.6), then $\lambda^* = \lambda_1$ and $\varphi^* = c\varphi_1$ for some positive constant c.

(iii) If $\Omega_1 \subset \Omega_2$, then $\lambda_1(\Omega_1) \geq \lambda_1(\Omega_2)$.

Proof. First consider part (i). Let $p \in (k - \frac{1}{2}, k)$ and let $c_0 > 0$ be a large constant. By Theorem 6.1, there is an admissible solution $u_p \in \Phi_0^k(\Omega)$ to the problem

$$S_k[u] = c_0|u|^p,$$

which is a minimizer of the associated functional. Namely $J(u_p) = \inf J(u)$, where

$$J(u) = \frac{1}{k+1} \int_\Omega (-u) S_k[u] - \frac{c_0}{p+1} \int_\Omega |u|^{p+1}.$$

Let $v_p = u_p/m_p$, where $m_p = \sup |u_p|$. Then v_p satisfies

$$S_k[v] = c_0 m_p^{p-k} |v|^p.$$

If $m_p^{p-k} \to 0$ as $p \to k$, the right hand side converges to zero uniformly, which contradicts with the fact that $\inf v_p = -1$. If $m_p^{p-k} \to \infty$, then $m_p \to 0$ uniformly, which implies $J(u_p) = \inf J(u) \to 0$. But if we choose $c_0 > 0$ large, $J(u_p) = \inf J(u) \to -\infty$ as $p \to k$. The contradiction implies that m_p is uniformly bounded. Hence by the a priori estimates in §4, we see that $(c_0 m_p^{p-k}, v_p)$ sub-converges to (λ_1, φ_1), and (λ_1, φ_1) is a solution of (6.6). By Theorem 4.4, $\varphi_1 \in C^{1,1}(\overline{\Omega}) \cap C^\infty(\Omega)$.

Next we consider (ii). If (λ^*, φ^*) is also a solution of (6.6), we may assume that $\lambda^* > \lambda_1$ and $\varphi^* < \varphi_1$ by multiplying a constant to φ^*. Denote $a_{ij} = \frac{\varphi}{\varphi u_{ij}} [S_k[u]]^{1/k}$ at $u = \varphi_1$. Then λ_1 and φ_1 are respectively the eigenvalue and eigenfunction of the elliptic operator $L = \sum a_{ij} \partial^2_{ij}$. By the concavity of $S_k^{1/k}[u]$, and noting that ψ^* and ψ_1 are negative in Ω, we deduce that

$$L(\varphi^* - \varphi_1) \geq S_k^{1/k}[\varphi^*] - S_k^{1/k}[\varphi_1] = -(\lambda^*)^{1/k}\varphi^* + \lambda_1^{1/k}\varphi_1 > \lambda_1^{1/k}(\varphi^* - \varphi_1)$$

in Ω, which contradicts the fact that $\lambda_1^{1/k}$ is the first eigenvalue of L. Hence we have $\lambda^* = \lambda_1$ and $\varphi_1 = \varphi^*$.

Part (iii) was proved in Step 3 of the proof of Theorem 5.1.

6.3 Superlinear Growth Case

We say f is superlinear with respect to the k-Hessian operator if

$$\lim_{u \to -\infty} |u|^{-k} f(x, u) \to \infty \qquad (6.7)$$

uniformly for $x \in \Omega$.

Theorem 6.3 *Suppose that $f(x, z) > 0$ for $z < 0$,*

$$\lim_{z \to 0^-} f(x, z)/|z|^k < \lambda_1, \qquad (6.8)$$

$$\lim_{z \to -\infty} f(x, z)/|z|^k > \lambda_1, \qquad (6.9)$$

where λ_1 is the eigenvalue of the k-Hessian operator. Suppose there exist constants $\theta > 0$ and M large such that

$$\int_z^0 f(x, s)\, ds \le \frac{1 - \theta}{k + 1} |z| f(x, z) \quad \forall\, z < -M. \qquad (6.10)$$

When $k \le \frac{n}{2}$, we also assume that there exists $p \in (1, k^* - 1)$ such that

$$\lim_{z \to -\infty} f(x, z)/|z|^p = 0. \qquad (6.11)$$

Then (6.1) has a non-zero k-admissible solution in $C^{3,\alpha}(\Omega) \cap C^{0,1}(\overline{\Omega})$, $\alpha \in (0, 1)$.

When $k = 1$, Theorem 6.3 is a typical result in semilinear elliptic equation. The solution in Theorem 6.3 is a min-max critical point of the functional J. As indicated before, we cannot use the variational theory directly, but by studying a descent gradient flow, we can use the underlying idea in the Mountain Pass Lemma. The main difficulty is to prevent blowup of solutions near the boundary for both the elliptic equation (6.1) and the parabolic equation (6.3), in the case $2 \le k < \frac{n}{2}$. It requires some new techniques.

Proof. For clarity we divide the proof into four steps.

Step 1. Let $f_{\delta, K}$ be a smooth, positive function given by

$$f_{\delta, K}(x, u) = \delta + \eta_{\delta_1} \hat{f}_{\delta, K}(x, u),$$

where $\eta_{\delta_1} \in C_0^\infty(\Omega)$ is a nonnegative function satisfying $\eta_{\delta_1}(x) = 1$ when $\operatorname{dist}(x, \varphi\Omega) > 2\delta_1$ and $\eta_{\delta_1}(x) = 0$ when $\operatorname{dist}(x, \varphi\Omega) < \delta_1$, and

$$\hat{f}_{\delta, K}(x, u) = \begin{cases} \delta & \text{if } |u| < \frac{1}{2}\delta, \\ f(x, u) & \text{if } \delta < |u| < K, \\ |u|^p & \text{if } |u| > 2K. \end{cases}$$

We will choose the constants $\delta, \delta_1 > 0$ small and $K > 1$ large.

Remark. Before continuing, let us explain why we make these modifications when $k \geq 2$, which are not needed when considering semilinear elliptic equations (the case $k = 1$). The introduction of δ is such that f is positive, so that we can apply the C^3 a priori estimates for the parabolic Hessian equation (6.3). We modify f for large $|u|$ (namely $f = |u|^p$ when $|u| > 2K$) is to use the gradient estimate for the parabolic Hessian equation. The purpose of introducing η_{δ_1} is to prevent the solution to the parabolic Hessian equation blow-up near the boundary. In the following we choose $\delta_1 = \delta$.

Consider the functional

$$J_{\delta,K}(x,u) = \frac{1}{k+1} \int_\Omega (-u)S_k[u] - \int_\Omega F_{\delta,K}(x,u), \qquad (6.12)$$

where $F_{\delta,K}(x,u) = \int_u^0 f_{\delta,K}(x,t)$. When $\delta > 0$ is sufficiently small, by assumption (6.8), there exists a smooth, k-admissible function $u_0 \in \Phi_0^k(\Omega)$ with small L^∞-norm, such that

$$S_k[u_0] > f_{\delta,K}(x,u_0). \qquad (6.13)$$

Consider the parabolic Hessian equation (6.3) with initial condition $u(\cdot,0) = su_0$, where $s > 0$ is a parameter. We choose the function μ in (6.3) such that

$$\mu(t) = \log t \quad \text{if } t < 1/8,$$
$$\mu(t) = t^{1/p} \quad \text{if } t > 8, \qquad (6.14)$$

and

$$(t-s)(\mu(t)-\mu(s)) \geq C(t-s)(t^{1/p}-s^{1/p}) \qquad (6.15)$$

for all $t, s > 0$, where C is an absolute constant, independent of s,t. Then equation (6.3) has a unique smooth solution u_s. By (6.11) and (6.14), $\mu(f(u))$ is of linear growth in u. Hence the solution exists for all time t.

Since $S_k[u_0] > 0$, u_0 is a sub-barrier for the solution u_s for small $s > 0$. That is when $s > 0$ is small, one has $0 > u_s(\cdot,t) > u_0$ for all t. Hence $J_{\delta,K}(x,u_s)$ is uniformly bounded,

$$J_{\delta,K}(u_s(\cdot,t)) > -\int_\Omega F_{\delta,K}(u_s(x,t))dx \geq -\frac{1}{2}$$

for all t, provided $\delta > 0$ is small.

On the other hand, when $s > 1$ is large, we have $J_{\delta,K}(su_0) < -1$. Hence $J_{\delta,K}(u_s(\cdot,t)) < -1$ for all $t > 0$ as (6.3) is a descent gradient flow. Let

$$s^* = \inf\{\bar{s} \mid \lim_{t\to\infty} J_{\delta,K}(u_s(\cdot,t)) < -1 \ \forall s > \bar{s}\}. \qquad (6.16)$$

Then s^* is positive. By the continuous dependence of the solution u_s in s, and the monotonicity (6.4), we see that $J_{\delta,K}(u_{s^*}(\cdot,t)) \geq -1$ for all time t. We also have

$$\sup |u_{s^*}(\cdot, t)| \geq C > 0 \qquad (6.17)$$

for some $C > 0$ independent of t. Indeed, if $\sup |u_{s^*}(\cdot, t)|$ is small at some time $\bar{t}$, then $\sup |u_s(\cdot, t)|$ is also small at $\bar{t}$ for $s > s^*$, close to s^*. Hence by (6.13), u_0 is a sub-barrier, and so by the comparison principle, $\sup |u_s(\cdot, t)|$ is small for all $t > \bar{t}$, which contradicts with the definition of (6.16).

Suppose for a moment that

$$|u_{s^*}(\cdot, t)| \leq C \qquad (6.18)$$

uniformly for $t \in (0, \infty)$. The constant C is allowed to depend on δ and K. Then by the global regularity of the parabolic Hessian equation, we conclude that $u_{s^*}(\cdot, t)$ sub-converges to a solution $u_{\delta,K}^*$ to the equation

$$S_k[u] = f_{\delta,K}(x, u). \qquad (6.19)$$

In Step 4 we show that $u_{\delta,K}^*$ is uniformly bounded in δ and K, and so it sub-converges as $\delta \to 0, K \to \infty$ to a solution $u \in C^{0,1}(\overline{\Omega}) \cap C^3(\Omega)$ in $\Phi_0^k(\Omega)$ of (6.1). From (6.17), $u \neq 0$.

Step 2. In the following we prove (6.18). For brevity we will write u_{s^*} as u, dropping the subscript s^*. Recall that $J_{\delta,K}(u(\cdot, t)) \geq -1$ for all time t. Hence the set

$$K^0 = \{t \in (0, \infty) \mid \frac{d}{dt} J_{\delta,K}(u(\cdot, t)) < -\sigma\} \qquad (6.20)$$

has finite measure, where $\sigma > 0$ is a small constant. For any $t \notin K^0$, first we show that

$$\int_\Omega (-u(\cdot, t)) S_k[u(\cdot, t)] \leq C, \qquad (6.21)$$

$$\int_\Omega F_{\delta,K}(x, u(\cdot, t)) \leq C, \qquad (6.22)$$

where C is a constant independent of t, δ and K. Indeed, if $t \notin K^0$, we have

$$\int_\Omega \{S_k[u] - f_{\delta,K}(x, u)\}\{\mu(S_k[u]) - \mu(f_{\delta,K}(x, u))\}$$
$$= \int_\Omega \partial_t u \{S_k[u] - f_{\delta,K}(x, u)\} = -\frac{d}{dt} J_{\delta,K}(u(\cdot, t)) \leq \sigma.$$

Hence by (6.15),

$$\int_\Omega \{S_k[u] - f_{\delta,K}(x, u)\}\{(S_k[u])^{1/p} - (f_{\delta,K}(x, u))^{1/p}\} \leq C\sigma.$$

Denote $\alpha = \left(S_k[u]\right)^{1/p}$, $\beta = \left(f_{\delta,K}(x,u)\right)^{1/p}$. We obtain

$$\int_\Omega |\alpha - \beta|^{p+1} \le C \int_\Omega (\alpha^p - \beta^p)(\alpha - \beta) \le C\sigma.$$

We have

$$\left| \int_\Omega u(\alpha^p - \beta^p) \right| \le C \int_\Omega |u|\,|\alpha - \beta|\,(\alpha^{p-1} + \beta^{p-1})dx$$

$$\le C\left[\int_\Omega |\alpha - \beta|^{p+1}\right]^{\frac{1}{p+1}} \left[\int_\Omega |u|^{p+1}\right]^{\frac{1}{p(p+1)}} \left[\int_\Omega |u|\,|\alpha^p + \beta^p|\right]^{\frac{p-1}{p}}$$

$$\le C\sigma^{1/(p+1)} \|u\|_{L^{p+1}}^{1/p} \left[\int_\Omega |u|\,|\alpha^p + \beta^p|\right]^{\frac{p-1}{p}}.$$

On the other hand,

$$J_{\delta,K}(s^* u_0) = J_{\delta,K}(u(\cdot,0)) \ge J_{\delta,K}(u(\cdot,t)) \qquad (6.23)$$

$$= \frac{1}{k+1}\int_\Omega (-u)S_k[u] - \int_\Omega F_{\delta,K}(x,u) \ge -1.$$

By (6.10),

$$F_{\delta,K}(x,u) \le \delta|u| + \frac{1-\theta}{k+1}|u|f_{\delta,K}(x,u) + C.$$

Hence

$$J_{\delta,K}(u(\cdot,t)) = \frac{1}{k+1}\int_\Omega (-u)S_k[u] - \int_\Omega F_{\delta,K}(x,u)$$

$$\ge \frac{1}{k+1}\int_\Omega (-u)S_k[u] - \int_\Omega \left[\delta|u| + \frac{1-\theta}{k+1}|u|f_{\delta,K}(x,u) + C\right]$$

$$= \frac{1}{k+1}\int_\Omega (-u)(\alpha^p - \beta^p) + \frac{\theta}{k+1}\int_\Omega |u|f_{\delta,K}(x,u) - \int_\Omega (C+\delta|u|).$$

It follows that, by the Sobolev inequality (Theorem 5.1),

$$\int_\Omega |u|f_{\delta,K}(x,u) \le C\int_\Omega |u|\,|\alpha^p - \beta^p| + J_{\delta,K}(s^* u_0) + \int_\Omega (C + \delta|u|)$$

$$\le C\sigma^{1/(p+1)}\|u\|_{L^{p+1}}^{1/p}\left[\int_\Omega |u|\,|\alpha|^p + \int_\Omega |u|\,|\beta^p|\right]^{\frac{p-1}{p}} + \int_\Omega (C+\delta|u|)$$

$$\le C\sigma^{1/(p+1)}\|u\|_{\Phi_0^k(\Omega)}^{1/p}\left[\|u\|_{\Phi_0^k(\Omega)}^{k+1} + \int_\Omega |u|\,|\beta^p|\right]^{\frac{p-1}{p}} + \int_\Omega (C+\delta|u|).$$

By the Sobolev inequality again, $\|u\|_{L^1} \leq C\|u\|_{\Phi_0^k}$. We obtain

$$\int_\Omega |u| f_{\delta,K}(x,u) \leq C_1 \varepsilon_{\sigma,\delta} \int_\Omega |u| |\alpha|^p + C_2$$

with $\varepsilon_{\sigma,\delta} \to 0$ as $\sigma, \delta \to 0$. Inserting the estimate into (6.23) we obtain (6.21) and (6.22).

Step 3. Now we use (6.21) and (6.22) to establish (6.18). If $k > \frac{n}{2}$, in view of (5.12), (6.18) follows readily from (6.21). We need only to consider the cases $k \leq \frac{n}{2}$. Let $M_t = \sup_{\Omega_\delta} |u(\cdot,t)|$, $\widetilde{M}_t = \sup_\Omega |u(\cdot,t)|$. If M_t is not uniformly bounded, there exists a sequence $t_j \to \infty$ such that $M_{t_j} \to \infty$ and

$$M_t \leq M_{t_j} \text{ for all } t < t_j. \tag{6.24}$$

Let

$$w(x) = \frac{-d_x + K d_x^2}{-\delta + K \delta^2} M_{t_j},$$

where $d_x = \text{dist}(x, \varphi\Omega)$. We choose δ small and $K \in (1, \delta^{-1/2})$ large such that $S_k[w] > \delta$ in $\Omega - \Omega_\delta$, where $\Omega_\delta = \{x \in \Omega \mid d_x > \delta\}$. Then $w = M_{t_j}$ on $\varphi\Omega_\delta$. Recall that $f_{\delta,K} = \delta$ in $\Omega - \Omega_\delta$. By the comparison principle we have $u(x,t) \geq w(x)$ for any $x \in \Omega - \Omega_\delta$, $t \in (0, t_j)$. It follows that $\widetilde{M}_t \leq M_{t_j}$ for $t \in (0, t_j)$. By (6.11) and (6.14), the right hand side of the parabolic equation (6.3) is of linear growth. Hence we have

$$M_t \geq M_{t_j} e^{C(t_j - t)} \ \forall \ t < t_j.$$

Hence $M_t \geq C M_{t_j}$ for $t \in (t_j - 2, t_j)$. Since the set K^0 has finite measure, we may assume that $t_j \notin K^0$ for all j and $M_t \leq C M_{t_j}$ for all $t < t_j$.

Suppose the maximum M_{t_j} of $|u(\cdot, t_j)|$ is attained at the point y_j. By the interior gradient estimate (10.10) below, we have

$$u(x, t_j) \leq -\frac{1}{2} M_{t_j} \ \forall \ x \in B_r(y_j),$$

where $r = c_1 M_{t_j}^\beta$ and $c_1 > 0$ is independent of j, and

$$\beta = 1 - \frac{p+k}{2k} = \frac{k-p}{2k}.$$

By (6.21) (where the constant C is independent of δ, K) and the Sobolev inequality (5.9) and (5.11), we have

$$\|u(\cdot, t_j)\|_{L^q(B_r(y_j))} \leq \|u(\cdot, t_j)\|_{L^q(\Omega)} \leq C\|u\|_{\Phi_0^k(\Omega)} \leq C,$$

where $q = k^*$ if $k < \frac{n}{2}$ and $q > p + 1$ is any sufficiently large constant if $k = \frac{n}{2}$. On the other hand, we have

$$\|u(\cdot, t_j)\|_{L^q(B_r(y_j))} \geq C r^n M_{t_j}^q \geq C M_{t_j}^{q+b\beta}.$$

Since $p < k^* - 1$, we have $q + b\beta > 0$. Hence when M_{t_j} is sufficiently large, we reach a contradiction. Therefore (6.18) is proved.

  Step 4. We have therefore obtained a solution $u_{\delta,M}$ to (6.19) which satisfies (6.21) and (6.22), with the constant C in (6.21) and (6.22) independent of δ, K. If $k > \frac{n}{2}$, by (5.12) we obtain

$$\sup_{\Omega} |u_{\delta,K}| \leq C \qquad (6.25)$$

for a different C independent of δ, K. Sending $\delta \to 0$ and $K \to \infty$, we obtain a solution $u \in C^{0,1}(\overline{\Omega}) \cap C^3(\Omega)$ in $\Phi_0^k(\Omega)$ of (6.1).

  If $k < \frac{n}{2}$, denote $\psi = f_{\delta,K}(c, u_{\delta,K})$. By (6.21) and (6.11), and the Sobolev inequality, we have $\psi \in L^\beta(\Omega)$ for some $\beta > \frac{n}{2k}$. Hence applying Theorem 5.5 to equation $S_k[u] = \psi$ in Ω we obtain again (6.25). In all the cases, we obtain a solution to (6.1).

Remarks
(i) In Step 4 above, if $k = \frac{n}{2}$, by the Sobolev embedding (5.11), the right hand side of (6.19) belongs to $L^\beta(\Omega)$ for any $\beta > 1$. Write equation (6.19) in the form

$$\sum_{i,j} a_{ij} u_{ij} = [S_k[u]]^{1/k} = f_{\delta,K}^{1/k}. \qquad (6.26)$$

where $a_{ij} = \frac{1}{k}[S_k[u]]^{1/k-1} S_k^{ij}[u]$. By inequality (x) in Section 2, the determinant $|a_{ij}| \geq C_{n,k} > 0$. Hence by Aleksandrov's maximum principle,

$$\sup_{\Omega} |u_{\delta,K}| \leq C \int \frac{1}{|a_{ij}|} f_{\delta,K}^{n/k} \, dx \leq C$$

for some C independent of δ, K. We also obtain (6.25).

  (ii) If f is independent of x and the domain Ω is convex, by the method of moving planes, the maximum point of $u_{\delta,K}$ will stay away from the boundary. Hence we can obtain (6.25) by a usual blow-up argument. We don't need to use Theorem 5.5.

  (iii) Let u_1 be a k-admissible function with small L^∞ norm, u_2 be a k-admissible function such that $J(u_2) < -1$, where J is the functional in (6.2). Let Γ denote the set of paths in $\Phi_0^k(\Omega)$ connecting u_1 to u_2. Let

$$c_0 = \inf_{\gamma \in \Gamma} \sup_{s \in (0,1)} J(\gamma(s)). \qquad (6.27)$$

Then the assumptions (6.8)–(6.11) and the Sobolev inequality (Theorem 5.1) implies that $c_0 > 0$. The above proof implies that there is a solution $u \in \Phi_0^k(\Omega)$ to (6.1) such that $J(u) = c_0$.

6.4 The Critical Growth Case

In this section we extend Theorem 6.3 to the critical growth case. Consider the problem

$$\begin{cases} S_k(D^2u) = |u|^{k^*-1} + f(x,u) & \text{in } \Omega, \\ u = 0 & \text{on } \varphi\Omega, \end{cases} \tag{6.28}$$

where $1 < k < n/2$ and f is a lower order term of $|u|^{k^*-1}$. For simplicity we will consider the case

$$f(x,u) = \lambda|u|^q, \tag{6.29}$$

where $q \in (k, k^* - 1), \lambda > 0$. Denote

$$J(u) = \frac{1}{k+1}\int_\Omega (-u)S_k[u] - \frac{1}{k^*}\int_\Omega |u|^{k^*} - \frac{\lambda}{q+1}\int_\Omega |u|^{q+1}dx.$$

$$c_0 = \inf_{u\in\Phi_0^k(\Omega)} \sup_{s>0} J(su). \tag{6.30}$$

By the Sobolev inequality (5.9), we have $c_0 > 0$.

We also denote

$$J^*(u) = \frac{1}{k+1}\int_\Omega (-u)S_k[u] - \frac{1}{k^*}\int_\Omega |u|^{k^*}.$$

$$c^* = \inf_{\Phi_0^k(\Omega)} \sup_{t>0} J^*(tu). \tag{6.31}$$

The following theorem extends the existence of positive solutions to semilinear elliptic equations in [BN] to the k-Hessian equation. Our proof is completely different, due to the lack of a gradient estimate near the boundary for equation (6.40).

Theorem 6.4 *Suppose*

$$c_0 < c^*. \tag{6.32}$$

Then (6.28) has a non-zero k-admissible solution.

Proof. For any $p \in (q, k^* - 1)$, by Theorem 6.3, there exists a solution $u_p \in C^3(\Omega) \cap C^{0,1}(\overline{\Omega})$ of

$$\begin{cases} S_k(D^2u) = \psi_p(x,u) =: |u|^p + \lambda|u|^q & \text{in } \Omega \\ u = 0 & \text{on } \varphi\Omega \end{cases} \tag{6.33}$$

with

$$J_p(u_p) = c_p,$$

where

$$J_p(u) = \frac{-1}{k+1}\int_\Omega uS_k(D^2u)dx - \int_\Omega \Psi_p(x,u)dx,$$

$\Psi_p(x,u) = \int_u^0 \psi_p(x,t)dt$, and

$$c_p = \inf_{u \in \Phi_0^k(\Omega)} \sup_{s>0} J_p(su) > 0.$$

From equation (6.33) we have

$$\int_\Omega u_p S_k(D^2 u_p) dx = \int_\Omega u_p \psi_p(x, u_p) dx,$$

which, together with $J_p(u_p) = c_p$, implies that

$$\|u_p\|_{\Phi_0^k} \le C \quad \text{and} \quad \|u_p\|_{L^{p+1}(\Omega)} \le C. \tag{6.34}$$

We want to prove that

$$M_p = \sup_{x \in \Omega} |u_p(x)|$$

is uniformly bounded for $p < k^* - 1$ and close to $k^* - 1$. If this is true then by the regularity in §4, there exists a subsequence of $u_p(x)$ which converges to a solution u_0 of (6.28). Moreover, one can prove

$$J(u_0) = \lim_{p \to k^*-1} J_p(u_p) = \lim_{p \to k^*-1} c_p > 0.$$

Hence $u_0 < 0$ in Ω.

Suppose to the contrary that there is a subsequence p_j so that $M_j =: M_{p_j} \to \infty$ as $p_j \to k^* - 1$. Suppose the supremum M_j is attained at x_j. Let

$$v_j(y) = M_j^{-1} u(R_j^{-1} y + x_j), \quad y \in \Omega_j,$$

where $R_j = M_j^{(p_j-k)/2k}$, $\Omega_j = \{y \mid R_j^{-1} y + x_j \in \Omega\}$. Then $v_j(0) = -1$, $-1 \le v_j \le 0$ for $y \in \Omega_j$, and v_j satisfies

$$S_k(D_y^2 v) = \widetilde{\psi}_j(y), \tag{6.35}$$

where

$$\widetilde{\psi}_j(y) = |v_j|^{p_j} + \lambda M_j^{q-p_j} |v_j|^q.$$

Moreover,

$$\int_\Omega |u_j|^{p_j+1} dx = M_j^{\delta_j} \int_{\Omega_j} |v_j|^{p_j+1} dy, \tag{6.36}$$

$$\int_\Omega |u_j| S_k(D^2 u_j) dx = M_j^{\delta_j} \int_{\Omega_j} |v_j| S_k(D^2 v_j) dy,$$

where $\delta_j = p_j + 1 - \frac{n}{2k}(p_j - k) \ge 0$. Hence $\|v_j\|_{L^{p_j+1}}$ and $\|v_j\|_{\Phi_0^k(\Omega_j)}$ are uniformly bounded.

By passing to a subsequence we assume that $x_j \to x_\infty \in \overline{\Omega}$. Denote $d_j = \text{dist}(x_j, \varphi\Omega)$. If

$$d_j R_j \to \infty, \tag{6.37}$$

then for any $R > 0$, $B_R(0) \subset \Omega_j$ provided j is large enough. By the interior gradient estimate (Theorem 4.1) and the interior second derivative estimate (Theorem 4.2), we may suppose, by passing to a subsequence if necessary,

$$v_j(y) \to v_\infty \text{ in } C^2_{loc}(\mathbb{R}^n),$$

and v_∞ satisfies the equation

$$S_k(D^2 v) = |v|^{k^*-1} \text{ in } \mathbb{R}^n.$$

Note that to apply Theorem 4.2, we may choose the function w in (4.11) as $w = \frac{\varepsilon}{R_1^2}(|x|^2 - R_1^2)$ for large R_1. Hence

$$-\int_{\mathbb{R}^n} v_\infty S_k(D^2 v_\infty)dx = \int_{\mathbb{R}^n} |v_\infty|^{k^*} dx,$$

and so

$$J^*(v_\infty) = \sup_{t>0} J^*(tv_\infty) \geq c^*. \tag{6.38}$$

On the other hand, let $\widetilde{\Psi}_j(y,u) = \int_u^0 \widetilde{\psi}_j(y,t)dt$. We have

$$\widetilde{\Psi}_j(y,u) \to \frac{1}{k^*}|v_\infty(y)|^{k^*} \text{ in } L^\infty_{loc}(\mathbb{R}^n)$$

as $j \to \infty$. Note that by equation (6.35),

$$\frac{-1}{k+1}v_j S_k(D^2 v_j) - \widetilde{\Psi}_j(y,v_j) \geq 0.$$

By Fatou's lemma we obtain

$$\begin{aligned}
J^*(v_\infty) &\leq \underline{\lim}_{j\to\infty} \int_{\Omega_j} \left[\frac{-1}{k+1}v_j S_k(D^2 v_j) - \widetilde{\Psi}_j(y,v_j)\right]dx \\
&= \underline{\lim}_{j\to\infty} M_j^{-\delta_j} \int_\Omega \left[\frac{-1}{k+1}u_{p_j} S_k^2(D^2 u_{p_j}) - \Psi_{p_j}(x,u_{p_j})\right]dx \\
&\leq \underline{\lim}_{j\to\infty} c_{p_j} \leq c_0 < c^*,
\end{aligned}$$

which contradicts with (6.38). Hence M_p is uniformly bounded.

Next we consider the case that $d_j R_j$ is uniformly bounded. We may suppose

$$d_j R_j \to \alpha \geq 0. \tag{6.39}$$

By the interior gradient estimate (Theorem 4.1), v_j converges locally uniformly to a function v_∞, and v_∞ satisfies

$$S_k(D^2 v_\infty) = |v_\infty|^{k^*-1} \text{ in } \Omega_\infty \tag{6.40}$$

in a weak or the viscosity sense (see §9 for definition of the weak solution), where by a rotation of axes, $\Omega_\infty = \{y_n > -\alpha\}$. Moreover we have $-1 \le v_\infty \le 0$ in Ω_∞.

We note that the argument in Case 1 doesn't work at the current situation, as we don't have uniform gradient estimate near the boundary $\varphi\Omega_\infty$, we don't know whether $v_\infty = 0$ on $\varphi\Omega_\infty$.

Let $D_j = \{y \in \mathbb{R}^n \mid v_j(y) \le -\frac{1}{2}\}$. By (6.34) and (6.36) we have $\|v_j\|_{L^{p_j}} \le C$ and so $\mathrm{mes}(D_j) \le C$. Applying Theorem 5.5 to the equation (6.35) and noticing that $\inf v_j = -1$, we have

$$\mathrm{mes}(D_j) \ge C_1 \tag{6.41}$$

for some $C_1 > 0$ independent of j. Let v_j^* be the (usual) rearrangement of v_j. Namely v_j^* is a radially symmetric, monotone increasing function, satisfying $|\{v_j^* < a\}| = |\{v_j < a\}|$ for any $a \in \mathbb{R}$, where $|\cdot|$ denotes the Lebesgue measure in $\mathbb{R}^n$. Let $v^* = \lim_{j\to\infty} v_j^*$. Then $v^* \not\equiv 0$ because of (6.41). By Fatou's lemma,

$$\int_{\mathbb{R}^n} |v^*|^{k^*} dy \le \underline{\lim}_{j\to\infty} \int_{\mathbb{R}^n} |v_j^*|^{p_j+1} dy$$

$$= \underline{\lim}_{j\to\infty} \int_{\mathbb{R}^n} |v_j|^{p_j+1} dy \le C.$$

Therefore $v^*(r) = o(\frac{1}{r})$ as $r \to \infty$. It follows that for any given $\varepsilon > 0$, there exists $\delta = \delta_\varepsilon > 0$, with $\delta \to 0$ at $\varepsilon \to 0$, such that

$$\delta \cdot \mathrm{mes}\{y \in \mathbb{R}^n \mid |v^*(y)|^{k^*} > \delta\} < \varepsilon.$$

Hence

$$\delta \cdot \mathrm{mes}\{y \mid |v_j(y)|^{k^*} > \delta\} = \delta \cdot \mathrm{mes}\{y \mid |v_j^*(y)|^{k^*} > \delta\} < \varepsilon \tag{6.42}$$

for sufficiently large j.

Denote

$$\widetilde{v}_j(y) = v_j(y) + \delta^{1/k^*},$$
$$\Omega_{j,\delta} = \{y \in \Omega_j \mid \widetilde{v}_j(y) < 0\}.$$

We have

$$\sup_{t>0} J^*(tv_j \mid \Omega_{j,\delta}) = \sup_{t>0} \int_{\Omega_{j,\delta}} \left[\frac{-t^{k+1}}{k+1} v_j S_k(D^2 v_j) - \frac{t^{k^*}}{k^*} |v_j|^{k^*} \right] dy$$

$$\ge \sup_{t>0} \int_{\Omega_{j,\delta}} \left[\frac{-t^{k+1}}{k+1} \widetilde{v}_j S_k(D^2 \widetilde{v}_j) - \frac{t^{k^*}}{k^*} |v_j|^{k^*} \right] dy.$$

By (6.42),

$$\int_{\Omega_{j,\delta}} |v_j|^{k^*} = \int_{\Omega_{j,\delta}} |\tilde{v}_j + \delta^{1/k^*}|^{k^*}$$

$$\leq \int_{\Omega_{j,\delta}} [|\tilde{v}_j|^{k^*} + C\delta^{1/k^*}|\tilde{v}_j|^{k^*-1} + C\delta]$$

$$\leq \int_{\Omega_{j,\delta}} [|\tilde{v}_j|^{k^*} + (\delta|\Omega_{j,\delta}|)^{1/k^*} \left(\int_{\Omega_{j,\delta}} |\tilde{v}_j|^{k^*} \right)^{\frac{k^*-1}{k^*}} + C\delta|\Omega_{j,\delta}|$$

$$\leq (1 + C\varepsilon^{1/k^*}) \int_{\Omega_{j,\delta}} |\tilde{v}_j|^{k^*},$$

where we have used the fact that

$$\|\tilde{v}_j\|_{L^{k^*}} \geq C > 0,$$

which follows from (6.41). We obtain

$$\sup_{t>0} J^*(tv_j \mid \Omega_{j,\delta}) \geq \sup_{t>0} \int_{\Omega_{j,\delta}} \left[\frac{-t^{k+1}}{k+1} \tilde{v}_j S_k(D^2\tilde{v}_j) - \frac{t^{k^*}}{k^*}(1 + C\varepsilon^{1/k^*})|\tilde{v}_j|^{k^*} \right] dy.$$

Hence we obtain

$$\sup_{t>0} J^*(tv_j \mid \Omega_{j,\delta}) \geq (1 - C\varepsilon^{1/k^*})c^*.$$

On the other hand, since $f(x,u) = \lambda|u|^q$ is a lower order term of $|u|^{k^*-1}$ and $\text{mes}(\Omega_{j,\delta})$ are uniformly bounded for fixed δ, we see that when M_j is large enough,

$$\sup_{t>0} J_j(tv_j, \Omega_{j,\delta}) \geq \sup_{t>0} J^*(tv_j, \Omega_{j,\delta}) - \varepsilon_j \geq (1 - C\varepsilon^{1/k^*})c^* - \varepsilon_j$$

with $\varepsilon_j \to 0$ as $M_j \to \infty$, where

$$J_j(u, \Omega_{j,\delta}) = J_{p_j}(u, \Omega_{j,\delta}) = \int_{\Omega_{j,\delta}} \left[\frac{-1}{k+1} u S_k(D^2u) - \tilde{\Psi}_j(x,u) \right] dx.$$

For any subdomain $D \subset \Omega_j$, since

$$\int_D v_j S_k(v_j) dy = \int_D v_j(|v_j|^{k^*-1} + \lambda M_j^{q-p_j}|v_j|^{q+1}) dy,$$

we have

$$J_j(v_j, D) = \sup_{t>0} J_j(tv_j, D) \geq 0.$$

Hence

$$J_j(v_j, \Omega_j) \geq J_j(v_j, \Omega_{j,\delta}) \geq (1 - C\varepsilon^{1/k^*})c^* - \varepsilon_j.$$

We reach a contradiction with (6.32) when ε and ε_j are sufficiently small. This completes the proof.

The technique in the treatment of the case (6.39) is new. Moreover, it also applies to the case (6.37). By carefully examining the argument, one sees that the function f in (6.29) can be replaced by a more general $f(x, u)$, provided

$$\lim_{t \to 0} f(x, t)|t|^{-k} = 0,$$
$$\lim_{t \to 0} f(x, t)|t|^{-k^*+1} = 0.$$

In the case the constant c_0 in (6.30) should be replaced by

$$c_0 = \inf_{\gamma \in \Gamma} \sup_{s \in [0,1]} J(\gamma(s)),$$

where Γ denotes the set of all paths in $\Phi_0^k(\Omega)$ connecting $U \equiv 0$ to a function u_0 satisfying $J(u_0) < 0$.

The verification of (6.32) can be carried out in a similar way as [BN], and the computation is also similar. Let

$$w_\varepsilon(x) = C_n^k \left(\frac{n - 2k}{k} \right)^k \left(\frac{\varepsilon^{\frac{1}{k+1}}}{\varepsilon + |x|^2} \right)^{\frac{n-2k}{2k}}, \tag{6.43}$$

where C_n^k is the binary coefficient. Then w_ε satisfies the equation

$$S_k(D^2 u) = |u|^{k^* - 1} \quad \text{in } \mathbb{R}^n, \tag{6.44}$$

and the constant c^* in (9.3) is attained by w_ε when $\Omega = \mathbb{R}^n$.

Lemma 6.1 *Suppose there exists a ball $B_r(x_0) \subset \Omega$ and*

$$q > \max \left\{ k, \frac{(k+1)(nk - n + 2k)}{k(n - 2k)} - 1 \right\}$$

so that

$$f(x, u) \geq \lambda |u|^q \quad \text{for } x \in B_r(x_0) \tag{6.45}$$

for some $\lambda > 0$, then $c_0 < c^*$.

Proof. Let

$$B = \int_{R^n} w_\varepsilon^{k^*} dx = -\int_{\mathbb{R}^n} w_\varepsilon S_k(D^2 w_\varepsilon) dx.$$

B is independent of ε. By a translation we may suppose x_0 is the origin. Let $\varphi(x)$ be a radial cut-off function so that $\varphi = 1$ in $B_{r/2}(0)$ and $\varphi = 0$ outside $B_r(0)$. We may choose φ so that $u_\varepsilon =: \varphi w_\varepsilon \in \Phi_0^k$ for $\varepsilon > 0$ small. Direct computations show that

$$\int_\Omega u_\varepsilon^{k^*} dx = B + O(\varepsilon^{(n-2k)/2k}),$$

$$\int_\Omega (-u_\varepsilon) S_k(D^2 u_\varepsilon) dx = B + O(\varepsilon^{(n-2k)/2k}),$$

If $f(x, u)$ satisfies (6.45), we have

$$\int_\Omega F(x, u_\varepsilon) dx \geq C \int_\Omega |\mathring{u}_\varepsilon|^{q+1} dx \geq C\varepsilon^{\frac{n}{2} - \frac{n-2k}{2(k+1)}q},$$

where $F(x, u) = \int_u^0 f(x, t) dt$. If $q > \frac{(k+1)(nk-n+2k)}{k(n-2k)} - 1$, we have $\frac{n}{2} - \frac{n-2k}{2(k+1)}q < \frac{n-2k}{2k}$. Hence if ε is small enough we have $c_0 < c^*$.

We refer the reader to [CGY] for more details on radially symmetric solutions to the k-Hessian equations in the critical growth case.

7 Hessian Integral Estimates

In this section we establish some local integral estimates for k-admissible functions [TW2]. The main estimates include (7.2) and (7.5) below. Our estimates are based on the divergence structure of the k-Hessian operator. As shown as the beginning of Section 5, the k-Hessian operator can also be written as

$$\begin{aligned} S_k[u] &= [D^2 u]_k \qquad\qquad (7.1)\\ &= \frac{1}{k} \sum u_{ij} S_k^{ij}[u] \\ &= \frac{1}{k} \sum \partial_{x_i}(u_{x_j} S_k^{ij}[u]). \end{aligned}$$

As before we denote by $\Phi^k(\Omega)$ the set of all smooth k-admissible functions in Ω, and by $\Phi_0^k(\Omega)$ the set of all smooth k-admissible functions vanishing on $\varphi\Omega$.

7.1 A Basic Estimate

Here we establish the following basic estimate.

Theorem 7.1 *Let $u \in \Phi^k(\Omega)$. Suppose $u \leq 0$ in Ω. Then for any subdomain $\Omega' \subset\subset \Omega$,*

$$\int_{\Omega'} S_k[u] \leq C\left(\int_\Omega |u|\right)^k, \qquad\qquad (7.2)$$

where C is a constant depending on n, k, Ω and Ω'.

Proof. It suffices to consider the case $\Omega = B_R(y)$, $\Omega' = B_r(y)$, for some $y \in \mathbb{R}^n$ and $r < R$. Let $\eta \in C^\infty(\overline{\Omega})$ satisfy $0 \leq \eta \leq 1$, $\eta = 1$ in $B_r(y)$, and $\eta = 0$ when $|x - y| \geq (R + 2r)/3$. Let $\widetilde{u} \in C^\infty(\overline{\Omega})$ be the unique k-convex solution of the Dirichlet problem

$$S_k[\widetilde{u}] = \eta S_k[u] + \delta^n \ \text{ in } \ \Omega,$$
$$\widetilde{u} = 0 \ \text{ on } \ \varphi\Omega.$$

Then $S_k[u + \frac{\delta}{2}|x - y|^2] \geq S_k[\widetilde{u}]$. By the comparison principle, we have $\frac{\delta}{2}|x - y|^2 + u \leq \widetilde{u} \leq 0$ in Ω, so that

$$\int_\Omega |\widetilde{u}| \leq \int_\Omega |u| + C\delta.$$

Let $\zeta \in C_0^\infty(\Omega)$ be a cut-off function. Then, by integration by parts,

$$\begin{aligned}
\int_\Omega \zeta S_k[\widetilde{u}] &= \frac{1}{k} \int \zeta S_k^{ij}[\widetilde{u}] D_{ij}\widetilde{u} = \frac{1}{k} \int \widetilde{u} S_k^{ij}[\widetilde{u}] D_{ij}\zeta \\
&\leq \frac{1}{k} \max\left(|D^2\zeta| \, |\widetilde{u}|\right) \int_{\mathrm{supp} D^2\zeta} S^{ii}[\widetilde{u}] \\
&= \frac{n - k + 1}{k} \max\left(|D^2\zeta| \, |\widetilde{u}|\right) \int_{\mathrm{supp} D^2\zeta} S_{k-1}[\widetilde{u}].
\end{aligned}$$

Choose $\zeta = 1$ in $B_{(R+r)/2}(y)$, $\zeta = 0$ for $|x - y| \geq (5R + r)/6$, $|D^2\zeta| \leq C(R - r)^{-2}$. By the Harnack inequality (4.10), we have

$$|\widetilde{u}(x)| \leq \int_\Omega |\widetilde{u}| \qquad \forall \, |x - y| \geq \frac{1}{2}(R + r).$$

Hence by sending $\delta \to 0$, we obtain

$$\int_{\Omega'} S_k[u] \leq C \int_\Omega S_{k-1}[u] \int_\Omega (-u)$$

for some constant C depending on n, k, r, R. By iteration we obtain (7.2).

7.2 Local Integral Gradient Estimates

First we prove

Theorem 7.2 *Let* $u \in \Phi^k(\Omega)$, $k = 1, \cdots, n$, *satisfy* $u \leq 0$ *in* Ω. *Then for any sub-domain* $\Omega' \subset\subset \Omega$, *we have the estimates*

$$\int_{\Omega'} |Du|^q S_l[u] \leq C\left(\int_\Omega |u|\right)^{q+l} \tag{7.3}$$

for all $l = 0, \cdots, k-1$, $0 \le q < \frac{n(k-l)}{n-k}$, where C is a constant depending on Ω, Ω', n, k, l and q.

Corollary 7.1 Let $u \in \Phi^k(\Omega)$. Then $\forall \; \Omega' \subset\subset \Omega$,

$$\|Du\|_{L^q(\Omega')} \le C \int_\Omega |u| \qquad (7.4)$$

for $q < \frac{nk}{n-k}$, where C depends on n, k, q, Ω, and Ω'.

Inequality (7.4) follows from (7.3) by taking $l = 0$. When $k = 1$, (7.4) can be found in [H]. Corollary 7.1 asserts that a k-admissible function is in the local Sobolev space $W^{1,q}_{loc}(\Omega)$. When $k > n/2$, we have $q > n$, and by the Sobolev imbedding theorem, $u \in C^\alpha_{loc}(\Omega)$ with $\alpha \in (0, 2 - \frac{n}{k})$. But recall that in Theorem 5.4 we have shown that $u \in C^\alpha_{loc}(\Omega)$ with $\alpha = 2 - \frac{n}{k}$.

To prove Theorem 7.2, let us denote, for a real $n \times n$ matrix $\mathcal{A} = [a_{ij}]$, not necessarily symmetric,

$$S_k(\mathcal{A}) = [\mathcal{A}]_k, \qquad (7.5)$$
$$S_k^{ij}(\mathcal{A}) = \frac{\varphi}{\partial a_{ij}} [\mathcal{A}]_k.$$

Then for any vector field $g = (g_1, \cdots, g_n)$, $g_i \in C^1(\Omega)$, $i = 1, \cdots, n$, it follows that

$$D_i S_k^{ij}(Dg) = 0, \quad j = 1, \cdots, n, \qquad (7.6)$$
$$S_k^{ij}(Dg) D_i g_j = k S_k(Dg).$$

Now we introduce a broader class of operators, namely, the p-k-Hessian operators, given for $k = 1, \cdots, n$, $p \ge 2$, $u \in C^2(\Omega)$, by

$$S_{k,p}[u] = \left[D(|Du|^{p-2} Du) \right]_k. \qquad (7.7)$$

where

$$D(|Du|^{p-2} Du) = |Du|^{p-2} \left(I + (p-2) \frac{Du \otimes Du}{|Du|^2} \right) D^2 u.$$

When $k = 1$, it is the well-known p-Laplacian operator,

$$S_{1,p}[u] = \mathrm{div}\, (|Du|^{p-2} Du). \qquad (7.8)$$

One can verify by direct computation that the p-k-Hessian operator is invariant under rotation of coordinates.

Let us call a function $u \in C^2(\Omega)$, p-k-admissible in Ω if $S_{l,p}[u] \ge 0$ for all $l = 1, \cdots, k$. We then have the following relation between k-admissibility and p-k-admissibility.

Lemma 7.1 If u is k-admissible, then u is p-l-admissible for $l = 1, \cdots, k-1$ and $p - 2 \le \frac{n(k-l)}{l(n-k)}$.

Proof. At a point $y \in \Omega$, where $Du(y) \neq 0$, we fix a coordinate system so that the x_1 axis is directed along the vector $Du(y)$ and the remaining axes are chosen so that the reduced Hessian $[D_{ij}u]_{i,j=2,\cdots,n}$ is diagonal. It follows then that the p-Hessian is given by

$$D_i\big(|Du|^{p-2}D_j u\big) = |Du|^{p-2} \begin{cases} (p-1)D_{i1}u & \text{if } j=1, i \geq 1, \\ D_{1j}u & \text{if } i=1, j>1, \\ D_{ii}u & \text{if } j=i>1, \\ 0 & \text{otherwise.} \end{cases} \tag{7.9}$$

Hence by calculation, we obtain for $l = 1, \cdots, k-1$ at the point y, setting $\widetilde{\lambda}_i = D_{ii}u(y)$, $i = 1, \cdots, n$,

$$|Du|^{l(2-p)} S_{l,p}[u] = (p-1)\widetilde{\lambda}_1 \sigma_{l-1;1}(\widetilde{\lambda}) + \sigma_{l;1}(\widetilde{\lambda}) - (p-1)\sum_{i=2}^{n} \sigma_{l-2;1i}(\widetilde{\lambda})(D_{i1}u)^2. \tag{7.10}$$

From the k-admissibility of u, we have

$$S_k[u] = \widetilde{\lambda}_1 \sigma_{k-1;1}(\widetilde{\lambda}) + \sigma_{k;1}(\widetilde{\lambda}) - \sum_{i=2}^{n} \sigma_{k-2;1i}(\widetilde{\lambda})(D_{i1}u)^2 \geq 0$$

so that using Newton's inequality, in the form

$$\frac{\sigma_{k;1}}{\sigma_{k-1;1}} \leq \frac{l(n-k)}{k(n-l)} \frac{\sigma_{l;1}}{\sigma_{l-1;1}}, \tag{7.11}$$

we have, for $p - 1 \leq \frac{k(n-l)}{l(n-k)}$, the inequality

$$\frac{1}{p-1}|Du|^{l(2-p)} S_{l,p}[u] \geq \widetilde{\lambda}_1 \sigma_{l-1;1}(\widetilde{\lambda}) + \frac{\sigma_{k;1}}{\sigma_{k-1;1}}\sigma_{l-1;1}(\widetilde{\lambda}) - \sum_{i=2}^{n} \sigma_{l-2;1i}(\widetilde{\lambda})(D_{i1}u)^2$$

$$\geq \frac{\sigma_{l-1;1}}{\sigma_{k-1;1}} \sum_{i=2}^{n} \sigma_{k-2;1i}(D_{i1}u)^2 - \sum_{i=2}^{n} \sigma_{l-2;1i}(D_{i1}u)^2$$

$$= \frac{1}{\sigma_{k-1;1}} \sum_{i=2}^{n} \Big(\sigma_{l-1;1}\sigma_{k-2;1i} - \sigma_{k-1;1}\sigma_{l-2;1i}\Big)(D_{i1}u)^2$$

$$= \frac{1}{\sigma_{k-1;1}} \sum_{i=2}^{n} \Big(\sigma_{l-1;1i}\sigma_{k-2;1i} - \sigma_{k-1;1i}\sigma_{l-2;1i}\Big)(D_{i1}u)^2$$

$$\geq 0.$$

Note that when applying Newton's inequality to the last inequality, the coefficient in $(\sigma_{l-1;1i}\sigma_{k-2;1i} - \sigma_{k-1;1i}\sigma_{l-2;1i})$ is better than we need.

Let $\lambda = (\lambda_1, \cdots, \lambda_n) \in \Gamma_k$. Suppose $\lambda_1 \geq \cdots \geq \lambda_n$. Let $u = \frac{1}{2}\lambda_i x_i^2$. By Lemma 7.1, we have $\Delta_p u \geq 0$ for $p \leq 2 + \frac{n(k-1)}{n-k}$. Hence

$$\sum_i \lambda_i + \frac{n(k-1)}{n-k}\lambda_n \geq 0. \tag{7.12}$$

Proof of Theorem 7.2. Setting

$$p^* = 1 + \frac{k(n-l)}{l(n-k)}, \quad k < n, \; l < k,$$

we obtain from Lemma 7.1 and the formula (7.10), for $2 < p < p^*$ and $u \in \Phi^k(\Omega)$,

$$|Du|^{l(2-p)} S_{l,p}[u] = \frac{p^* - p}{p^* - 2} S_l[u] + \frac{p-2}{p^* - 2}|Du|^{l(2-p^*)} S_{l,p^*}[u]$$
$$\geq \frac{p^* - p}{p^* - 2} S_l[u],$$

and hence, for $q = (p-2)l < \frac{n(k-l)}{n-k}$, we have the estimate

$$|Du|^q S_l[u] \leq \frac{p^* - 2}{p^* - p} S_{l,p}[u]. \tag{7.13}$$

Theorem 7.2 will follow by estimation of $S_{l,p}[u]$ in $L^1_{\mathrm{loc}}(\Omega)$. For any non-negative cut-off function $\eta \in C^2_0(\Omega)$, we obtain

$$\int_\Omega \eta S_{l,p}[u] = \int_\Omega \eta S_l(D(|Du|^{p-2}Du)) \tag{7.14}$$
$$= \frac{1}{l}\int_\Omega \eta S_l^{ij} D_i(|Du|^{p-2}D_j u)$$
$$= -\frac{1}{l}\int_\Omega |Du|^{p-2} S_l^{ij} D_i\eta D_j u.$$

From (7.9), we have

$$S_l^{ij} D_j u = |Du|^{(l-1)(p-2)} S_l^{ij}(D^2 u)D_j u$$
$$= |Du|^{(l-1)(p-2)} S_l^{ij}[u]D_j u,$$

so that, by substituting in (7.14), we obtain

$$\int_\Omega \eta S_{l,p}[u] = -\frac{1}{l}\int_\Omega |Du|^{l(p-2)} S_l^{ij} D_i\eta D_j u$$
$$\leq \frac{1}{l}\int_\Omega |Du|^{q+1}|D\eta| S_{l-1}[u],$$

and hence, replacing η by η^l and using (7.13), we obtain

$$\int_\Omega |Du|^q \eta^l S_l[u] \le C \max |D\eta| \int_\Omega |Du|^{q+1} \eta^{l-1} S_{l-1}[u].$$

Consequently,

$$\int_\Omega |Du|^q \eta^l S_l[u] \le (C \max |D\eta|)^l \int_\Omega |Du|^{q+l}, \qquad (7.15)$$

so that the estimate (7.3) is reduced to the case $l = 0$. To handle this case, we take $l = 1$ in (7.15) with

$$q = q(1) < \frac{n(k-1)}{n-k}.$$

If u is k-admissible for $k \ge 2$, we have

$$S_2[u] = \frac{1}{2}\big((\Delta u)^2 - |D^2 u|^2\big) \ge 0$$

and hence

$$|D^2 u| \le \Delta u. \qquad (7.16)$$

Therefore we obtain from (7.15)

$$\int_\Omega \eta |Du|^q |D^2 u| \le C \max |D\eta| \int_\Omega |Du|^{1+q}$$

so that

$$\int_\Omega \eta D(|Du|^{1+q}) \le C \max |D\eta| \int_\Omega |Du|^{1+q}.$$

Thus by an appropriate choice of η, we obtain for any subdomain $\Omega' \subset\subset \Omega$,

$$\||Du|^{1+q}\|_{L^{n/(n-1)}(\Omega')} \le C d_{\Omega'}^{-1} \int_\Omega |Du|^{1+q}, \qquad (7.17)$$

where $d_{\Omega'} = \operatorname{dist}(\Omega', \varphi\Omega)$, C is a constant depending on k, q and n. The estimate (7.3) now follows by interpolation. $\qquad \square$

From Theorem 7.2 we may derive corresponding estimates for the k-admissible function themselves.

Theorem 7.3 *Let u be a nonpositive k-admissible function in Ω, $k \le n/2$. Then for any subdomain $\Omega' \subset\subset \Omega$, we have*

$$\int_{\Omega'} |u|^q S_l[u] \le C \left(\int_\Omega |u| \right)^{l+q} \qquad (7.18)$$

for all $l = 0, \cdots, k-1$, $0 \le q < \frac{n(k-l)}{n-2k}$, where C is a constant depending on Ω, Ω', n, k, l and q.

Proof. With $\eta \geq 0, \in C_0^1(\Omega)$, we estimate

$$\int_\Omega \eta^2 (-u)^q S_l[u] = \frac{q}{l} \int_\Omega \eta^2 (-u)^{q-1} S_l^{ij} D_i u D_j u - \frac{1}{l} \int_\Omega (-u)^q S_l^{ij} D_i u D_j \eta^2$$

$$\leq \frac{q(n-l+1)}{l} \int_\Omega \eta^2 (-u)^{q-1} S_{l-1} |Du|^2$$

$$+ \frac{2(n-l+1)}{l} \int_\Omega \eta (-u)^q S_{l-1} |Du| |D\eta|$$

$$\leq \frac{(q+1)(n-l+1)}{l} \int_\Omega \eta^2 (-u)^{q-1} S_{l-1} |Du|^2$$

$$+ \frac{n-l+1}{l} \int_\Omega |D\eta|^2 (-u)^{q+1} S_{l-1}$$

Now, for any $p < \frac{n(k-l+1)}{n-k}$, we have

$$\int_\Omega \eta^2 (-u)^{q-1} S_{l-1} |Du|^2 \leq \left(\int_\Omega \eta^2 S_{l-1} |Du|^p \right)^{2/p} \left(\int_\Omega \eta^2 (-u)^{\frac{p(q-1)}{p-2}} S_{l-1} \right)^{1-2/p}$$

so that if $q < \frac{n(k-l)}{n-2k}$, we may choose p so that $q^* = \frac{p(q-1)}{p-2} < \frac{n(k-l+1)}{n-2k}$, and the estimate (7.18) follows from Theorem 7.2 by induction on l.

8 Hessian Measures

In this section we extend the notion of k-admissible functions to nonsmooth functions. We assign a measure $\mu_k[u]$ to a k-admissible function u and prove the weak continuity of μ_k. The proof of the weak continuity of μ_k in [TW2] involves delicate integral estimates and is based on the estimates in §7. Here we provide a simpler proof, using ideas from [TW1, TW5]. As an application we prove the existence of a weak solution to the Dirichlet problem of the k-Hessian equation.

8.1 Non-Smooth k-Admissible Functions

Observe that a C^2 function u is k-admissible if and only if for any matrix $A = \{a_{ij}\}$ with eigenvalues in the cone

$$\Gamma_k^* = \{\lambda^* \in \mathbb{R}^n \mid \lambda^* \cdot \lambda \leq 0 \ \forall \ \lambda \in \Gamma_k\},$$

there holds

$$\sum a_{ij} D_{ij}^2 u \leq 0. \tag{8.1}$$

Note that a matrix A with eigenvalues in Γ_k^* must be negative definite. From (8.1) we can extend the notion of k-admissibility to non-smooth functions as follows.

Definition 8.1 *A function u in Ω is k-admissible if*
(i) it is upper semi-continuous and the set $\{u = -\infty\}$ has measure zero; and
(ii) for any matrix $A = \{a_{ij}\}$ with eigenvalues in Γ_k^*,

$$\int_\Omega u a_{ij} D_{ij}^2 \varphi \le 0 \quad \forall\, \varphi \in C_0^\infty(\Omega), \varphi \ge 0. \tag{8.2}$$

Note that when $k = 1$, Γ_k^* contains only the vector $-(1, \cdots, 1)$, and (8.2) becomes $\int_\Omega u(-\Delta\varphi) \le 0$ for any $\varphi \ge 0, \in C_0^\infty(\Omega)$. The above definition implies that an upper semi-continuous function u is k-admissible if it is subharmonic with respect to the operator $L = \sum a_{ij} D_{ij}^2$ for any matrix A with eigenvalues in Γ_k^*.

From (8.2) we see that if u is k-admissible, so is its mollification u_ε, given by

$$u_\varepsilon(x) = \int_\Omega u(x - \varepsilon y)\rho(y)\, dy = \int_\Omega \varepsilon^{-n} \rho(\frac{x - y}{\varepsilon}) u(y), \tag{8.3}$$

where ρ is a mollifier, namely ρ is a smooth, nonnegative function with support in the unit ball $B_1(0)$, and $\int_{B_1(0)} \rho = 1$. Observe that if u is k-admissible, it is also subharmonic. Hence its mollification u_ε converges to u monotone decreasingly. Therefore by Corollary 7.1, a k-admissible function is locally in the Sobolev space $W_{loc}^{1,q}(\Omega)$ for any $q < \frac{nk}{n-k}$.

Lemma 8.1 *Let u_j be a sequence of k-admissible functions which converges to u almost everywhere. Suppose u is upper semi-continuous and the set $\{u = -\infty\}$ has measure zero. Then u is k-admissible and u_j converges to u pointwise.*

Proof. The first assertion follows readily from the definition. The second one is due to that u is upper semi-continuous and u_j is subharmonic and so it satisfies the mean value inequality below.

Recall that a k-admissible function u is subharmonic, it satisfies the mean value inequality

$$u(y) \le \frac{1}{|B_r(y)|} \int_{B_r(y)} u \quad \forall\, B_r(y) \subset \Omega. \tag{8.4}$$

Therefore if u_j and u are k-admissible and $\{u_j\}$ converges to u almost everywhere, u_j is locally uniformly bounded in $L^1(\Omega)$. Conversely, if a sequence of k-admissible functions $\{u_j\}$ is uniformly bounded in $L_{loc}^1(\Omega)$, then by Corollary 7.1, the set $\{u = -\infty\}$ has measure zero.

We also have the following comparison principle for k-admissible functions.

Lemma 8.2 *Suppose u and v are k-admissible and v is smooth in Ω. Suppose $S_k[v] = 0$ in Ω and for any point $y \in \varphi\Omega$, $\underline{\lim}_{x \to y}[v(x) - u(x)] \geq 0$. Then $v \geq u$ in Ω.*

Proof. If there is an interior point $x_0 \in \Omega$ such that $v(x_0) < u(x_0)$, by adding a positive constant $\delta = \frac{1}{2}(u(x_0) - v(x_0))$ to v we may suppose that for any $y \in \varphi\Omega$, $\lim_{x \to y}[v(x) - u(x)] \geq \delta > 0$, so that $v > u$ in a neighborhood of the boundary $\varphi\Omega$. Therefore for $\varepsilon > 0$ small, we have $v > u_\varepsilon$ near $\varphi\Omega$, where u_ε is the mollification of u. By the comparison principle for smooth k-admissible functions, we conclude that $v \geq u_\varepsilon$, which is in contradiction with $v(x_0) < u(x_0) \leq u_\varepsilon(x_0)$.

Lemma 8.3 *Suppose u, v are two k-admissible functions. Then $w = \max(u, v)$ is also k-admissible.*

Proof. Let $u_\varepsilon, v_\varepsilon$ be the mollification of u and v, respectively. Then it suffices to show that $w_\varepsilon = \max(u_\varepsilon, v_\varepsilon)$ is k-admissible. For brevity we drop the subscript ε. Since the function w is semi-convex (i.e., $w + C|x|^2$ is convex for sufficiently large constant C), w is twice differentiable almost everywhere and the eigenvalues of $D^2 w$ lies in Γ_k. Let $w_{\varepsilon'}$ be the mollification of w. By integration by parts in (8.3), we have

$$D^2 w_{\varepsilon'}(x) \geq \int_\Omega D^2 w(x - \varepsilon'y)\rho(y)\, dy.$$

Hence for any matrix A with eigenvalues in Γ_k^*,

$$a_{ij} D^2_{ij} w_{\varepsilon'}(x) \leq \int_\Omega a_{ij} D^2_{ij} w(x - \varepsilon'y)\rho(y)\, dy \leq 0.$$

Hence $w_{\varepsilon'}$, and so also w, is k-admissible.

8.2 Perron Lifting

Let u be a k-admissible function in Ω and $\omega \Subset \Omega$ be a subdomain of Ω. The Perron lifting of u in ω, u^ω, is the upper semicontinuous regularization of the function $\hat{u}$,

$$u^\omega(x) = \lim_{t \to 0} \sup_{B_t(x)} \hat{u}, \qquad (8.5)$$

where

$$\hat{u}(x) = \sup\{v(x) \mid v \text{ is } k\text{-admissible in } \Omega \text{ and } v \leq u \text{ in } \Omega - \omega\}.$$

Obviously we have $u^\omega \geq \hat{u}$, and $\hat{u}$ and u^ω coincide in Ω except possibly on $\partial\omega$.

Lemma 8.4 *Assume $\varphi\omega$ is $C^{3,1}$ smooth. Then u^ω is a solution of*

$$S_k[w] = 0 \quad in \ \ \omega, \tag{8.6}$$
$$w = u \quad on \ \ \varphi\omega,$$

in the sense that there is a sequence of smooth k-admissible functions w_ε which satisfies $S_k[w_\varepsilon] = 0$ in ω and $w_\varepsilon \to u$ on $\overline{\omega}$ pointwise.

Proof. Let u_ε be a mollification of u, as given in (8.3). Let $u_{\varepsilon,j} = u_\varepsilon + 2^{-j}|x|^2$. Then $S_k[u_{\varepsilon,j}] \geq C2^{-kj}$. That is, $u_{\varepsilon,j}$ is a smooth sub-solution to the Dirichlet problem

$$S_k[w] = C2^{-kj} \ \ in \ \ \omega,$$
$$w = u_{\varepsilon,j} \ \ on \ \ \varphi\omega.$$

Hence from [G], there is a unique global smooth solution $w_{\varepsilon,j} \in C^3(\overline{\omega})$, monotone in j. By (3.5) we have $\sup_\omega |Dw_{\varepsilon,j}| \leq \sup_{\varphi\omega} |Dw_{\varepsilon,j}|$. On the boundary $\varphi\omega$, we have $u_{\varepsilon,j} \leq w_{\varepsilon,j} \leq \overline{u}_{\varepsilon,j}$, where $\overline{u}_{\varepsilon,j}$ is the harmonic extension of $u_{\varepsilon,j}$ in ω. Hence

$$\sup_{\varphi\omega} |Dw_{\varepsilon,j}| \leq \sup_{\varphi\omega} |Du_{\varepsilon,j}| \leq \sup_{\varphi\omega} |Du_\varepsilon| + C2^{-j}.$$

Therefore by passing to a subsequence, $w_{\varepsilon,j}$ converges as $j \to \infty$ to a solution w_ε of (8.6) which satisfies the boundary condition $w_\varepsilon = u_\varepsilon$ on $\varphi\omega$.

Let $u_\varepsilon^\omega = w_\varepsilon$ in ω and $u_\varepsilon^\omega = u_\varepsilon$ in $\Omega - \omega$. It is easy to see that u_ε^ω is the Perron lifting of u_ε in ω. The proof of Lemma 8.3 implies that u_ε^ω is k-admissible. Since u_ε is monotone decreasing in ε, so is u_ε^ω. By the comparison principle (Lemma 8.2), we have $u_\varepsilon^\omega \geq u^\omega$ in Ω. Hence $u_0 := \lim_{\varepsilon\to 0} u_\varepsilon^\omega \geq u^\omega$.

On the other hand, let u_0^ω be the upper semicontinuous regularization of u_0. Then by Lemma 8.1, u_0^ω is k-admissible in Ω. Obviously $u_0^\omega = u$ in $\Omega - \overline{\omega}$. Hence by definition of u^ω, $u_0^\omega \leq u^\omega$. We obtain $u_0 \geq u^\omega \geq u_0^\omega \geq u_0$. Hence

$$\lim_{\varepsilon\to 0} u_\varepsilon^\omega = u^\omega.$$

Lemma 8.1 implies the convergence is pointwise. The interior gradient estimate implies that u^ω is locally uniformly Lipschitz continuous in ω.

Below we will consider the Perron lifting in an annulus $\omega_t = B_{r+t}(x_0) - B_{r-t}(x_0)$. Let us fix r and let t vary. Then u^{ω_t} is monotone in t, namely

$$\lim_{t\to\delta^-} u^{\omega_t}(x) \leq u^{\omega_\delta}(x) \leq \lim_{t\to\delta^+} u^{\omega_t}(x) \ \ \forall \, x \in \Omega.$$

It follows that $\|u^{\omega_t}\|_{L^1(\Omega)}$, as a function of t, is monotone and bounded. Hence, $\|u^{\omega_t}\|_{L^1(\Omega)}$ is continuous at almost all t. It follows that for almost all $t > 0$

$$\lim_{s\to t} u^{\omega_s}(x) = u^{\omega_t}(x). \tag{8.7}$$

234

Lemma 8.5 *Suppose u_j, u are k-admissible and $u_j \to u$ a.e. in Ω. Suppose (8.7) holds at t. Then we have $u_j^{\omega_t} \to u^{\omega_t}$ a.e. in Ω as $j \to \infty$.*

Proof. Since $u_j^{\omega_t}$ and u^{ω_t} are locally uniformly Lipschitz continuous in ω_t, by passing to a subsequence, we may assume that $u_j^{\omega_t}$ is convergent. Let $w' = \lim u_j^{\omega_t}$ and w be the upper semicontinuous regularization of w'. Then by Lemma 8.1, w is k-admissible and $w = u$ in $\Omega - \overline{\omega}_t$. Hence by the definition of the Perron lifting, we have $u^{\omega_t} \geq w$.

Next we prove that for any $\delta > 0$, $w \geq u^{\omega_{t-\delta}}$. Once this is proved, we have $u^{\omega_t} \geq w \geq u^{\omega_{t-\delta}}$. Sending $\delta \to 0$, we obtain $u^{\omega_t} = w$ by (8.7).

To prove $w \geq u^{\omega_{t-\delta}}$, it suffices to prove that for any given $\varepsilon > 0$, $u_j^{\omega_t} \geq u - \varepsilon$ on $\varphi\omega_{t-\delta}$ for sufficiently large j. By the interior gradient estimate, $u_j^{\omega_t}$ is locally uniformly Lipschitz continuous in ω_t. If there exists a point $x_0 \in \varphi B_{r-\delta/2}$ such that $u(x_0) > u_j^{\omega_r}(x_0) + \varepsilon$ for all large j, by (8.4), there is a Lebesgue point $x_1 \in B_{\delta/4}(x_0)$ of u such that $u(x_1) > u_j^{\omega_r}(x_1) + \frac{1}{2}\varepsilon$ for all large j. It follows that the limit function w' is strictly less than u a.e. near x_1. We reach a contradiction as $w' = \lim_{j\to\infty} u_j^{\omega_r} \geq \lim_{j\to\infty} u_j = u$.

8.3 Weak Continuity

Denote $\mu_k[u] = S_k[u]dx$. It is a nonnegative measure if u is a C^2 smooth, k-admissible function. First we prove the following monotonicity formula.

Lemma 8.6 *Let u, v be two smooth k-admissible function in Ω. Suppose $u = v$ on $\varphi\Omega$ and $u(x) > v(x)$ for $x \in \Omega$, near $\varphi\Omega$. Then*

$$\int_\Omega S_k[u] \leq \int_\Omega S_k[v]. \tag{8.8}$$

Proof. We may assume that $\varphi\Omega$ is smooth, otherwise it suffices to prove (8.8) in $\{u - \delta > v\}$ and send $\delta \to 0$. We have

$$\frac{d}{dt}\int_\Omega S_k[u + t(v-u)] = \int_\Omega S_k^{ij}[u + t(v-u)](v-u)_{ij}$$

$$= \int_{\varphi\Omega} (v-u)_i \gamma_j S_k^{ij}[u + t(v-u)].$$

It is easy to see that the integrand on the right hand side is nonnegative.

Lemma 8.7 *Let $u_j \in C^2(\Omega)$ be a sequence of k-admissible functions which converges to a k-admissible function u in Ω almost everywhere. Then $\mu_k[u_j]$ converges to a measure μ weakly, namely for any smooth function φ with compact support in Ω,*

$$\int_\Omega \varphi \, d\mu_k[u_j] \to \int_\Omega \varphi \, d\mu. \tag{8.9}$$

Proof. For any open set $\Omega' \Subset \Omega$, by Theorem 7.1, $\mu_k[u_j](\Omega')$ is uniformly bounded. Hence there is a subsequence of $\mu_k[u_j]$ which converges weakly to a measure μ. We need to prove that μ is independent of the choice of subsequences of $\{u_j\}$.

Let $\{u_j\}, \{v_j\}$ be two sequences of k-admissible functions. Suppose both sequences $\{u_j\}$ and $\{v_j\}$ converge to u almost everywhere in Ω. Suppose that

$$\mu_k[u_j] \to \mu, \qquad \mu_k[v_j] \to \nu \tag{8.10}$$

weakly as measures. To prove that $\mu = \nu$, it suffices to prove that for any ball $B_r(x_0) \Subset \Omega$, $\mu(B_r) = \nu(B_r)$, or equivalently, for any small $t > 0$,

$$\mu(B_{r-2t}) \le \nu(B_{r+2t}), \tag{8.11}$$

$$\nu(B_{r-2t}) \le \mu(B_{r+2t}). \tag{8.12}$$

Let ε_j be a sequence of small positive constants converging to zero. Let $\hat{u}_j = \frac{1}{2}\varepsilon_j|x|^2 + u_j$ and $\hat{v}_j = \frac{1}{2}\varepsilon_j|x|^2 + v_j$. Then

$$S_k[\hat{u}_j] = S_k[u_j] + \sum_{i=1}^{k} C_{k,i}\varepsilon_j^i S_{k-i}[u] \ge C\varepsilon_j^k.$$

By Theorem 7.1, $S_{k-i}[u_j]$ is locally uniformly bounded in $L^1(\Omega)$. Hence by (8.10), $\mu_k[\hat{u}_j] \to \mu$ weakly. Therefore we may assume directly that $S_k[u_j] \ge \varepsilon_j > 0$ and $S_k[v_j] \ge \varepsilon_j > 0$.

We prove (8.11) and (8.12) in two steps. In the first one we assume that $u_j, v_j \in C^2(\Omega)$, $u \in C^0(\Omega)$, and $u_j, v_j \to u$ locally uniformly in Ω.

Let $\hat{v}_j = v_j + \delta_j[|x - x_0|^2 - r^2]$. Since $|u_j - v_j|$ converges to zero uniformly, there exists $\delta_j \to 0$ such that $\hat{v}_j < u_j$ in $B_{r-\frac{1}{2}t}$ and $\hat{v}_j > u_j$ on $\varphi B_{r+\frac{1}{2}t}$. Let $A = \{x \in \Omega \mid \hat{v}_j(x) < u_j(x)\}$. Then $B_{r-\frac{1}{2}t}(x_0) \subset A \subset B_{r+\frac{1}{2}t}(x_0)$. Hence by Lemma 8.6,

$$\int_{B_{r-2t}(x_0)} S_k[u_j] \le \int_A S_k[u_j] \le \int_A S_k[\hat{v}_j]$$

$$\le \int_A S_k[v_j] + O(\delta_j) \le \int_{B_{r+2t}} S_k[v_j] + O(\delta_j).$$

Sending $j \to \infty$ we obtain (8.11). Similarly we can prove (8.12).

The second step essentially repeats the first step. From the first step we see that for any continuous k-admissible function u, we can assign a measure $\mu_k[u]$ such that if a sequence of smooth k-admissible functions u_j converges to u uniformly, then $\mu_k[u_j]$ converges to $\mu_k[u]$ weakly as measure. In particular it means $\mu_k[u_j^t]$ and $\mu_k[v_j^t]$ are well defined, where we denote by u_j^t, v_j^t and u^t the Perron lifting of u_j, v_j and u in $\omega_t = B_{r+t}(x_0) - B_{r-t}(x_0)$. By Lemma 8.5, we have

$$u_j^t, v_j^t \to u^t \quad \text{in} \quad \Omega.$$

By the interior gradient estimate, u_j^t, v_j^t are locally uniformly Lipschitz continuous in ω_t. But u_j and v_j may not be C^2 in ω_t. To avoid such situation, we replace u_j^t (and v_j^t) in ω_t by the solution of $S_k[u] = \varepsilon_j'$ in ω_t satisfying the boundary condition $u_j^t = u_j$ (and $v_j^t = v_j$) on $\varphi\omega_t$, for sufficiently small ε_j'.

Let $\hat{v}_j^t = v_j^t + \delta_j[|x - x_0|^2 - r^2]$. Since $|u_j^t - v_j^t|$ converges to zero uniformly in $B_{r+\frac{3}{4}t} - B_{r-\frac{3}{4}t}$, there exists $\delta_j \to 0$ such that $\hat{v}_j^t < u_j^t$ on $\partial B_{r-\frac{1}{2}t}$ and $\hat{v}_j^t > u_j^t$ on $\varphi B_{r+\frac{1}{2}t}$. Let A' be the component of $\{u_j^t < \hat{v}_j^t\}$ which contains $\partial B_{r-\frac{1}{2}t}$. Let $\varphi' A'$ be the boundary of A' in the annulus $B_{r+\frac{3}{4}t} - B_{r+\frac{1}{4}t}$. Let A be the domain enclosed by $\varphi' A'$. Then $B_{r-\frac{1}{2}t} \subset A \subset B_{r+\frac{1}{2}t}$. Hence as above,

$$\int_{B_{r-2t}(x_0)} S_k[u_j] \le \int_A S_k[u_j^t] \le \int_A S_k[\hat{v}_j^t]$$

$$\le \int_A S_k[v_j^t] + O(\delta_j) \le \int_{B_{r+2t}} S_k[v_j] + O(\delta_j).$$

Sending $j \to \infty$ we obtain (8.11). Similarly we can prove (8.12).

Therefore for any k-admissible function u, we can assign a measure $\mu_k[u]$ to u, and μ_k is weakly continuous in u.

Theorem 8.1 *For any k-admissible function u, there exists a Radon measure $\mu_k[u]$ such that*

(i) $\mu_k[u] = S_k[u]dx$ if $u \in C^2(\Omega)$; and
(ii) *if $\{u_j\}$ is a sequence of k-admissible functions which converges to u a.e., then $\mu_k[u_j] \to \mu_k[u]$ weakly as measure.*

As an application, we compute the k-Hessian measure for the function

$$w_k(x) = \begin{cases} |x - y|^{2-n/k} & k > n/2, \\ \log|x - y| & k = n/2, \\ -|x - y|^{2-n/k} & k < n/2. \end{cases}$$

We have

$$\mu_k[w_k] = \begin{cases} \left(2 - \frac{n}{k}\right)\left[\binom{n}{k}\omega_n\right]^{1/k} \delta_y & \text{if } k \ne \frac{n}{2}, \\ \left[\binom{n}{k}\omega_n\right]^{1/k} \delta_y & \text{if } k = \frac{n}{2}, \end{cases}$$

where ω_n is the area of the unit sphere, and δ_y is the Dirac measure at y.

8.4 The Dirichlet Problem

As another application of Theorem 8.1, we consider the Dirichlet problem

$$S_k[u] = \nu \text{ in } \Omega, \tag{8.13}$$

$$u = \varphi \text{ on } \varphi\Omega.$$

When u is not smooth, the Hessian operator $S_k[u]$ in (8.13) is understood as $\mu_k[u]$, and u is called a *weak solution*. The following theorem was included in [TW2]. Here we give a different proof.

Theorem 8.2 *Let Ω be a $(k-1)$-convex domain with smooth boundary. Let φ be a continuous function on $\varphi\Omega$ and ν be a nonnegative Radon measure. Suppose that ν can be decomposed as*

$$\nu = \nu_1 + \nu_2 \tag{8.14}$$

such that ν_1 is a measure with compact support in Ω, and $\nu_2 \in L^p(\Omega)$ for some $p > \frac{n}{2k}$ if $k \le \frac{n}{2}$, or $p = 1$ if $k > \frac{n}{2}$. Then there exists a k-admissible weak solution u to (8.13).

Proof. Let ν_j be a sequence of smooth, positive functions which converges to ν weakly as measure. By the decomposition (8.14), we may assume that ν_j is uniformly bounded in $L^p(N_{\delta_0})$, where $N_\delta = \{x \in \Omega, \text{ dist}(x, \varphi\Omega) < \delta\}$. Let φ_j be a sequence of smooth functions which converges monotone increasingly to φ. Let u_j be the solution of

$$S_k[u] = \nu_j \text{ in } \Omega, \tag{8.15}$$
$$u = \varphi_j \text{ on } \varphi\Omega.$$

If u_j is uniformly bounded in $L^1(\Omega)$, by Corollary 7.1, $\{u_j\}$ contains a convergent subsequence which converges to a k-admissible function u. By Lemma 8.7, u is a weak solution to (8.13). Therefore it suffices to prove that u_j is uniformly bounded in $L^1(\Omega)$ and the limit function u satisfies the boundary condition $u = \varphi$ on $\varphi\Omega$.

For $\delta > 0$, let $\eta_\delta \in C_0^\infty(\Omega)$ be a nonnegative function satisfying $\eta_\delta(x) = 1$ when $\text{dist}(x, \varphi\Omega) > \delta$ and $\eta_\delta(x) = 0$ when $\text{dist}(x, \varphi\Omega) < 3\delta/4$. Let

$$\nu_{j,\delta} = \nu_j \eta_\delta + \delta,$$
$$\nu'_{j,\delta} = \nu_j(1 - \eta_\delta) + \delta.$$

Then both $\nu_{j,\delta}$ and $\nu'_{j,\delta}$ are smooth, positive functions. Let $u_{j,\delta}$ be the solution of the

$$S_k[u] = \nu_{j,\delta} \text{ in } \Omega, \tag{8.16}$$
$$u = \varphi_j \text{ on } \varphi\Omega.$$

Let $u'_{j,\delta}$ be the solution of the

$$S_k[u] = \nu'_{j,\delta} \text{ in } \Omega,$$
$$u = 0 \text{ on } \varphi\Omega.$$

By Theorem 5.5, we have

$$\|u'_{j,\delta}\|_{L^\infty(\Omega)} \le C\|\nu'_{j,\delta}\|_{L^p(\Omega)}^{1/k} \to 0$$

uniformly in j, as $\delta \to 0$. By the concavity of $S_k^{1/k}[u]$,

$$S_k^{1/k}[u_{j,\delta} + u'_{j,\delta}] \geq S_k^{1/k}[u_{j,\delta}] + S_k^{1/k}[u'_{j,\delta}]$$

Hence $u_{j,\delta} + u'_{j,\delta}$ is a sub-barrier to the Dirichlet problem (8.15). Hence it suffices to prove that for any given $\delta > 0$, $u_{j,\delta}$ is uniformly bounded in $L^1(\Omega)$ and $u_\delta = \lim u_{j,\delta}$ satisfies the boundary condition $u_\delta = \varphi$ on $\varphi\Omega$.

For any fixed $\delta > 0$, we claim that $u_j = u_{j,\delta}$ is uniformly bounded in $N_{\delta/2}$ (in the following we drop the subscript δ). Indeed, if this is not true, for a fixed, sufficiently small $\varepsilon \in (0, \frac{1}{4}\delta)$, let $D_\varepsilon = \{x \in \mathbb{R}^n \mid \mathrm{dist}(x, \Omega) < \varepsilon\}$ be the ε-neighborhood of Ω. Let $\eta = K(\rho_{D_\varepsilon}(x) + K'\rho_{D_\varepsilon}^2(x))$, where $\rho_{D_\varepsilon}(x)$ is the distance from x to ∂D_ε. Then η is k-admissible when K' is large and $d_x < \delta_1$ for some $\delta_1 > 0$ depending only on n, k and $\varphi\Omega$. If u_j is not uniformly bounded in $N_{\delta/2}$, by the Harnack inequality (Theorem 4.2), $u_j(x) \to -\infty$ uniformly for $x \in \{x \in \Omega \mid \mathrm{dist}(x, \varphi\Omega) = 2\varepsilon\}$. Choose K large such that $\eta < u_j$ on $\varphi\Omega$ and $u_j < \eta$ on $\varphi\Omega_\varepsilon$. Let

$$\partial A_j = \{x \in D_\varepsilon \mid \rho_{D_\varepsilon}(x) \in (0, 2\varepsilon), \eta(x) = u_j(x)\}.$$

Then $\varphi A_j \subset N_{\delta/2}$. Let A_j be the domain enclosed by ∂A_j. By Lemma 8.6,

$$\nu_{j,\delta}(\Omega) \geq \nu_{j,\delta}(A_j) = \int_{A_j} S_k[u_j] \geq \int_{A_j} S_k[\eta] \to \infty$$

as $K \to \infty$. But the left hand side is uniformly bounded. The contradiction implies that u_j is uniformly bounded in $N_{\delta/2}$. Since u_j is sub-harmonic, by the mean value inequality (8.4) it follows that u_j is uniformly bounded in $L^1_{loc}(\Omega)$.

To show that $u = \lim u_j$ satisfies the boundary condition $u = \varphi$ on $\varphi\Omega$, extend φ to a harmonic function in Ω. Since u is sub-harmonic, by the comparison principle we have $u \leq \varphi$ in Ω. Hence for any $y \in \varphi\Omega$, $\overline{\lim}_{x \to y} u(x) \leq \varphi(y)$. Next we prove that $\underline{\lim}_{x \to y} u(x) \geq \varphi(y) \ \forall \ y \in \varphi\Omega$. Let w_j be the solution to the Dirichlet problem

$$S_k[w] = K \ \text{in} \ \Omega$$
$$w = \varphi_j \ \text{on} \ \varphi\Omega.$$

Since u_j is uniformly bounded in $N_{\delta/2}$, we can fix a sufficiently large K, which may depend on δ but is independent of j, such that the solution $w_j < -K$ on $\Omega \cap \varphi N_{\delta/2}$. Recall that $\nu_{j,\delta} = \delta < 1$ in $N_{\delta/2}$. By the comparison principle we have $w_j \leq u_j$ in $N_{\delta/2}$. But when K is fixed, we have $\lim_{j \to \infty} \underline{\lim}_{x \to y} w_j(x) = \varphi(y)$ uniformly.

The uniqueness is a more complicated issue. It is proved in [TW3] that if $\nu \in L^1$, the solution in Theorem 8.2 is unique.

9 Local Behavior of Admissible Functions

In this section we prove a Wolff potential estimate and give a necessary and sufficient condition such that a weak solution is Hölder continuous. Results in this section are due to D. Labutin [Ld].

9.1 The Wolff Potential Estimate

Given a Radon measure μ on Ω, we denote

$$W_k^\mu(x, r) = \int_0^r \left(\frac{\mu(B_t(x))}{t^{n-2k}} \right)^{\frac{1}{k}} \frac{dt}{t}. \tag{9.1}$$

$W_k^\mu(x, r)$ is called Wolff potential.

Lemma 9.1 *Let $u \leq 0$ be k-admissible in $B_R(0)$. Then*

$$\left[\frac{\mu_k[u](B_{9R/10})}{R^{n-2k}} \right]^{\frac{1}{k}} \leq C \inf_{\partial B_{R/2}} (-u). \tag{9.2}$$

If furthermore $\mu_k[u] = 0$ in $(B_{5R/8} - B_{3R/8}) \cup (B_{11R/10} - B_{9R/10})$, then

$$\inf_{\partial B_R} u - \inf_{\partial B_{R/2}} u \leq C \left(\frac{\mu[u](B_R)}{R^{n-2k}} \right)^{\frac{1}{k}}, \tag{9.3}$$

where C is independent of R and u.

Proof. First we prove (9.2). Let ψ be the solution of $S_k[\psi] = 0$ in $B_R - \overline{B}_{9R/10}$, $\psi = 0$ on ∂B_R, and $\psi = u$ on $\partial B_{9R/10}$. By replacing u by ψ in the annulus $B_R - \overline{B}_{9R/10}$, we may assume that $u = 0$ on ∂B_R and $\mu_k[u] = 0$ in $B_R - \overline{B}_{9R/10}$. By the Harnack type inequality,

$$\sup_{\partial B_{19R/20}} (-u) \leq C \inf_{\partial B_{19R/20}} (-u). \tag{9.4}$$

Let φ be a radial k-admissible function satisfying $S_k[\varphi] = 0$ in $B_R - B_{19R/20}$, $\varphi = 0$ on ∂B_R, and $\varphi = \inf_{\partial B_{19R/20}} u$ on $\partial B_{19R/20}$. Then by the comparison principle we have $u \geq \varphi$ in $B_R - \overline{B}_{9R/10}$. By Lemma 8.6, it follows that

$$\mu_k[u](B_R) \leq \mu_k[\varphi](B_R) = C_{n,k} R^{n-2k} \left| \inf_{\partial B_{19R/20}} u \right|^k.$$

We obtain by the Harnack inequality (9.4)

$$\left[\frac{\mu_k[u](B_{9R/10})}{R^{n-2k}}\right]^{\frac{1}{k}} \leq C \inf_{\partial B_{19R/20}} |u|.$$

Note that u is subharmonic, $\inf_{\partial B_{R/2}} |u| \leq \inf_{\varphi B_{19R/20}} |u|$. We obtain (9.2).

To prove (9.3), let w be the solution of $S_k[w] = \mu_k[u]$ in B_R and $w = \inf_{\partial B_R} u$ on ∂B_R. The solution w should be obtained as the limit of the solution w_ε to $S_k[w] = \mu_k[u_\varepsilon]$ in B_R and $w_\varepsilon = u_\varepsilon$ on ∂B_R. Note that by assumption, $S_k[u] = 0$ near ∂B_R, so u is Lipschitz continuous near ∂B_R. It follows that $w \leq u$ and so

$$\inf_{\partial B_R} u - \inf_{\partial B_{R/2}} u \leq \inf_{\partial B_R} w - \inf_{\partial B_{R/2}} w.$$

Therefore to prove (9.3), we may assume that $u =$ constant on φB_R. By subtracting we may assume that $u = 0$ on $\varphi\Omega$.

Let φ be a radial k-admissible function satisfying $S_k[\varphi] = 0$ in $B_R - B_{R/2}$, $\varphi = 0$ on ∂B_R, and $\varphi = \sup_{\partial B_{R/2}} u$. Then by the comparison principle we have $u \leq \varphi$ in $B_R - B_{R/2}$. It follows that

$$\mu_k[u](B_R) \geq \mu_k[\varphi](B_R) = C_{n,k} R^{n-2k} \left| \sup_{\partial B_{R/2}} u \right|^k.$$

By the Harnack inequality, $|\sup_{\partial B_{R/2}} u| \geq C |\inf_{\varphi B_{R/2}} u|$. We obtain (9.3).

Theorem 9.1 *Let $u \leq 0$ be a k-admissible function in $B_{2R}(0)$. Then we have*

$$C^{-1} W_k^\mu(0, R) \leq -u(0) \leq C\{W_k^\mu(0, 2R) + |\sup_{\partial B_R} u|\}, \tag{9.5}$$

where $\mu = \mu[u]$, and C is independent of u and R.

Proof. First we prove the left inequality, namely $|u(0)| \geq C^{-1} W_k^\mu(0, R)$. For any $r \in (0, \frac{1}{2}R)$, let $\omega = B_{9r/8} - B_{3r/4}$, let u^ω be the Perron lifting of u over ω, and let $\tilde{u}$ be the Perron lifting of u^ω over $B_{7r/8}$. By (9.2) we have

$$\left[\frac{\mu_k[\tilde{u}](B_{9r/10})}{r^{n-2k}}\right]^{1/k} \leq C\left(\sup_{\partial B_{9r/8}} \tilde{u} - \sup_{\partial B_{7r/8}} \tilde{u}\right)$$

$$\leq C\left(\sup_{\partial B_{3r/2}} u - \sup_{\partial B_{3r/4}} u\right).$$

Observing that

$$\mu_k[u](B_{r/2}) = \mu_k[u^\omega](B_{r/2}) \leq \mu_k[u^\omega](B_{9r/10}) = \mu_k[\tilde{u}](B_{9r/10}),$$

we obtain,

$$\left[\frac{\mu_k[u](B_{r/2})}{r^{n-2k}}\right]^{1/k} \leq C\left(\sup_{\partial B_{3r/2}} u - \sup_{\partial B_{3r/4}} u\right). \tag{9.6}$$

For $j = 0, 1, \cdots$, let $R_j = 2^{-j}R$. We have,

$$C^{-1} \sum_{j=0}^{\infty} \left[\frac{\mu_k(B_{R_j})}{R_j^{n-2k}} \right]^{1/k} \leq W_k^{\mu}(0, R) \leq C \sum_{j=0}^{\infty} \left[\frac{\mu_k(B_{R_j})}{R_j^{n-2k}} \right]^{1/k}. \qquad (9.7)$$

Hence letting $r = R_j$ in (9.6) and summing up we obtain the first inequality of (9.5).

To prove the second inequality, we may suppose $\mu_k[u] = 0$ in $B_{2R} - B_R$. Let $R_j = 2^{-j}R$, $\omega = \bigcup_{j=1}^{\infty}(B_{5R_j/4} - B_{3R_j/4})$, and let u^{ω} be the Perron lifting of u over ω. Then $u = u^{\omega}$ in $B_{2R} - \omega$, $\mu_k[u^{\omega}] = 0$ in ω. Since $\mu_k[u]$ depends on u locally, we have, for any $r > 0$,

$$\mu_k[u](B_r) \leq \mu_k[u^{\omega}](B_{2r}) \leq \mu_k[u](B_{4r}).$$

Hence to prove the second inequality we may suppose directly that $\mu_k[u] = 0$ in ω.

Let $u_j = u^{B_{R_j}}$, the Perron lifting of u over B_{R_j}. Then $u_j \searrow u$ pointwise. In particular $u_j \searrow u$ at the origin. Hence to prove the second inequality it suffices to show that for all $s \geq 1$,

$$\left| \inf_{\partial B_{R_s}} u \right| \leq C \sum_{j=0}^{s-1} \left[\frac{\mu_k[u](B_{R_j})}{R_j^{n-2k}} \right]^{1/k} + C \left| \sup_{\partial B_R} u \right| \qquad (9.8)$$

and send $s \to \infty$.

By (9.3) we have

$$\left| \inf_{\partial B_{R_s}} u \right| \leq \left| \inf_{\partial B_{R_{s-1}}} u \right| + C \left[\frac{\mu_k[u](B_{R_{s-1}})}{R_{s-1}^{n-2k})} \right]^{1/k}.$$

Applying (9.3) repeatedly, we obtain, for $0 \leq j \leq s$,

$$\left| \inf_{\partial B_{R_s}} u \right| \leq \left| \inf_{\partial B_{R_j}} u \right| + C \sum_{i=j}^{s-1} \left[\frac{\mu_k[u](B_{R_j})}{R_j^{n-2k}} \right]^{1/k}.$$

Letting $j = 0$, we obtain (9.8). Since $\mu_k[u] = 0$ in $B_{5R/4} - B_{3R/4}$, we have $|\inf_{\partial B_R} u| \leq C |\sup_{\varphi B_R} u|$ by the Harnack inequality. This completes the proof.

9.2 Hölder Continuity of Weak Solutions

From Theorem 9.1 we obtain a necessary and sufficient condition for a weak solution to be Hölder continuous.

Theorem 9.2 *A k-admissible function u in Ω is Hölder continuous if and only if there exists a constant $\varepsilon > 0$ such that for any $x \in \Omega$, $r \in (0, 1)$, the measure $\mu_k[u]$ satisfies*

$$\mu_k[u](B_r \cap \Omega) \leq Cr^{n-2k+\varepsilon}. \tag{9.9}$$

Proof. If u is Hölder continuous with exponent $\alpha \in (0, \frac{1}{2})$, from the first inequality in (9.5) we obtain

$$\int_0^r \left(\frac{\mu(B_t(x))}{t^{n-2k}} \right)^{\frac{1}{k}} \frac{dt}{t} \leq Cr^\alpha.$$

Hence

$$\frac{\mu(B_{r/2}(x))}{t^{n-2k}} \leq Cr^{k\alpha}.$$

We obtain (9.9) with $\varepsilon = k\alpha$.

Next assume that (9.9) holds. Consider the function u in $B_R(0)$. We want to prove that $|u(x) - u(0)| < Cr^\alpha$ for $|x| < r = \frac{1}{2}R^2$. Replacing u by the Perron lifting u^ω, where $\omega = B_R - B_{R/2}$, we may assume that $\mu_k[u] = 0$ in ω. Let w_1 be the solution of $S_k[w] = 0$ in B_R and $w = u$ on ∂B_R. Let w_2 be the solution of $S_k[w] = \mu_k[u]$ in B_R and $w = 0$ on ∂B_R. Then

$$w_1 \geq u \geq w_1 + w_2.$$

Hence

$$u(x) - u(0) \leq w_1(x) - [w_1(0) + w_2(0)] \leq [w_1(x) - w_1(0)] + w_2(0).$$

By (9.9) and the second inequality in (9.5), we have $w_2(0) \leq CR^{\varepsilon/k}$. By the interior gradient estimate, w_1 is Lipschitz continuous. Hence

$$|w_1(x) - w_1(0)| \leq \frac{C}{R}|x| \leq C|x|^{1/2}.$$

We obtain

$$u(x) - u(0) \leq C|x|^{1/2} + CR^{\varepsilon/k}.$$

Similarly we have $u(0) - u(x) \leq C|x|^{1/2} + CR^{\varepsilon/k}$. Hence u is Hölder continuous at the origin with exponent $\varepsilon/2k$.

From Theorem 9.2, we obtain

Corollary 9.1 *Let u be a k-admissible solution ($k \leq \frac{n}{2}$) to*

$$S_k[u] = f. \tag{9.10}$$

Suppose $f \in L^p(\Omega)$ for some $p > \frac{n}{2k}$. Then u is Hölder continuous.

From Theorem 9.1, one can also prove that a k-admissible function u is continuous at x if and only if $W_k^\mu(x, r) \to 0$ as $r \to 0$. One can also introduce the notion of capacity, and establish various potential theoretical results, such as quasi-continuity of k-admissible functions and the Wiener criterion for the continuity of k-admissible functions at the boundary, just as in the Newton potential theory. We refer the reader to [TW3, Ld] for more details. More applications of the Wolff potential estimate can be found in [PV1, PV2].

10 Parabolic Hessian Equations

This section includes the a priori estimates and existence of solutions for the parabolic Hessian equations used before. We refer the reader to [Lg] for more general fully nonlinear parabolic equations of parabolic type.

Consider the initial boundary value problem

$$F[u] - u_t = f(x, t, u) \quad \text{in } \Omega \times [0, \infty) \tag{10.1}$$
$$u(\cdot, 0) = u_0,$$
$$u = 0 \quad \text{on } \varphi\Omega \times [0, \infty)$$

where

$$F[u] = \mu(S_k[u]). \tag{10.2}$$

We assume that μ is a smooth function defined on $(0, \infty)$, satisfying $\mu'(t) > 0$, $\mu''(t) < 0$ for all $t > 0$, and

$$\mu(t) \to -\infty \quad \text{as } t \to 0, \tag{10.3}$$
$$\mu(t) \to +\infty \quad \text{as } t \to +\infty. \tag{10.4}$$

Furthermore we assume that $\mu(\sigma_k(\lambda))$ is concave in λ, which implies that $F[u]$ is concave in $D^2 u$. A natural candidate for μ is $\mu(t) = \log t$, such as in (5.17). But we have also used different function μ, such as in (6.3) (6.14).

We say a function $u(x, t)$ is k-admissible with respect to the parabolic equation (10.1) if for any given $t \geq 0$, $u(\cdot, t)$ is k-admissible. Equation (10.1) is parabolic when u is k-admissible. Condition (10.3) is to ensure that $\sigma_k(\lambda) > 0$ so that the admissibility can be kept at all time.

Theorem 10.1 *Assume $u_0 \in C^4(\overline{\Omega})$ is k-admissible, and satisfies the compatibility condition*

$$F[u_0] = f \quad on \quad \varphi\Omega \times \{t = 0\}. \tag{10.5}$$

Assume $\varphi\Omega \in C^{3,1}$ and Ω is $(k-1)$-convex, $f \in C^{2,1}_{x,t}(\overline{Q}_T)$, and μ satisfies the conditions above. Let u be an k-admissible solution of (10.1). Then we have the a priori estimate

$$\|u\|_{C^{2,1}_{x,t}(\overline{Q}_T)} \leq C, \tag{10.6}$$

where $Q_T = \Omega \times (0,T]$, C depends on n, k, $\varphi\Omega$, $\|u_0\|_{C^4(\overline{\Omega})}$, $\sup_{Q_T} |u|$, and $\|f\|_{C^{1,1}(\overline{Q}_T)}$.

To prove Theorem 10.1, one first establish an upper bound for $\sup_{Q_T} |u_x|$ and $\sup_{Q_T} |u_t|$, then prove $\sup_{Q_T} |u_{xx}|$ is bounded. The estimates for $\sup_{Q_T}(|u_x| + |u_t|)$ will be given in the proof of Theorem 10.2 below. The estimate for $\sup |u_{xx}|$ is similar to that for the elliptic equation (3.1) and is omitted here. We refer the reader to [Ch1, W2] for details. See also [Lg].

Note that when applying Theorem 10.1 to equation (5.17), the a priori bound for $\sup_{Q_T} |u|$ is guaranteed by our truncation of $|u|^p$, namely the function $f(u)$ in (5.15). In equation (5.17), the right hand side involves an integration $\beta(u)$, which satisfies the estimate $C_1 \leq \beta(u) \leq C_2$ for two absolute positive constants C_1, C_2. This integration does not affect the a priori estimate for $\sup_{Q_T} |u_t|$. See the proof of (10.11) below. Once u_t is bounded, $\beta(u)$ is positive and Lipschitz continuous in t.

Estimate (10.6) implies that equation (10.1) is uniformly parabolic. Therefore by Krylov's regularity theory for uniformly parabolic equation [K1], we obtain higher order derivative estimates. By the a priori estimates, one can then prove the local existence of smooth solutions by the contraction mapping theorem. In particular, if $\sup_{\Omega \times [0,T]} |u| < \infty$ for any $T > 0$, the smooth solution exists at all time $t > 0$.

In Step 3 of the proof of Theorem 6.3, we need a special interior gradient estimate, namely (10.10) below, for solutions to equation (10.1) with μ given in (6.14). We provide a proof for it below. See also [CW1]. Estimate for higher order derivatives and existence of solutions can be obtained similarly as above.

Theorem 10.2 Let μ be the function in (6.14). Assume $u_0 \in C^4(\overline{\Omega})$ is k-admissible, and satisfies the compatibility condition (10.5). Assume that f is C^2 in x and u, C^1 in t, and satisfies,

$$f(x,t,u) \leq C_0(1+|u|) \quad \forall\, (x,t,u) \in \overline{\Omega} \times \mathbb{R}. \tag{10.7}$$

Suppose $u \in C^{4,2}_{x,t}(\overline{\Omega} \times [0,\infty))$ is a k-admissible solution of (10.1). Then we have, for $0 < t < T$,

$$u(x,t) \geq -e^{C_1 t} \sup_{\Omega} |u_0(x)|, \tag{10.8}$$

$$|\nabla_x u(x,t)| \leq C_2(1 + \frac{1}{r} M_t^{(p+k)/2k}), \tag{10.9}$$

$$|u_t(x,t)| \leq C_3(1 + M_t), \tag{10.10}$$

where $M_t = \sup_{Q_t} |u|$, $r = dist(x, \varphi\Omega)$. The constant C_1 depends only on n, k, p and C_0; C_2 and C_3 depends additionally on u_0 and the gradient of f.

Proof. Estimate (10.8) is obvious as the right hand side is a lower barrier. To prove (10.9) and (10.10) we assume for simplicity that $M_t \geq 1$. First we prove (10.10). Let

$$G = \frac{u_t}{M - u},$$

where $M = 2M_t$. If G attains its minimum on the parabolic boundary of Q_t, we have $u_t \geq -C$ for some $C > 0$ depending on the initial value u_0. Hence we may suppose G attains its minimum at an interior point in Q_t. At this point we have

$$u_{tt} + (M - u)^{-1}u_t^2 \leq 0,$$
$$u_{jt} + (M - u)^{-1}u_t u_j = 0, \quad j = 1, 2, \cdots, n,$$

and the matrix

$$\{u_{ijt} + (M - u)^{-1} \ (u_{it}u_j + u_{jt}u_i + u_t u_{ij}) + 2(M - u)^{-2}u_i u_j u_t\}$$
$$= \{u_{ijt} + (M - u)^{-1}u_t u_{ij}\} \geq 0.$$

Differentiating the equation (10.1) we get

$$F_{ij}u_{ijt} - u_{tt} = f_t + f_u u_t,$$
$$F_{ij}u_{rij} - u_{rt} = f_r + f_u u_r,$$

where $F_{ij} = \frac{\varphi}{\partial u_{ij}}F[u]$. We may suppose $u_t \leq 0$ at this point. From the above formulae we obtain

$$(M - u)^{-1}u_t^2 \leq -F_{ij}u_{ijt} + f_t + f_u u_t$$
$$\leq (M - u)^{-1}u_t F_{ij}u_{ij} + f_t + f_u u_t$$
$$\leq f_t + f_u u_t.$$

Hence $u_t \geq -C$ for some C depending on $\sup_{Q_t} f_t$ and $\inf_{Q_t} f_u$.

Similarly let

$$G = \frac{u_t}{M + u}.$$

If G attains its maximum on the parabolic boundary of Q_t, we have $u_t \leq C$. If it attains its maximum at some point in Q_t. At this point we have

$$u_{tt} - (M + u)^{-1}u_t^2 \geq 0,$$
$$u_{jt} - (M + u)^{-1}u_t u_j = 0, \quad j = 1, 2, \cdots, n,$$
$$\{u_{ijt} - (M + u)^{-1}u_t u_{ij}\} \leq 0.$$

Hence as above we obtain

$$
\begin{aligned}
(M+u)^{-1}u_t^2 &\leq F_{ij}u_{ijt} - f_t - f_u u_t \\
&\leq (M+u)^{-1}u_t F_{ij}u_{ij} - f_t - f_u u_t \\
&= (M+u)^{-1}ku_t\mu' S_k[u] - f_t - f_u u_t.
\end{aligned}
$$

If $S_k[u] \leq 10$ at the point, by the equation (10.1) we have

$$
u_t = F[u] - f \leq C.
$$

Otherwise we have $\mu(t) = t^{1/p}$ and so

$$
\mu' S_k[u] = \frac{1}{p}\mu(S_k[u]) = \frac{1}{p}(u_t + f).
$$

It follows that

$$
(M+u)^{-1}u_t^2 \leq (M+u)^{-1}ku_t(u_t + f)/p - f_t - f_u u_t.
$$

That is

$$
\frac{p-k}{p}\frac{u_t^2}{M+u} \leq \frac{kf u_t}{p(M+u)} - f_t - f_u u_t.
$$

We obtain $u_t \geq -C$ for some C depending on $\inf_{Q_t} f_t$ and $\inf_{Q_t} f_u$.

Next we prove (10.9). For simplicity let us take $t = T$. The proof below is similar to that of the interior gradient estimate in §4. Assume that $B_r(0) \subset \Omega$. Consider

$$
G(x,t,\xi) = \rho(x)\varphi(u)u_\xi,
$$

where $\rho(x) = 1 - |x|^2/r^2$, $\varphi(u) = (M-u)^{-1/4}$. Suppose

$$
\sup\{G(x,t,\xi) \mid x \in B_r(0), t \in (0,T], |\xi| = 1\}
$$

is attained at (x_0, t_0) (with $t_0 > 0$) and $\xi_0 = (1, 0, \cdots, 0)$. Then at the point we have

$$
0 = (\log G)_i = \frac{\rho_i}{\rho} + \frac{\varphi_i}{\varphi} + \frac{u_{1i}}{u_1},
$$

$$
\begin{aligned}
0 &\geq F_{ij}(\log G)_{ij} - (\log G)_t \\
&= F_{ij}\left(\frac{\rho_{ij}}{\rho} - \frac{\rho_i\rho_j}{\rho^2}\right) - \frac{\rho_t}{\rho} + F_{ij}\left(\frac{\varphi_{ij}}{\varphi} - \frac{\varphi_i\varphi_j}{\varphi^2}\right) - \frac{\varphi_t}{\varphi} + F_{ij}\left(\frac{u_{1ij}}{u_1} - \frac{u_{1i}u_{1j}}{u_1^2}\right) - \frac{u_{1t}}{u_1} \\
&\geq F_{ij}\left(\frac{\rho_{ij}}{\rho} - 3\frac{\rho_i\rho_j}{\rho^2}\right) - \frac{\rho_t}{\rho} + F_{ij}\left(\frac{\varphi_{ij}}{\varphi} - 3\frac{\varphi_i\varphi_j}{\varphi^2}\right) - \frac{\varphi_t}{\varphi} + \frac{1}{u_1}(F_{ij}u_{1ij} - u_{1t}) \\
&\geq -\frac{C}{\rho^2}\mathcal{F} + \left(\frac{\varphi''}{\varphi} - 3\frac{\varphi'^2}{\varphi^2}\right)F_{11}u_1^2 + \frac{\varphi'}{\varphi}(F_{ij}u_{ij} - u_t) + \frac{f_1}{u_1},
\end{aligned}
$$

where $\mathcal{F} = \sum F_{ii}$. By our choice of φ,

$$\frac{\varphi''}{\varphi} - 3\frac{\varphi'^2}{\varphi} = \frac{1}{8(M-u)^2}$$

Note that $F_{ij}u_{ij} \geq 0$ and $\varphi' \geq 0$. We obtain

$$0 \geq \frac{F_{11}u_1^2}{32M^2} - \frac{C}{\rho^2}\mathcal{F} - \frac{\varphi'}{\varphi}u_t + \frac{f_1}{u_1}.$$

Therefore we have either

$$F_{11}u_1^2 \leq \frac{CM^2}{\rho^2}\mathcal{F}, \tag{10.11}$$

or

$$F_{11}u_1^2 \leq CM^2(\frac{\varphi'}{\varphi}u_t - \frac{f_1}{u_1}) \leq CM^2. \tag{10.12}$$

In (10.12) we have used the estimate (10.10).
    Recall that

$$u_{11} = -u_1(\frac{\rho_1}{\rho} + \frac{\varphi'}{\varphi}u_1).$$

We may assume that $u_1 \geq CM/r$, for otherwise (10.9) is readily verified. Hence we have

$$u_{11} \leq -\frac{\varphi'}{2\varphi}u_1^2 \leq -\frac{C}{M}u_1^2.$$

From the proof of Theorem 4.1 we then have

$$S_k^{11}[u] \geq C \sum S_k^{ii}[u],$$

$$S_k^{11}[u] \geq C\frac{u_1^{2k-2}}{M^{k-1}},$$

Therefore in the case (10.11), we obtain $\rho u_1 \leq CM$ and (10.9) follows. In the case (10.12), we observe by equation (10.1) and estimate (10.10) that $S_k[u] \leq CM^p$ at (x_0, t_0). Hence $\mu'(S_k[u]) \geq CM^{-p+1}$. We therefore obtain

$$F_{11} \geq \frac{Cu_1^{2k-2}}{M^{p+k-2}}.$$

Inserting into (10.12) we obtain $u_1(x_0, t_0) \leq CM^{(p+k)/2k}$. Hence at the center $x = 0$ we have

$$|Du(0,t)| \leq \frac{\rho\varphi(x_0, t_0)}{\rho\varphi(0, t)}u_1(x_0, t_0) \leq CM^{(p+k)/2k}.$$

This completes the proof.

11 Examples of Fully Nonlinear Elliptic Equations

This is the notes for my lectures under the title *Fully nonlinear elliptic equations*, given in a workshop at C.I.M.E., Italy. So it is appropriate to give more examples of fully nonlinear elliptic equations here.

(i) One of the most important fully nonlinear equations is the *Monge-Ampère equation*

$$\det D^2 u = f(x, u, Du). \tag{11.1}$$

The Monge-Ampère equation finds many applications in geometry and applied sciences. A special case of the Monge-Ampère equation is the *prescribing Gauss curvature equation*

$$\frac{\det D^2 u}{(1 + |Du|^2)^{(n+2)/2}} = \kappa(x), \tag{11.2}$$

where κ is the Gauss curvature of the graph of u.

(ii) A related equation is the *complex Monge-Ampère equation*

$$\det u_{z_i \bar{z}_j} = f, \tag{11.3}$$

which plays an important role in complex geometry.

(iii) The *k-Hessian equation*

$$S_k[u] = f(x), \tag{11.4}$$

studied in previous sections.

(iv) The *k-curvature equation*

$$H_k[u] = f(x), \tag{11.5}$$

where $1 \leq k \leq n$ and $H_k[u] = \sigma_k(\kappa)$, is a class of prescribing Weingarten curvature equations, where $\kappa = (\kappa_1, \cdots, \kappa_n)$ are the principal curvatures of the graph of u. The k-curvature equation is just the mean curvature equation when $k = 1$, and the Gauss curvature equation when $k = n$.

Related to the k-Hessian and k-curvature equations are the Hessian quotient and curvature quotient equations, that is

$$\frac{S_k[u]}{S_l[u]} = f(x), \tag{11.6}$$

$$\frac{H_k[u]}{H_l[u]} = f(x), \tag{11.7}$$

where $0 \le l < k \le n$, and S_k and H_k are respectively the k-Hessian and k-curvature operator. A special case of (11.7) is the prescribing harmonic curvature equation, that is when $k = n, l = n - 1$.

(v) The *special Lagrangian equation*

$$\arctan\lambda_1 + \cdots \arctan\lambda_n = c \qquad (11.8)$$

is a fully nonlinear equation arising in geometry. If u is a solution, the graph $(x, \nabla u(x))$ is a minimal surface in $\mathbb{R}^n \times \mathbb{R}^n$. When $n = 3$ and $c = k\pi$, equation (11.8) can be written as

$$\Delta u = \det D^2 u. \qquad (11.9)$$

(vi) In stochastic control theory there arises the *Bellman equation*

$$F[u] = \inf_{\alpha \in V} \{L_\alpha[u] - f_\alpha(x)\}, \qquad (11.10)$$

or more generally the Bellman-Isaacs equation

$$F[u] = \sup_{\alpha \in U} \inf_{\beta \in V} \{L_{\alpha,\beta}[u] - f_{\alpha,\beta}(x)\}, \qquad (11.11)$$

where α, β are indexes and $L_{\alpha,\beta}$ are linear elliptic operators.

(vii) Another well-known fully nonlinear equation is Pucci's equation [GT], which is a special Bellman equation. For $\alpha \in (0, \frac{1}{n}]$, let $\mathcal{L}_\alpha$ denote the set of linear uniformly elliptic operator of the form $L[u] = a_{ij}(x)\partial_{ij}u$ with bounded measurable coefficients a_{ij} satisfying $a_{ij}\xi_i\xi_j \ge \alpha|\xi|^2$, $\Sigma a_{ii} = 1$ for all $\xi \in \mathbb{R}^n, x \in \Omega$. Pucci's operators are defined by

$$P_\alpha^+[u] = \sup_{L \in \mathcal{L}_\alpha} L[u], \qquad (11.12)$$
$$P_\alpha^-[u] = \inf_{L \in \mathcal{L}_\alpha} L[u].$$

By direct calculation [GT],

$$P_\alpha^+[u] = \alpha\Delta u + (1 - n\alpha)\lambda_1(D^2 u), \qquad (11.13)$$
$$P_\alpha^-[u] = \alpha\Delta u + (1 - n\alpha)\lambda_n(D^2 u),$$

where $\lambda_1(D^2 u)$ and $\lambda_n(D^2 u)$ denote the maximum and minimum eigenvalues of $D^2 u$.

(viii) Equation (11.1) is the standard Monge-Ampère equation. In many applications one has the Monge-Ampère equation of general form,

$$\det\{D^2 u - A(x, u, Du)\} = f(x, u, Du), \qquad (11.14)$$

where A is an $n \times n$ matrix. Similarly one has an extension of the k-Hessian equation (11.3), that is

$$S_k\{\lambda(D^2u - A(x, u, Du))\} = f. \qquad (11.15)$$

Equation (11.14) arises in applications such as reflector design, optimal transportation, and isometric embedding. Equation (11.15) is related to the so-called k-Yamabe problem in conformal geometry.

Some of the above equations may not be elliptic in general, such as the Monge-Ampère equation (11.1) and the k-Hessian equation (11.4). But they are elliptic when restricted to an appropriate class of functions.

References

[B]   I.J. Bakelman, Variational problems and elliptic Monge-Ampère equations, J. Diff. Geo. 18(1983), 669–999.

[BN]   H. Brezis and L. Nirenberg, Positive solutions of nonlinear elliptic equations involving critical Sobolev exponents, Comm. Pure Appl. Math., 36(1983), 437–477.

[Ca]   L.A. Caffarelli, Interior $W^{2,p}$ estimates for solutions of Monge-Ampère equations, Ann. Math., 131 (1990), 135–150.

[CNS1]   L. Caffarelli, L. Nirenberg, J. Spruck, Dirichlet problem for nonlinear second order elliptic equations I, Monge-Ampere equations, Comm. Pure Appl. Math., 37(1984), 369–402.

[CNS2]   L. Caffarelli, L. Nirenberg, J. Spruck, Dirichlet problem for nonlinear second order elliptic equations III, Functions of the eigenvalues of the Hessian, Acta Math., 155(1985), 261–301.

[Ch1]   K.S. Chou, On a real Monge-Ampère functional, Invent. Math., 101(1990), 425–448.

[Ch2]   K.S. Chou, Remarks on the critical exponents for the Hessian operators, Ann. Inst. H. Poincaré Anal. Non linéaire, 7(1990), 113–122.

[Ch3]   K.S. Chou, On symmetrization and Hessian equations, J. d'Analyse Math., 52(1989), 94–116.

[CGY]   K.S. Chou, D. Geng, Di, and S.S.Yan, Critical dimension of a Hessian equation involving critical exponent and a related asymptotic result. J. Diff. Eqns 129 (1996), 79–110.

[CW1]   K.S. Chou and X.-J. Wang, Variational theory for Hessian equations, Comm. Pure Appl. Math., 54(2001), 1029–1064.

[CW2]   K.S. Chou and X-J. Wang, Variational solutions to Hessian equations, preprint, 1996. Available at wwwmaths.anu.edu.au/research.reports/.

[CS]   A. Colesanti and P. Salani, Generalized solutions of Hessian equations, Bull. Austral. Math. Soc. 56(1997), 459–466.

[D]   P. Delanoe, Radially symmetric boundary value problems for real and complex elliptic Monge-Ampère equations, J. Diff. Eqns 58(1985), 318–344.

[Do]   H. Dong, Hessian equations with elementary symmetric functions, Comm. Partial Diff. Eqns, 31(2006), 1005–1025.

[E]   L.C. Evans, Classical solutions of fully nonlinear, convex, second order elliptic equations, Comm. Pure Appl. Math., 25(1982), 333–362.

[FZ]   D. Faraco and X. Zhong, Quasiconvex functions and Hessian equations. Arch. Ration. Mech. Anal. 168(2003), 245–252.

[GT]   D. Gilbarg and N.S. Trudinger, Elliptic partial differential equations of second order, Springer, 1983.

[G1]   B. Guan, The Dirichlet problem for a class of fully nonlinear elliptic equations, Comm. PDE, 19(1994), 399–416.

[G2]   B. Guan, The Dirichlet problem for Hessian equations on Riemannian manifolds, Calc. Var. PDE, 8(1999), 45–69.

[H]   L. Hörmander, Notions of convexity, Birkhauser, 1994.

[I]   N.M. Ivochkina, Solutions of the Dirichlet problem for certain equations of Monge-Ampère type (in Russian), Mat. Sb., 128(1985), 403–415; English translation in Math. USSR Sb., 56(1987).

[ITW]   N.M. Ivochkina, N.S. Trudinger, X.-J. Wang, The Dirichlet problem for degenerate Hessian equations, Comm. Partial Diff. Eqns 29(2004), 219–235.

[J]   H.Y. Jian, Hessian equations with infinite Dirichlet boundary value, Indiana Univ. Math. J. 55(2006), 1045–1062.

[K1]   N.V. Krylov, Nonlinear elliptic and parabolic equations of the second order, Moscow, Nauka 1985 (in Russian); English translation: Dordrecht, Reidel 1987.

[K2]   N.V. Krylov, Lectures on fully nonlinear second order elliptic equations, Lipschitz Lectures, Univ. Bonn, 1994.

[KT]   H.J. Kuo and N.S. Trudinger, New maximum principles for linear elliptic equations, Indiana Univ. Math. J., to appear.

[Ld]   D. Labutin, Potential estimates for a class of fully nonlinear elliptic equations, Duke Math. J., 111(2002), 1–49.

[Lg]   G. Lieberman, Second order parabolic differential equations, World Scientific, 1996.

[Lp]   P.L. Lions, Two remarks on Monge-Ampère equations, Ann. Mat. Pura Appl. (4) 142(1985), 263–275.

[LT]   M. Lin and N.S. Trudinger, On some inequalities for elementary symmetric functions, Bull. Austral. Math. Soc. 50(1994), 317–326.

[PV1]   N.C. Phuc and I.E. Verbitsky, Local integral estimates and removable singularities for quasilinear and Hessian equations with nonlinear source terms, Comm. Partial Diff.Eqns 31(2006), 1779–1791.

[PV2]   N.C. Phuc and I.E. Verbitsky, Quasilinear and Hessian equations of Lane-Emden type, To appear in Ann. Math.

[P]   A.V. Pogorelov, The multidimensional Minkowski problem, New York, Wiley 1978.

[R]   R.C. Reilly, On the Hessian of a function and the curvatures of its graph, Michigan Math. J. 20(1973–74), 373–383.

[S]   P. Salani, Boundary blow-up problems for Hessian equations, Manuscripta Math. 96(1998), 281–294.

[STW]   W.M. Sheng, N.S. Trudinger, X.-J. Wang, The Yamabe problem for higher order curvatures, J. Diff. Geom., 77(2007), 515–553.

[T]   N.S. Trudinger, On the Dirichlet problem for Hessian equations, Acta Math., 175(1995), 151–164.

[TW1]   N.S. Trudinger and X.-J. Wang, Hessian measures I, Topol. Methods Nonlin. Anal., 10(1997), 225–239.

[TW2]   N.S. Trudinger and X.-J. Wang, Hessian measures II, Ann. Math., 150(1999), 579–604.

[TW3]   N.S. Trudinger and X.-J. Wang, Hessian measures III, J. Funct. Anal., 193(2002), 1–23.

[TW4]   N.S. Trudinger and X.-J. Wang, A Poincaré type inequality for Hessian integrals, Calc. Var. and PDE, 6(1998), 315–328.

[TW5]   N.S. Trudinger and X.-J. Wang, The weak continuity of elliptic operators and applications in potential theory, Amer. J. Math., 551(2002), 11–32.

[TW6]   N.S. Trudinger and X.-J. Wang, Boundary regularity for the Monge-Ampère and affine maximal surface equations, Ann. Math., to appear.

[U1]   J. Urbas, On the existence of nonclassical solutions for two class of fully nonlinear elliptic equations, Indiana Univ. Math. J., 39(1990), 355–382.

[U2]   J. Urbas, An interior second derivative bound for solutions of Hessian equations, Calc. Var. PDE, 12(2001), 417–431.

[U3]   J. Urbas, The second boundary value problem for a class of Hessian equations, Comm. Partial Diff. Eqns, 26(2001), 859–882.

[W1]   X.-J. Wang, Existence of multiple solutions to the equations of Monge-Ampère type, J. Diff. Eqns 100(1992), 95–118.

[W2]   X.-J. Wang, A class of fully nonlinear elliptic equations and related functionals, Indiana Univ. Math. J., 43(1994), 25–54.

[W3]   X.-J. Wang, Some counterexamples to the regularity of Monge-Ampère equations, Proc. Amer. Math. Soc. 123(1995), 841–845.

[WY]   M. Warren and Y. Yuan, Hessian estimates for the σ_2-equation in dimension three, preprint.

Minimal Surfaces in CR Geometry

Paul Yang

In this lecture course, I plan to introduce the subject of minimal surfaces in pseudo-hermitian geometry from an elementary point of view. The topics to be covered are:

1. Introduction to pseudo-hermitian structure.
2. Area of surfaces in 3-D pseudo-hermitian manifold, the p-mean curvature equation.
3. Structure of singular sets.
4. Some applications of the $\mathcal{C}^2$-structure theory.
5. Weak solution and condition for minimizer.
6. Existence and Uniqueness of Boundary value problem.
7. Regularity results.

1 Pseudo-Hermitian Structure

We are given a 3-dim contact structure (M^2, θ) where θ is a one-form with non-degeneracy condition $\theta \wedge d\theta \neq 0$. When (M^3, θ) is given an additional almost complex structure

$$J : \ker \theta \to \ker \theta$$

satisfying $J^2 = -I$, then we say (M^3, θ, J) is a 3-dimensional pseudo-hermitian manifold.

P. Yang
Department of Mathematics, Princeton University, Fine Hall, Washington Road, Princeton, New Jersey 08544
e-mail: yang@math.princeton.edu

S.-Y.A. Chang et al. (eds.), *Geometric Analysis and PDEs*,
Lecture Notes in Mathematics 1977, DOI: 10.1007/978-3-642-01674-5_6,
© Springer-Verlag Berlin Heidelberg 2009

The basic examples come from several complex variables where $\Omega \subset \mathbb{C}^2$ is a smooth strictly pseudo-convex domain given by a defining function $\Omega = \{p|\varphi(p) < 0\}$ and $\partial\Omega = \{p|\varphi(p) = 0\}$ and φ is strictly pluri-subharmonic i.e.,

$$\sum \frac{\partial^2 \varphi}{\partial z_i \partial \bar{z}_j} v^i \bar{v}^i > 0 \quad \text{for all } v \text{ satisfying} \sum \frac{\partial \varphi}{\partial z^i} v^i = 0.$$

Then it is easy to verify that $\frac{-\partial\varphi + \bar\partial\varphi}{\sqrt{-1}}$ is a contact form on $\partial\Omega$, and J is given by the natural complex structure from $\mathbb{C}^2$.

A particular example is the Heisenberg group $\mathbb{H}$ which may be represented as the hyper-surface $\varphi = (Im)\, w - |z|^2 = 0$ in the (z, w) space.

Check $\frac{-\partial\varphi + \bar\partial\varphi}{\sqrt{-1}} = du + 2(xdy - ydx)$ where $w = u + iv$, $z = x + iy$.

The pseudo-hermitian connection of Tanaka and Webster defines covariant differentiation and the associated curvature and torsion.

We begin by a normalizing choice of vector field T called the Reeb vector field satisfying

$$\begin{cases} \theta(T) = 1 \\ \mathcal{L}_T\theta = d\theta(T, \cdot) = 0. \end{cases}$$

We then choose a complex v.f Z_1 to be an eigenvector of J with eigenvalue i, and a complex 1-form θ^1 s.t.

$$\{\theta, \theta^1, \theta^{\bar 1}\} \text{ is dual to } \{T, Z_1, Z_{\bar 1}.\}$$

It follows that

$$d\theta = ih_{1\bar 1}\theta^1 \wedge \theta^{\bar 1} \text{ for real } h_{1\bar 1} > 0$$

then we can normalize further by choosing Z_1 so that $h_{1\bar 1} = 1$

$$d\theta = i\theta^1 \wedge \theta^{\bar 1}.$$

The pseudo-hermitian connection ∇ is given by

$$\nabla Z_1 = \omega_1^1 \otimes Z_1, \, \nabla Z_{\bar 1} = \omega_{\bar 1}^{\bar 1} \otimes Z_{\bar 1}, \, \nabla T = 0.$$

The connection form ω_1^1 is uniquely determined by

$$\begin{cases} d\theta^1 = \theta^1 \wedge \omega_1^1 + A_{\bar 1}^1 \theta \wedge \theta^{\bar 1} \\ \omega_1^1 + \omega_{\bar 1}^{\bar 1} = 0 \end{cases}$$

Then we have

$$d\,\omega_1^1 = W\,\theta^{\bar{1}} \wedge \theta^{\bar{1}} + 2i\,Im\,(A_{11,\bar{1}})\,\theta^{\bar{1}} \wedge \theta$$

$A_{\bar{1}}^1$ — Torsion

W — Webster scalar curvature.

Converting to real forms: $\theta^1 = e^1 + \sqrt{-1}e^2$, $Z_1 = \frac{1}{2}(e_1 - ie_2)$, $\omega_1^1 = iw$

$$d\theta = 2e^1 \wedge e^2$$
$$\nabla e_1 = \omega \otimes e_2 \qquad \nabla e_2 = -\omega \otimes e_1$$

$$de^1 = -e^2 \wedge \omega + \theta \wedge (\Re A_{\bar{1}}^1 e^1 + \Im A_{\bar{1}}^1 e^2)$$
$$de^2 = e^1 \wedge \omega + \theta \wedge (\Im A_{\bar{1}}^1 e^1 - \Re A_{\bar{1}}^1 e^2)$$

$$d\omega(e_1, e_2) = -2W$$

$$[e_1, e_2] = \quad -2T - \omega(e_1)e_1 - \omega(e_2)e_2$$

$$[e_1, T] = (\Re A_{11})\,e_1 - [(\Im A_{11}) + \omega(T)]e_2$$

$$[e_2, T] = [\Im A_{11} + \omega(T)]\,e_1 + (\Re A_{11})e_2.$$

Extend J to all TM by requiring $J(T) = 0$ so that

$$J^2 x = -x + \theta(x)T \quad \forall x \in TM$$

Then the connection is characterized by the conditions

1) $\nabla_x V \in \xi \quad \forall V \in \xi.$
2) $\nabla T = 0$, $\nabla J = 0$, $\nabla(\theta) = 0.$
3) $Tor\,(x, y) = -d\theta(x, y)T$
   $Tor\,(T, Jy) = -JTor\,(T, y)$
   $Tor\,(T, x) = -\frac{1}{2}\,J\,\mathcal{L}_T\,Jx \quad$ for $x \in \xi.$

Remark 1.1. Vanishing torsion condition means that T is an infinitesimal CR transformation.

Equation of Geodesics It turns out that, if $x = \dot{\gamma}(t)$ then

$$\begin{cases} \triangle_x x = \alpha Jx \\ \\ \dot{\alpha} = \langle Tor\,(T, x), x \rangle. \end{cases}$$

This can be checked considering the energy functional

$$E[\gamma] = \int_0^\ell \| \dot{\gamma}(t) \|^2 \, dt.$$

Letting $v = \frac{\partial}{\partial u}\gamma$, then

$$\frac{\partial}{\partial u} E[\gamma_n]\big|_{u=0} = \int_0^\ell v\langle x, x\rangle\, dt$$

$$= 2\int_0^\ell \langle \nabla_v x, x\rangle\, dt$$

$$= 2\int_0^\ell \langle \nabla_x v + (Tor\,(v,x)), x\rangle\, dt$$

$$= 2\int_0^\ell x\langle v, x\rangle - \langle v, \nabla_x x\rangle + \langle Tor\,(v,x), x\rangle\, dt$$

$$= 2\int_0^\ell - \langle v, \nabla_x x\rangle + \langle Tor\,(v,x), x\rangle\, dt$$

$$(\text{write } \nabla_x x = \alpha J x + y) = 2\int_0^\ell -\alpha\langle v, Jx\rangle + \langle v, y\rangle + \theta(v)\langle Tor\,(T,x), x\rangle\, dt$$

$$\begin{pmatrix} d\theta(v,x) = -\langle v, Jx\rangle \\ \| \\ -x\theta(v) \end{pmatrix} = 2\int_0^\ell - \alpha \cdot (x\theta(v)) + \langle v, y\rangle + \theta(v)\langle Tor\,(Tx), x\rangle\, dt$$

$$= 2\int_0^\ell \langle \theta(v) x\alpha(t) + Tor\,(T,x), x\rangle + \langle v, y\rangle\, dt.$$

Thus: $\langle v, y\rangle = 0$ and $x\alpha(t) + \langle Tor\,(T,x), x\rangle = 0$ which is the required Euler equation.

Remark 1.2. Under the assumption $Tor\,(T,x) = 0$ the equation of geodesics becomes

$$\begin{cases} \nabla_x x = \alpha J x \\ \dot{\alpha} = 0, \end{cases}$$

a second order system.

Example 1.3. In the Heisenberg group $\quad \theta = dz + x\,dy - y\,dx$
Then the geodesics γ issuing from 0 have constant curvature α. The projection of γ to the $x\,y$ plane are planar circles of radius $\frac{1}{\alpha}$

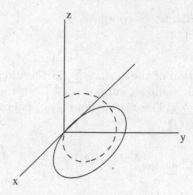

2 Area of a Surface in (M^3, θ, J)

Given a smooth surface $\sum \subset (M^3, \theta, J)$, there is at each point p a (unique up to sign) vector $e \in T_p \sum \cap \xi_p$ if the latter intersection is one-dimension, the generic situation.

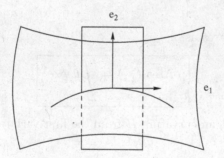

The line field defines a foliation consisting of integral curves of e_1, which we shall call characteristic curves.

Definition 2.1. $p \in \sum$ is a singular point if $T_p \sum = \xi_p$.

Let us define the p-area as variation of volume:

$$\mathsf{Vol}(\Omega) = \frac{1}{2} \int_\Omega \theta \wedge d\theta \quad \text{for a domain } \Omega \subset M.$$

Now suppose $\sum = \partial\Omega$, let us compute

$$\delta_{fe_2} v(\Omega)$$

by considering a variation of the surface $\sum$ in the direction fe_2, where f is a smooth function with compact support in $\sum$, we then extend fe_2 locally to a neighborhood of $\sum$. The vector field fe_2 generates a $1 - p$-group of diffeomorphisms φ_t near $\sum$:

$$\delta_{fe_2} \text{Vol}(\Omega) = \tfrac{d}{dt}\big|_{t=0} \text{Vol}(\varphi_t(\Omega)) = \tfrac{d}{dt}\big|_{t=0} \int_\Omega \varphi_t^*(\theta \wedge d\theta)$$

$$= \int_\Omega \mathcal{L}_{fe_2}(\theta \wedge d\theta)$$

$$= \int_\Omega (d \circ i_{fe_2} + i \circ d_{fe_2})\theta \wedge d\theta$$

$$= \int_{\partial\Omega} i_{fe_2} \theta \wedge d\theta$$

$$= \int_{\partial\Omega} f\theta \wedge e^1$$

Hence we define p-area as

$$\boxed{p\,\text{Area}(\textstyle\sum) = \int_\Sigma \theta \wedge e^1}$$

Next we define p-mean curvature H from the first variation of the p-area:

$$\delta_{fe_2} \int_\Sigma \theta \wedge e^1 = \int_\Sigma \mathcal{L}_{fe_2}(\theta \wedge e^1) = \int_\Sigma i_{fe_2} \circ d(\theta \wedge e^1)$$

$$= \int_\Sigma i_{fe_2}(\theta \wedge e^2 \wedge \omega)$$

$$= \int_\Sigma -f\theta \wedge \omega$$

$$= \int_{\Sigma} -f \underbrace{\omega(e_1)}_{\substack{\| \triangle \\ H}} \theta \wedge e^1 - f\omega(e_2)\theta \wedge e^2$$

$$= \int_{\Sigma} -f \underbrace{\omega(e_1)}_{\substack{\| \triangle \\ H}} + 0.$$

Thus we define the mean curvature to be

$$H = \omega(e_1).$$

Computation Let ψ be a defining function of $\sum$ then $\nabla_b\psi = e_1(\psi)e_1 + (e_2\psi)e_2 = (e_2\psi)e_2$.

$\therefore \frac{\nabla_b\psi}{|\nabla_b\psi|} = e_2$ - called the horizontal normal

$$\mathrm{div}_b\, e_2 = \langle \nabla_{e_1}e_2, e_1 \rangle + \langle \nabla_{e_2}e_2, e_2 \rangle$$

$$= -\langle e_2, \nabla_{e_1}e_1 \rangle$$

$$= -H$$

Example 2.2. In the Heisenberg group

$$H = D^{-3} \left\{ (u_y + x)^2\, u_{xx} - 2(u_y + x)\, u_{xy} + (u_x - y)^2\, u_{yy} \right\}.$$

where

$$D = \left[(u_x - y)^2 + (u_y + x)^2 \right]^{1/2}.$$

Often we will also denote the horizontal normal as N.

3 Structure of the Singular Set

In this section we study the singular set of a C^2-surface in the Heisenberg. So let us consider a surface represented as a graph $\sum : z = u(x,y)$ over a domain $\Omega \subset \mathbb{R}^2$. The singular set of $\sum$ is given by

$$S[u] = \{ (x,y) \in \Omega \,|\, u_x - y = 0,\, u_y + x = 0 \}$$

We have the following dichotomy for $S[u]$:

Theorem A ([CHMY]) *Let $u \in C^2(\Omega)$ satisfy*

$$\begin{cases} \mathrm{div}N(u) = H & \text{in } \Omega \smallsetminus S[u] \\[2mm] |H(p)| = O\left(\frac{1}{|p-p_0|}\right) \end{cases}$$

where p_0 is a singular point. Then either

(a) *p_0 is an isolated singular point of $S[u]$ and*

$$d(N^\perp D)_{p_0} = I$$

or

(b) *$S[u] \cap Bp_0(\in)$ is a C^1-curve and the characteristic curves meet $S[u] \cap Bp_0(t)$ in a C^1 manner transversely, so that the e_1-line field has a continuous extension over $S[u]$.*

Remark 3.1. In case (a), the condition $d(N^\perp D)_{p_0} = I$ means that the e^1-field has index one at an isolated singular point. In case (b), the assertion is that the line field given by e_1 extends continuously across the singular curve

Proof of Theorem A:

Given any unit vector (a,b) at p_0, consider the function $F_{ab} = a(u_x - y) + b(u_y + x)$ and the set

$$\Gamma_{a,b} = \{(x,y) \mid F_{ab}(x,y) = 0\}$$

$$= S[u] \cup \{(x,y) \mid N_u = \pm(b,-a)\}.$$

We claim that if p_0 is not an isolated point, then $\Gamma_{a,b}$ is a C^1 curve passing through p_0 for all but one choice of (a,b). Consider the matrix

$$U = \begin{pmatrix} u_{xx} & u_{xy} - 1 \\ u_{xy} + 1 & u_{yy} \end{pmatrix}.$$

If rank $(U(p)) = 2$, then p_0 is an isolated singular point. On the other hand, if p_0 is not an isolated singular point, we have

$$\mathrm{rank}\,(U(p_0)) = 1.$$

Therefore there exists at most one direction (a_0, b_0) s.t.

$$\nabla F_{a_0 b_0} = ((a_0, b_0)\, U)\,(p_0) = 0.$$

Hence for all other choices of $(a,b) \neq \pm(a_0, b_0)$, The $\Gamma_{a,b}$'s are C^1 curve passing through p, and have a common tangent direction at p_0.

If $S[u]$ is not a $\mathcal{C}^1$-curve near p_0 then for different choices of direction say (a,b) and (a',b'), such a picture must hold:

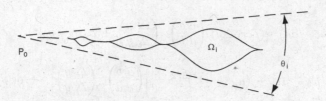

Let v be the unit normal to the ith region enclosed by Γ and Γ'. We have in polar coordinates centered at p_0,

$$\left| \oint_{\partial\Omega_i} N(u) \cdot v \, ds \right| = \left| \iint_{\Omega_i} H \, dx \, dy \right|$$

$$\leq \int_{r(e_i)}^{r(\bar{e}_i)} \int_0^{\theta_i} \left(\frac{C}{r} \, d\theta \right) r \, dr$$

$$\leq C \, \theta_i | r(\bar{e}_i) - r(e_1) |.$$

But

$$\text{LHS} \geq (|(-b,a) \cdot \nu_0 - (-b',a') \cdot \nu_0| - \delta_i) \, | r(\bar{e}_i) - r(e_i) |$$

$$\geq (|c - c'| - \delta_i) \, | r(\bar{e}_i) - r(e_i) |$$

where
$$c \neq c', \theta_i \text{ and } \delta_i \longrightarrow 0 \text{ as } i \longrightarrow \infty.$$

This is a contradiction.

Next we consider case (a) when p_0 is an isolated singular point. Let $N^{\perp} D = (u_y + x, -u_x + y)$. We consider the first order Taylor expansion near p_0:

$$(u_y + y)(p) = (u_{yx} + 1)(p_0) \triangle_x + u_{yy}(p_0) \triangle_y + o(|p - p_0|)$$
$$(u_x - y)(p) = u_{xx}(p_0) \triangle_x + (u_{xy} - 1)(p_0) \triangle_y + o(|p - p_0|).$$

Set $\begin{pmatrix} u_{xx} & u_{xy}+1 \\ u_{xy}-1 & u_{yy} \end{pmatrix} (p_0) = \begin{pmatrix} c \ a \\ d \ b \end{pmatrix}.$

Then

$$P = (u_y + x)^2 u_{xx} - 2(u_y + x)(u_x - y) u_{xy} + (u_x - y)^2 u_{yy}$$

$$= (bc - ad) \cdot (\triangle x, \triangle y) \begin{pmatrix} c & a \\ d & b \end{pmatrix} \begin{pmatrix} \triangle x \\ \triangle y \end{pmatrix} + o(|p - p_0|^2)$$

$$D^3 = ((u_x - y)^2 + (u_y + x)^2)^{3/2} = \left| \begin{pmatrix} a & b \\ c & d \end{pmatrix} \begin{pmatrix} \triangle y \\ \triangle y \end{pmatrix} \right|^3 + o(|p - p_0|^3).$$

Hence the condition $H = \frac{P}{D^3} = O\left(\frac{1}{r}\right)$ implies:
Choosing $\triangle y = 0$:

$$H \sim \frac{(bc - ad) c(\triangle x)^2}{((a^2 + c^2)(\triangle x)^2)^{3/2}} = \frac{(bc - ad)c}{(a^2 + c^2)^{3/2} \triangle x}, \quad \text{we find } c = 0.$$

Choosing $\triangle x = 0$:

$$H \sim O\left(\frac{1}{|p - p_0|}\right) \implies b = 0.$$

Hence,

$$H \sim \frac{-ad(a + d) \triangle x \triangle y}{(a^2(\triangle x)^2 + d^e(\triangle y)^2)^{3/2}} = O\left(\frac{1}{|p - p_0|}\right)$$

Hence $a + d = 0$.

Thus $\qquad d(N^\perp \triangle) \big|_{p_0} = \begin{bmatrix} 1 & 0 \\ 0 & 1 \end{bmatrix}.$

This finishes the proof in the case (a).

In case (b), it remains to show that characteristic curves cross singular curves transversely in C^1-manner. Without loss of generality let us assume Γ is a singular curve, locally represented as a graph over the y-axis. Thus given c near y_0, there is x_c s.t. $(x_c, c) \in \Gamma$. Let us consider the ratio

$$\frac{(u_y + x)(x,c)}{(u_x - y)(x,c)} = \frac{(u_y + x)(x,c) - (u_y + x)(x_c,c)}{(u_x - y)(x,c) - (u_x - y)(x_c,c)}$$

$$= \frac{(x - x_c)(u_{xy} + 1)(x'_c,c)}{(x - x_c)(u_{xx})(x''_c \cdot c)}$$

$$= \frac{(u_{xy} + 1)(x^1_c,c)}{u_{xx}(x''_c,c)}$$

where in the second line, the mean value theorem gives the existence of x'_c and x''_c between x and x_c. Now letting (x, c) tends to (x_0, y_0) we find

$$\lim_{p > p_0} \frac{u_y + x}{u_x - y} = \frac{u_{xy} + 1}{u_{xx}} (p_0).$$

Therefore the right-handed limit of $N(u)$ must agree with the left-handed limit of $N(u)$ up to *sign*.

4 Some Applications of Theorem A

It is always important to understand the blow-up limits of minimal surfaces. In case of graphs, the classification is known as Berstein's theorem. In the setting of entire p-minimal graphs, we have

Theorem 1 ([CHMY]) *The entire C^2-solutions of equation*

(4.1) $(u_y + x)^2 u_{xx} - 2(u_y + x)(u_x - y) u_{xy} + (u_x - y)^2 u_{yy} = 0$ *are either*

(a) *plane or*
(b) *of the form*

$$u = -abx^2 + a^2 - b^2xy + aby^2 + g(-bx + ay)$$

where g is a C^2-function of a single variable, and $a, b \in \mathbb{R}$ are such that $a^2 + b^2 = 1$.

Remark 4.1. 1. ([CH]) Cheng and Hwang classified entire, properly embedded simply connected p-minimal surface. These include the following family of surfaces:

$$(x - x_0) \cos \theta(t) + (y - y_0) \sin \theta(t) = 0$$

where $t = z - y_0x + x_0y$, and θ is a C^2- function of t.

2. Later on we shall prove that the following surface are entire $C^{1,1}$ minimizing solutions

$$u(x, y) = \begin{cases} -xy + (\cot \theta) y^2, \ y \geq 0 \\ \\ -xy + (\cot \eta) y^2, \ y \leq 0 \end{cases}$$

where $\theta + \eta = 2\pi$.

Proof of Theorem 1:

Geometrically $H = 0$ is the condition $\omega_1^1(e_1) = 0$. This is the same as saying the integral curve of e_1-field are contact geodesics, which are lines everywhere tangent to the contact plane. Therefore the graph of u is a ruled surfaces over regular points. Consider the projection of the rulings to the xy plane. There are two possibilities according to whether the rulings project onto parallel lines. If so, we may by rotating the coordinates if necessary write

$$0 = e_1^2 u = u_{xx}$$

so that $N_u = \partial_y$ hence

$$u = xf(y) + g(y) \implies$$

$$u = xy + g(y) \text{ for some } C^2 \text{ function } g.$$

In the other possibility, some rulings intersect at some p_0. Then p_0 is a singular point. Applying Theorem A, it follows that all the rulings must pass through p_0. Reasoning as above, but using polar coordinates, one can check that hence the surface is a plane.

A second application is a topological obstruction to the existence of C^2 embedded surfaces of bounded mean curvature in a general pseudo-hermitian manifold.

Theorem 2 *Let (M^3, θ, J) be a 3-dimensional pseudo-hermitian structure, $\sum \subset M$ be a closed C^2 embedded surface of bounded p-mean curvature, then genus $(\sum) \leq 1$.*

Proof

The index of the line field-e_1 at isolated singular point is 1. The e_1-field can be extended continuously across the singular curves. It follows from Hopf index theorem that the Euler characteristic of such a surface is non-negative.

The classification of singular set also has application to the Isoperimetric problem proposed by Pansu [P]. Consider the extremal problem of a domain $\Omega \subset \mathbb{H}^1$ with smooth boundary and of a given volume, and try to minimize the p-area of the boundary $\partial\Omega$. The optimal configuration, if smooth, will satisfy the condition

$$H = \lambda$$

where λ is the optimal isoperimetric constant. Thus the optimal surface will have rulings given by geodesics of curvature λ. The recent article [RR] uses Theorem A to prove that such a configuration must be congruent to the surface obtained by rotating the following geodesic γ_λ around the z-axis:

$$\begin{cases} x(s) = \frac{1}{2\lambda} \sin(2\lambda s) \\ y(s) = \frac{1}{2\lambda} (-1 + \cos(2\lambda s)) \\ z(s) = \frac{1}{2\lambda} (s - \frac{1}{2\lambda} \sin(2\lambda s)). \end{cases}$$

This surface has precisely two singular points $(0, 0, 0)$ and $(0, 0, \pi)$.

5 Weak Solutions and Condition for Minimizer

In this section we consider the functional

$$\mathcal{F}[u] = \int_\Omega |\nabla u + \mathbf{F}| + Hu \, dx$$

for a fixed vector valued function $F \in L^1(\Omega)$, and a prescribed $H \in L^\infty(\Omega)$ for all test functions $u \in W^{1,1}(\Omega)$. We make the following defintion:

Definition 5.1. (Weak solution): $u \in W^{1,1}(\Omega)$ is a weak solution to the equation

(5.1) $$\operatorname{div} \frac{\nabla u + F}{|\nabla u + F|} = H$$

in Ω if for all $\varphi \in C_0^\infty(\Omega)$ we have

(5.2) $$\int_{S[u]} |\nabla \varphi| + \int_{\Omega \smallsetminus S[u]} N(u) \cdot \nabla \varphi + \int_\Omega H\varphi \geq 0$$

where

$$S[u] = \{x \in \Omega \mid \nabla u(x) + F(x) = 0\}$$

$$N[u] = \frac{\nabla u + F}{|\nabla u + F|}.$$

In the general case, the singular set may be quite large. Even in the special case $\Omega \subset \mathbb{R}^{2n}$ and $F(x,y) = (-y,x)$ corresponding to the p-mean curvature equation, there exists examples (see [Ba]) where the singular set has positive measure.

We observe that when $F(x,y) = (-y,x)$ we have

(5.3) $$(N(u) - N(v)) \cdot (\nabla u - \nabla v) \geq 0.$$

Under such a condition we can characterize minimizers.

Theorem B *Suppose F satisfies (5.3), then $u \in W^{1,1}(\Omega)$ is a minimizer for*

$$\mathcal{F}[u] = \int_\Omega |\nabla u + \mathbf{F}| + Hu \, dx$$

if and only if u is a weak solution.

Remark 5.2. 1. If $H_{n-1}(S[u]) = 0$, and u is a smooth solution to (5.1) then u is a minimizer.
2. Thus a C^2-solution to the equation (4.1) need not be a weak solution. In fact, an elementary calculation shows that if $S[u]$ is a smooth singular curve Γ, then the characteristic curves meeting Γ must do so orthogonally.

Outline of Proof of Theorem B:

Given a test function $\varphi \in C_0^\infty(\Omega)$, consider the variation $u_\epsilon = u + \epsilon\varphi$. We will verify the following:

Claims:

(1) $\dfrac{d\mathcal{F}[u_\epsilon]}{d\epsilon}\Big|_{\hat\epsilon} = \pm \displaystyle\int_{S[u_{\hat\epsilon}]} |\nabla\varphi| + \int_{\Omega\smallsetminus S[u_{\hat\epsilon}]} N(u_{\hat\epsilon}) \cdot \nabla\varphi + \int_\Omega H\varphi.$

if $\hat\epsilon$ is regular (see (3) below).

(2) $\epsilon \longmapsto \mathcal{F}[u_\epsilon]$ is a Lipschitz function.

(3) $\{\epsilon|\ \text{meas}\ (S[u_\epsilon] \cap \{|\nabla\varphi| \neq 0\}) > 0\}$ is a countable set, such ϵ shall be called singular.

(4) If $\epsilon_2 > \epsilon_1$, are regular then

$$\frac{d\mathcal{F}(u_\epsilon)}{d\epsilon}\Big|_{\epsilon_2} - \frac{d\mathcal{F}(u_\epsilon)}{d\epsilon}\Big|_{\epsilon_1} \geq 0.$$

(5) For any sequence of regular $\epsilon_j \searrow \hat\epsilon$

$$\lim_{j\to\infty} \frac{d\mathcal{F}[u_\epsilon]}{d\epsilon}\Big|_{\epsilon_j} = \frac{d\mathcal{F}[u_\epsilon]}{d\epsilon}\Big|_{\hat\epsilon}.$$

It will then follow that a weak solution is a minimizer:

$$\mathcal{F}[u + \varphi] - \mathcal{F}[u] = \int_0^1 \frac{d\mathcal{F}[u_\epsilon]}{d\epsilon}\, d\epsilon \geq 0.$$

Proof

$$\mathcal{F}[u_\epsilon] = \int_{S[u_{\hat\epsilon}]} |\nabla u_\epsilon + F| + \int_{\Omega\smallsetminus S[u_{\hat\epsilon}]} |\nabla u_\epsilon + F| + \int_\Omega H u_{\hat\epsilon} + (\epsilon - \hat\epsilon)\, H\varphi$$

(1)

$$= \int_{S[u_{\hat\epsilon}]} |\epsilon - \hat\epsilon||\nabla\varphi| + \int_{\Omega\smallsetminus S[u_{\hat\epsilon}]} |\nabla u_\epsilon + F| + \int_\Omega H u_{\hat\epsilon} + (\epsilon - \hat\epsilon)\, H\varphi.$$

Comparing with

$$\mathcal{F}[u_{\hat\epsilon}] = \int_{\Omega\smallsetminus S[u_{\hat\epsilon}]} |\nabla u_{\hat\epsilon} + F| + \int_\Omega H u_{\hat\epsilon}$$

we find

$$\frac{\mathcal{F}[u_{\hat\epsilon}] - \mathcal{F}[u_\epsilon]}{\epsilon - \hat\epsilon} = \frac{|\epsilon - \hat\epsilon|}{\epsilon - \hat\epsilon}\int_{S[u_{\hat\epsilon}]} |\nabla\varphi| + \int_{\Omega\smallsetminus S[u_{\hat\epsilon}]} \frac{|\nabla u_\epsilon + F| - |\nabla u_{\hat\epsilon} + F|}{\epsilon - \hat\epsilon} + \int_\Omega H\varphi.$$

The integrand in the second term is bounded by $|\nabla\varphi|$, hence (2). Expanding the integrand: we find

$$\frac{|\nabla u_\epsilon + F| - |\nabla u_{\hat\epsilon} + F|}{\epsilon - \hat\epsilon} = \frac{2(\nabla u_{\hat\epsilon} + F) \cdot \nabla\varphi + (\epsilon - \hat\epsilon)|\nabla\varphi|^2}{|\nabla u_\epsilon + F| + |\nabla u_{\hat\epsilon} + F|}.$$

Taking limit $\epsilon \longrightarrow \hat\epsilon$ in case $\hat\epsilon$ is regular the first integral vanishes and we obtain (1).

(3) Follows from the simple observation that for $\epsilon \neq \epsilon'$ the sets $S[u_\epsilon] \cap \{\nabla\varphi \neq 0\}$ and $S[u_{\epsilon'}] \cap \{\nabla\varphi \neq 0\}$ are disjoint.

(4) For regular values $\epsilon_2 > \epsilon_1$

$$\frac{d\mathcal{F}[u_\epsilon]}{d\epsilon}\bigg|_{\epsilon_2} - \frac{d\mathcal{F}[u_\epsilon]}{d\epsilon}\bigg|_{\epsilon_1} = \int_{\Omega \cap S[u_{\epsilon_1}] \cup S[u_{\epsilon_2}]} (N(u_{\epsilon_2}) - N(u_{\epsilon_1})) \cdot \frac{1}{\epsilon_2 - \epsilon_1}(\nabla u_{\epsilon_2} - \nabla u_{\epsilon_1}) \geq 0$$

on account of (5.3).

(5) First observe that for each j,

$$\int_{S[u_{\epsilon_j}]} |\nabla\varphi| = 0 \text{ hence } \int_{\bigcup_1^\infty S[u_{\epsilon_j}]} |\nabla\varphi| = 0 \text{ and } \int_{\bigcup_1^\infty S[u_{\epsilon_j}]} N(u_{\epsilon_j}) \cdot \nabla\varphi = 0.$$

$$\therefore \frac{d\mathcal{F}[u_\epsilon]}{d\epsilon}\bigg|_{\epsilon_j} = \int_{\Omega \setminus S[u_{\epsilon_j}]} N(u_{\epsilon_j}) \cdot \nabla\varphi + \int_\Omega H\varphi$$

$$= \int_{\Omega \setminus \bigcup_1^\infty S[u_{\epsilon_j}]} N(u_{\epsilon_j}) \cdot \nabla\varphi + \int_\Omega H\varphi.$$

In $S[u_{\hat\epsilon}] \setminus \bigcup_1^\infty S[u_{\epsilon_j}]$ we have

$$N(u_{\epsilon_j}) = \frac{(\nabla u_{\hat\epsilon} + F) + (\epsilon_j - \hat\epsilon)\nabla\varphi}{|\nabla u_{\hat\epsilon} + F + (\epsilon_j - \hat\epsilon)\nabla\varphi|} = \frac{(\epsilon_j - \hat\epsilon)\nabla\varphi}{(\epsilon_j - \hat\epsilon)|\nabla\varphi|}$$

and in $\Omega \setminus (S[u_{\hat\epsilon}] \cup \bigcup_1^\infty S[u_{\epsilon_j}])$ $\lim_{j \longrightarrow \infty} N(u_{\epsilon_j}) = N(u_{\hat\epsilon})$. Hence

$$\int_{\Omega \setminus \bigcup_1^\infty S[u_{\epsilon_j}]} N(u_{\epsilon_j}) \cdot \nabla\varphi = \left\{ \int_{S[u_{\hat\epsilon}] \setminus \bigcup_1^\infty S[u_{\epsilon_j}]} + \int_{\Omega \setminus \left(S[u_{\hat\epsilon}] \cup \bigcup_1^\infty S[u_{\epsilon_j}]\right)} \right\} N(u_{\epsilon_j}) \cdot \nabla\varphi$$

$$\longrightarrow \int_{S[u_{\hat\epsilon}]} |\nabla\varphi| + \int_{\Omega \setminus S[u_{\hat\epsilon}]} N(u_{\hat\epsilon}) \cdot \nabla\varphi.$$

Therefore

$$\frac{d\mathcal{F}[u_\epsilon]}{d\epsilon}\Big|_{\epsilon_j} = \int_{\Omega \smallsetminus S[u_{\epsilon_j}]} N(u_{\epsilon_j}) \cdot \nabla\varphi + \int_\Omega H\,\varphi$$

$$\longrightarrow \int_{S[u_{\hat\epsilon}]} |\nabla\varphi| + \int_{\Omega \smallsetminus S[u_{\hat\epsilon}]} N(u_{\hat\epsilon}) \cdot \nabla\varphi + \int_\Omega H\varphi \cdot$$

$$= \frac{d\mathcal{F}[u_\epsilon]}{d\epsilon}\Big|_{\hat\epsilon}$$

6 Existence and Uniqueness for Boundary Value Problems

Let $\Omega \subset \mathbb{R}^2$ be a convex domain with smooth boundary. Given φ, we consider the problem to find a minimizer of the p-area functional so that:

$$(6.1) \qquad \operatorname{div} \frac{(u_x - y,\, u_y + x)}{((u_x - y)^2 + (u_y + x)^2)^{1/2}} = 0$$

with u agreeing with φ on $\partial\Omega$.

Theorem C *There exists $u \in C^{0,1}(\bar\Omega)$ which is a weak solution to (6.1) that $u = \varphi$ on $\partial\Omega$.*

Indication of Proof

We make use of the elliptic approximation to the equation (6.1):

$$(6.2) \qquad \operatorname{div} \frac{(u_x - y,\, u_y + x)}{\sqrt{\epsilon^2 + (u_x - y)^2 + (u_y + x)^2}} = 0.$$

and then let $\epsilon \longrightarrow 0$, making sure that the solution to the approximate equation converges with uniformly bounded Lipschitz constant. This is a somewhat routine application of known elliptic theory, we skip the details.

We discuss next the question of uniqueness. The following example was first discovered by S. Pauls [P].

Example 6.1.

$$\sum_u : u = \quad x^2 + xy$$

$$\sum_v : v = xy + 1 - y^2$$

are C^2 solutions of the equation (4.1). $\sum_u$ and $\sum_v$ coincide over the unit circle $x^2 + y^2 = 1$, thus represent distinct solutions to the boundary value problem over the unit disc. The singular set $S[u]$ is given by the y-axis while the singular set $S[v]$ is given by $\{x = y\}$.

It is easy to check that in each case the characteristic curves do not meet the singular curve orthogonally. It follows that neither surface is a minimizer. On the other hand, we have the following uniqueness criterion:

Theorem C ([CHMY]) *(Comparison principle for C^2 solutions). Suppose $\Omega \subset \mathbb{R}^2$ is a smoothly bounded domain, and that u, v satisfy*

$$\begin{cases} \operatorname{div} N(u) \geq \operatorname{div} N(v) & \text{in } \Omega \smallsetminus (S[u] \cup S[v]) \\ u \leq v & \text{on } \partial\Omega \end{cases}$$

and

$$H^1 - \operatorname{meas} (S[u] \cup S[v]) = 0;$$

then

$$u \leq v.$$

<u>Claim:</u> $N(u) = N(v)$ on the set $\Omega_+ = \{u > v\}$.

We observe that it follows that Ω_+ is empty: for, if not, Sard's theorem shows that there exists arbitrarily small $\epsilon > 0$ such that $\Omega_{+,\epsilon} = \{u - v \geq \epsilon\}$ has smooth boundary Γ_ϵ. Hence Γ_ϵ may be parametrized by a C^1 curve $(x(s), y(s), u(s))$

$$\frac{du}{ds} + x \frac{dy}{ds} - y \frac{dx}{ds} = 0$$

where $N^\perp(u)$ is tangent to Γ_ϵ: we compute

$$N^\perp(u) = N^\perp(u) \cdot \nabla u = \tfrac{1}{D}\{(\nabla u)^\perp + (x, y)\} \cdot \nabla u$$

$$= \tfrac{1}{D}(x, y) \cdot \nabla u$$

$$= \tfrac{1}{D}(x, y) \cdot \{(\nabla u) + (-y, x)\}$$

$$= (x, y) \cdot N(u).$$

Similarly

$$N^+(u) \cdot v = N^-(v) \cdot v = (x, y) \cdot N(v).$$

This leads to a contradiction:

$$0 = \oint_{\Gamma_\epsilon} \theta = \int_{\Omega_{\epsilon,+}} d\theta > 0.$$

Proof of Comparison Principle

We begin with the formula:

$$(6.3) \quad (\nabla u - \nabla v) \cdot (N(u) - N(v)) = \frac{D_u + D_v}{2} \cdot |N(u) - N(v)|^2.$$

For simplicity, we start by assuming there is no singular set. Let $\Omega_+ = \{u > v\}$.

Proof of Claim that $N(u) = N(v)$ on Ω_+

Again it suffices to show $N(u) = N(v)$ on $\Omega_{+,\delta} = \{u - v > \delta\}$ where $\partial\Omega_{+,\delta}$ is smooth. Consider the integral

$$I^\delta = \oint_{\partial\Omega_{+,\delta}} \tan^{-1}(u - v)\,(N(u) - N(v))\,v\,ds$$

$$= \int_{\Omega_{+,\delta}} \{\frac{1}{1 + (u - v)^2}\,(\nabla u - \nabla v)\,(N(u) - N(v))$$

$$+ \tan^{-1}(u - v)\,\mathrm{div}\,(N(u) - N(v))\}\,dx\,dy$$

$$\geq \int_{\Omega_{+,\delta}} \frac{1}{1 + (u - v)^2} \cdot \frac{D_u + D_v}{2}\,|N(u) - N(v)|^2\,dx\,dy.$$

Hence if $N(u) \neq N(v)$ somewhere, the RHS is bounded from below by a positive number for all δ small. But

$$I^\delta \leq \tan^{-1}\delta \oint_{\partial\Omega_{+,\delta}} (N(u) - N(v)) \cdot \left(\frac{-\nabla(u - v)}{|\nabla(u - v)|}\right)\,ds$$

where the integrand is non-positive. This leads to a contradiction.

When there is a singular set, we cover $S[u] \cup S[v]$ by small balls $\bigcup_{j=1}^{N} B_j$ s.t. $\sum |\partial B_j| < \frac{\epsilon}{2}$. Then apply the argument to the region $\Omega_{+,\delta}$ with these balls removed. Since ϵ can be made arbitrarily small, we still have a contradiction.

7 Regularity of the Regular Part

Our motivation in this section comes from the analytic description of sets of bounded perimeter by Franchi, Serapioni and Serra-Cassano ([FSS]). Their main result is the decomposition of any set $\sum$ of bounded perimeter - which includes in particular regions bounded by Lipschitz graphs: $\sum$ can be decomposed into a regular part which is a $\mathcal{C}^1_{\mathbb{H}}$-hypersurface, and a singular set of measure zero. Although such a surface needs not be C^1 in the ordinary sense, we are interested in improving the regularity of the C^1 regular part. It will be worthwhile to present some examples.

Example 7.1.

$$u = \begin{cases} xy, \ y \geq 0 \\ 0 \quad y \leq 0. \end{cases}$$

The characteristic lines are drawn as follows:

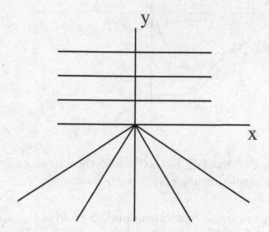

This is a piecewise $\mathcal{C}^1$ surface with folding angles along the x-axis.

Example 7.2. This example is the minimizer of the boundary value side considered in Example 6.1 in section VI.

$$x = s(\sin \eta(t)) + \alpha(t)$$

$$y = -s(\cos \eta(t)) + \beta(t)$$

$$z = s[\beta(t) \sin \eta(t) + \alpha(t) \cos \eta(t)] + \gamma(t)$$

where

$\eta(t) = \frac{\pi}{2} + \theta_2(t) - \delta(t)$

$\cos(\theta_2(t) - \frac{3}{2}\pi) = \frac{t}{\sqrt{2}}$

$\tan \delta(t) = t\sqrt{1 - t^2/2}\Big/(1 - t^2/\sqrt{2})$

$\alpha(t) = t \cos 3\pi/8$

$\beta(t) = t \sin 3\pi/8$

$\dot{\gamma}(t) = \cos^2 3\pi/8 + \cos 3\pi/8 \sin 3\pi/8.$

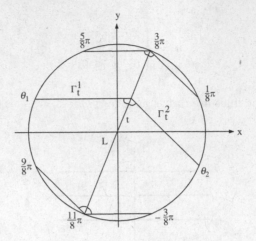

This surface is $C^{1,1}$ but fails to be C^2 along the singular curve which is the line through origin making angle $3\pi/8$ with x-axis.

Example 7.3. This example "smoothens out" the fold of example 7.1 by introducing a buffer region. Choose $\alpha, \beta = [0,1] \longrightarrow \mathbb{R}$ be C^∞ smooth function with the properties;

(7.1) $\alpha(0) = \alpha\left(\frac{1}{2}\right) = \alpha(1) = 0, \alpha'(0) = \alpha'(1) = 1$

$\alpha(t) > 0$ for $0 < t < 1/2$, $\alpha(t) < 0$ for $\frac{1}{2} < t < 1$; $|\alpha(t)| < \frac{\pi}{2}$

$\beta(0) = 0, \beta(\frac{1}{2}) = \frac{1}{2}, \beta(1) = 1, \beta'(t) = 0 \quad 0 < t < 1$

$\beta^{(n)}(0) = \beta^{(n)}(1) = 0$ for all $n = 1, 2 \ldots$

$x = s \cos(\alpha(t))$

$y = s \sin(\alpha(t)) + \beta(t)$

$z = s \beta(t) \cos(\alpha(t))$

for $(s,t) \in [0,\infty) \times [0,1]$. Then for $(s,t) \in (-\infty,0] \times [0,1]$ we take

$x(t) = s \cos(\alpha(t)1)$

$y(t) = -s \sin(\alpha(t)) + \beta(t)$

$z(t) = s \beta(t) \cos(\alpha(t))$

We then glue to the original surface

$$u = \begin{cases} 0 \text{ for } y \leq 0 \\ \\ x \text{ for } y \geq 1. \end{cases}$$

The diligent reader may check that this gives a minimizing surface which is a graph of a function $u \in \mathcal{C}^{1,\alpha}(\mathbb{R}^2)$ for every $\alpha < 1$. It only misses being $\mathcal{C}^2$ at the point $(0,0)$, $(0,\frac{1}{2})$ and $(0,1)$.

References

[B]   Z. Balogh; Size of rectifiable sets and functions of prescribed gradient, *J. Reine Angew. Math.* **564** (2003), 63–83.

[CHMY]   J.-H. Cheng, J.-F. Hwang, A. Malchiodi and P. Yang; "Minimal surfaces in Pseudo-hermitian geometry, " *Ann. Scuola Normal Sup.*, Pisa, I. Sci , **(5)** 2005, 129–177.

[CHY1]   J.-H. Cheng, J.-F Hwang and P. Yang; "Existence and Uniqueness for P-area minimizers in the Heisenberg group," *Math Ann.*, (2007), $337 = 253 - 293$.

[Ch.Y]   S. Chanillo and P. Yang; "Isoperimetric inequality and volume comparison on CR manifolds (WIP).

[P]   P. Pansu; "Une inegalité isoperimétrique sur le group de Heisenberg," C.R. **295**, 127–130.

[Pa]   S. Pauls, "Minimal surfaces in the Heisenberg group," *Geometry Dedicata*, **104** (2004), 201–231.

[T]   N. Tanaka, "A differetial geometric study on strongly pseudo-convex manifolds," Kinokuniya, Tokyo, (1975).

[W]   S. Webster, "Pseudo-hermitian structures of a real hyper-surface," **JDG 13** (1978), 25–41.

List of Participants

1. Abdellaoui Boumediene
   boumediene.abdellaoui@uam.es
   Univ. Tlemcen, Tlemcen, Algeria
2. Ambrosetti Antonio
   ambr@sissa.it
   SISSA/ISAS, Trieste, Italy (**editor**)
3. Amendola Maria Emilia
   emamendola@unisa.it
   Univ. Salerno, Salerno, Italy
4. Ana Primo
   ana.primo@uam.es
   Univ. Autonoma de Madrid, Madrid, Spain
5. Antonangeli Giorgio
   antonang@mat.uniroma1.it
   Univ. La Sapienza, Roma, Italy
6. Autuori Giuseppina
   autuori@math.unifi.it
   Univ. Firenze, Firenze, Italy
7. Bonforte Matteo
   bonforte@calvino.polito.it
   Univ. Autonoma de Madrid, Madrid, Spain
8. Borrello Francesco
   fborrello@dmi.unict.it
   Univ. Catania, Catania, Italy
9. Capogna Luca
   lcapogna@uark.edu
   Univ. Arkansas, Fayetteville, AR, USA
10. Caponigro Marco
    caponigro@sissa.it
    SISSA/ISAS, Trieste, Italy

11. Caravenna Laura
    laura.caravenna@hotmail.com
    SISSA/ISAS, Trieste, Italy
12. Catino Giovanni
    catino@mail.dm.unipi.it
    Univ. Pisa, Pisa, Italy
13. Chang Alice
    chang@math.princeton.edu
    Princeton Univ., Princeton NJ, USA (**editor**)
14. Charro Caballero Fernando
    fernando.charro@uam.es
    Univ. Autonoma de Madrid, Madrid, Spain
15. Coglitore Federico
    fedecogli@libero.it
    Univ. Roma Tre, Roma, Italy
16. Fall Mouhamed
    fall@sissa.it
    SISSA/ISAS, Trieste, Italy
17. Gazzini Marita
    gazzini@sissa.it
    SISSA/ISAS, Trieste, Italy
18. Ghannam Clara
    claraghannam@hotmail.com
    Univ. La Rochelle, La Rochelle, France
19. Ghergu Marius
    marius.ghergu@imar.ro
    Inst. of Math. Simion Stoiłow, Romanian Academy, Bucharest, Romania
20. Giuffrè Sofia
    sofia.giuffre@unirc.it
    Univ. Mediterranea di Reggio Calabria, Reggio Calabria, Italy
21. Gonzalez Maria del Mar
    mgonzale@math.utexas.edu
    The University of Texas at Austin, Austin, TX, USA
22. Grillo Gabriele
    gabriele.grillo@polito.it
    Politecnico di Torino, Torino, Italy
23. Gursky Matthew J.
    mgursky@nd.edu
    Univ. Notre Dame, Notre Dame, IN, USA (**lecturer**)
24. Han Xiaoli
    xhan@ictp.it
    ICTP, Trieste, Italy
25. Ianni Isabella
    ianni@sissa.it
    SISSA/ISAS, Trieste, Italy

26. Khmelynitskaya Alena
    Alena.Khmelynitskaya@ksu.ru
    Kazan State Univ., Kazan, Russia
27. Kokocki Piotr
    koksi@mat.uni.torun.pl
    Nicolaus Copernicus Univ., TORUN, Poland
28. Kruglikov Boris
    kruglikov@math.uit.no
    Univ. Tromso, Tromso, Norway
29. Loiudice Annunziata
    loiudice@dm.uniba.it
    Univ. Bari, Bari, Italy
30. Lanconelli Ermanno
    lanconel@dm.unibo.it
    Univ. Bologna, Bologna, Italy (**lecturer**)
31. Mahmoudi Fethi
    mahmoudi@sissa.it
    SISSA/ISAS, Trieste, Italy
32. Malchiodi Andrea
    malchiod@sissa.it
    SISSA/ISAS, Trieste, Italy (**editor, lecturer**)
33. Mantegazza Carlo
    c.mantegazza@sns.it
    Scuola Normale di Pisa, Pisa, Italy
34. Mercuri Carlo
    mercuri@sissa.it
    SISSA/ISAS, Trieste, Italy
35. Mircea Petrache
    m.petrache@sns.it
    Scuola Normale Superiore Pisa, Pisa, Italy
36. Monticelli Dario
    monticel@mat.unimi.it
    Univ. Milano, Milano, Italy
37. Montoro Luigi
    montoro@mat.unical.it
    Univ. Calabria, Cosenza Italy
38. Munoz Claudio
    claumuno@gmail.com
    Univ. Pierre et Marie Curie, Paris 6, Paris, France
39. Ndiaye Cheikh Birahim
    nadiaye@sissa.it
    SISSA/ISAS, Trieste, Italy
40. Olech Michal
    olech@math.uni.wroc.pl
    University Paris XI, Paris, France

41. Paniccia Irene
    `irene_paniccia@katamail.com`
    Univ. Roma La Sapienza, Roma, Italy
42. Rebai Yomna
    `yomna.rebai@gmail.com`
    Facultè des Sciences de Bizerte, Bizerte, Tunisia
43. Riey Giuseppe
    `riey@mat.unical.it`
    Univ. Calabria, Cosenza, Italy
44. Ruiz David
    `ruiz@sissa.it`
    SISSA/ISAS, Trieste, Italy
45. Scienza Matteo
    `matteo.scienza@sns.it`
    Scuola Normale Superiore Pisa, Pisa, Italy
46. Selvitella Alessandro
    `selvit@sissa.it`
    SISSA/ISAS, Trieste, Italy
47. Siciliano Gaetano
    `siciliano@dm.uniba.it`
    Univ. Bari, Bari, Italy
48. Solferino Viviana
    `solferino@mat.unical.it`
    Univ. Calabria., Cosenza, Italy
49. Stephane Pia
    `stephane.pia@abenavoli.191.it`
    Univ. Napoli, Napoli, Italy
50. Tarantello Gabriella
    `tarantel@mat.uniroma2.it`
    Univ. Roma Tor Vergata, Roma, Italy (**lecturer**)
51. Tran Vu Khanh
    `khanh@math.unipd.it`
    Univ. Padova, Padova, Italy
52. Vaira Giusi
    `vaira@sissa.it`
    SISSA/ISAS, Trieste, Italy
53. Wang Xu-Jia
    `wang@maths.anu.edu.au`
    Australian National Univ., Camberra, Australia (**lecturer**)
54. Yang Paul
    `yang@math.princeton.edu`
    Princeton Univ., Princeton, NJ, USA (**lecturer**)
55. Yashagin Eugene
    `Evgene.Yashagin@ksu.ru`
    State University, Kazan, Russia

LIST OF C.I.M.E. SEMINARS

Published by Ed. Cremonese, Firenze

1966   39. Calculus of variations
        40. Economia matematica
        41. Classi caratteristiche e questioni connesse
        42. Some aspects of diffusion theory

1967   43. Modern questions of celestial mechanics
        44. Numerical analysis of partial differential equations
        45. Geometry of homogeneous bounded domains

1968   46. Controllability and observability
        47. Pseudo-differential operators
        48. Aspects of mathematical logic

1969   49. Potential theory
        50. Non-linear continuum theories in mechanics and physics and their applications
        51. Questions of algebraic varieties

1970   52. Relativistic fluid dynamics
        53. Theory of group representations and Fourier analysis
        54. Functional equations and inequalities
        55. Problems in non-linear analysis

1971   56. Stereodynamics
        57. Constructive aspects of functional analysis (2 vol.)
        58. Categories and commutative algebra

1972   59. Non-linear mechanics
        60. Finite geometric structures and their applications
        61. Geometric measure theory and minimal surfaces

1973   62. Complex analysis
        63. New variational techniques in mathematical physics
        64. Spectral analysis

1974   65. Stability problems
        66. Singularities of analytic spaces
        67. Eigenvalues of non linear problems

1975   68. Theoretical computer sciences
        69. Model theory and applications
        70. Differential operators and manifolds

Published by Ed. Liguori, Napoli

1976   71. Statistical Mechanics
        72. Hyperbolicity
        73. Differential topology

1977   74. Materials with memory
        75. Pseudodifferential operators with applications
        76. Algebraic surfaces

Published by Ed. Liguori, Napoli & Birkhäuser

1978   77. Stochastic differential equations
        78. Dynamical systems

1979   79. Recursion theory and computational complexity
        80. Mathematics of biology

1980   81. Wave propagation
        82. Harmonic analysis and group representations
        83. Matroid theory and its applications

Published by Springer-Verlag

Lecture Notes in Mathematics

For information about earlier volumes
please contact your bookseller or Springer
LNM Online archive: springerlink.com

Vol. 1833: D.-Q. Jiang, M. Qian, M.-P. Qian, Mathematical Theory of Nonequilibrium Steady States. On the Frontier of Probability and Dynamical Systems. IX, 280 p, 2004.

Vol. 1834: Yo. Yomdin, G. Comte, Tame Geometry with Application in Smooth Analysis. VIII, 186 p, 2004.

Vol. 1835: O.T. Izhboldin, B. Kahn, N.A. Karpenko, A. Vishik, Geometric Methods in the Algebraic Theory of Quadratic Forms. Summer School, Lens, 2000. Editor: J.-P. Tignol (2004)

Vol. 1836: C. Năstăsescu, F. Van Oystaeyen, Methods of Graded Rings. XIII, 304 p, 2004.

Vol. 1837: S. Tavaré, O. Zeitouni, Lectures on Probability Theory and Statistics. Ecole d'Eté de Probabilités de Saint-Flour XXXI-2001. Editor: J. Picard (2004)

Vol. 1838: A.J. Ganesh, N.W. O'Connell, D.J. Wischik, Big Queues. XII, 254 p, 2004.

Vol. 1839: R. Gohm, Noncommutative Stationary Processes. VIII, 170 p, 2004.

Vol. 1840: B. Tsirelson, W. Werner, Lectures on Probability Theory and Statistics. Ecole d'Eté de Probabilités de Saint-Flour XXXII-2002. Editor: J. Picard (2004)

Vol. 1841: W. Reichel, Uniqueness Theorems for Variational Problems by the Method of Transformation Groups (2004)

Vol. 1842: T. Johnsen, A. L. Knutsen, K_3 Projective Models in Scrolls (2004)

Vol. 1843: B. Jefferies, Spectral Properties of Noncommuting Operators (2004)

Vol. 1844: K.F. Siburg, The Principle of Least Action in Geometry and Dynamics (2004)

Vol. 1845: Min Ho Lee, Mixed Automorphic Forms, Torus Bundles, and Jacobi Forms (2004)

Vol. 1846: H. Ammari, H. Kang, Reconstruction of Small Inhomogeneities from Boundary Measurements (2004)

Vol. 1847: T.R. Bielecki, T. Björk, M. Jeanblanc, M. Rutkowski, J.A. Scheinkman, W. Xiong, Paris-Princeton Lectures on Mathematical Finance 2003 (2004)

Vol. 1848: M. Abate, J. E. Fornaess, X. Huang, J. P. Rosay, A. Tumanov, Real Methods in Complex and CR Geometry, Martina Franca, Italy 2002. Editors: D. Zaitsev, G. Zampieri (2004)

Vol. 1849: Martin L. Brown, Heegner Modules and Elliptic Curves (2004)

Vol. 1850: V. D. Milman, G. Schechtman (Eds.), Geometric Aspects of Functional Analysis. Israel Seminar 2002-2003 (2004)

Vol. 1851: O. Catoni, Statistical Learning Theory and Stochastic Optimization (2004)

Vol. 1852: A.S. Kechris, B.D. Miller, Topics in Orbit Equivalence (2004)

Vol. 1853: Ch. Favre, M. Jonsson, The Valuative Tree (2004)

Vol. 1854: O. Saeki, Topology of Singular Fibers of Differential Maps (2004)

Vol. 1855: G. Da Prato, P.C. Kunstmann, I. Lasiecka, A. Lunardi, R. Schnaubelt, L. Weis, Functional Analytic Methods for Evolution Equations. Editors: M. Iannelli, R. Nagel, S. Piazzera (2004)

Vol. 1856: K. Back, T.R. Bielecki, C. Hipp, S. Peng, W. Schachermayer, Stochastic Methods in Finance, Bressanone/Brixen, Italy, 2003. Editors: M. Fritelli, W. Runggaldier (2004)

Vol. 1857: M. Émery, M. Ledoux, M. Yor (Eds.), Séminaire de Probabilités XXXVIII (2005)

Vol. 1858: A.S. Cherny, H.-J. Engelbert, Singular Stochastic Differential Equations (2005)

Vol. 1859: E. Letellier, Fourier Transforms of Invariant Functions on Finite Reductive Lie Algebras (2005)

Vol. 1860: A. Borisyuk, G.B. Ermentrout, A. Friedman, D. Terman, Tutorials in Mathematical Biosciences I. Mathematical Neurosciences (2005)

Vol. 1861: G. Benettin, J. Henrard, S. Kuksin, Hamiltonian Dynamics – Theory and Applications, Cetraro, Italy, 1999. Editor: A. Giorgilli (2005)

Vol. 1862: B. Helffer, F. Nier, Hypoelliptic Estimates and Spectral Theory for Fokker-Planck Operators and Witten Laplacians (2005)

Vol. 1863: H. Führ, Abstract Harmonic Analysis of Continuous Wavelet Transforms (2005)

Vol. 1864: K. Efstathiou, Metamorphoses of Hamiltonian Systems with Symmetries (2005)

Vol. 1865: D. Applebaum, B.V. R. Bhat, J. Kustermans, J. M. Lindsay, Quantum Independent Increment Processes I. From Classical Probability to Quantum Stochastic Calculus. Editors: M. Schürmann, U. Franz (2005)

Vol. 1866: O.E. Barndorff-Nielsen, U. Franz, R. Gohm, B. Kümmerer, S. Thorbjønsen, Quantum Independent Increment Processes II. Structure of Quantum Lévy Processes, Classical Probability, and Physics. Editors: M. Schürmann, U. Franz, (2005)

Vol. 1867: J. Sneyd (Ed.), Tutorials in Mathematical Biosciences II. Mathematical Modeling of Calcium Dynamics and Signal Transduction. (2005)

Vol. 1868: J. Jorgenson, S. Lang, $Pos_n(R)$ and Eisenstein Series. (2005)

Vol. 1869: A. Dembo, T. Funaki, Lectures on Probability Theory and Statistics. Ecole d'Eté de Probabilités de Saint-Flour XXXIII-2003. Editor: J. Picard (2005)

Vol. 1870: V.I. Gurariy, W. Lusky, Geometry of Müntz Spaces and Related Questions. (2005)

Vol. 1871: P. Constantin, G. Gallavotti, A.V. Kazhikhov, Y. Meyer, S. Ukai, Mathematical Foundation of Turbulent Viscous Flows, Martina Franca, Italy, 2003. Editors: M. Cannone, T. Miyakawa (2006)

Vol. 1872: A. Friedman (Ed.), Tutorials in Mathematical Biosciences III. Cell Cycle, Proliferation, and Cancer (2006)

Vol. 1873: R. Mansuy, M. Yor, Random Times and Enlargements of Filtrations in a Brownian Setting (2006)

Vol. 1874: M. Yor, M. Émery (Eds.), In Memoriam Paul-André Meyer - Séminaire de Probabilités XXXIX (2006)

Vol. 1875: J. Pitman, Combinatorial Stochastic Processes. Ecole d'Eté de Probabilités de Saint-Flour XXXII-2002. Editor: J. Picard (2006)

Vol. 1876: H. Herrlich, Axiom of Choice (2006)

Vol. 1877: J. Steuding, Value Distributions of L-Functions (2007)

Vol. 1878: R. Cerf, The Wulff Crystal in Ising and Percolation Models, Ecole d'Eté de Probabilités de Saint-Flour XXXIV-2004. Editor: Jean Picard (2006)

Vol. 1879: G. Slade, The Lace Expansion and its Applications, Ecole d'Eté de Probabilités de Saint-Flour XXXIV-2004. Editor: Jean Picard (2006)

Vol. 1880: S. Attal, A. Joye, C.-A. Pillet, Open Quantum Systems I, The Hamiltonian Approach (2006)

Vol. 1881: S. Attal, A. Joye, C.-A. Pillet, Open Quantum Systems II, The Markovian Approach (2006)

Vol. 1882: S. Attal, A. Joye, C.-A. Pillet, Open Quantum Systems III, Recent Developments (2006)

Vol. 1883: W. Van Assche, F. Marcellàn (Eds.), Orthogonal Polynomials and Special Functions, Computation and Application (2006)

Vol. 1884: N. Hayashi, E.I. Kaikina, P.I. Naumkin, I.A. Shishmarev, Asymptotics for Dissipative Nonlinear Equations (2006)

Vol. 1885: A. Telcs, The Art of Random Walks (2006)

Vol. 1886: S. Takamura, Splitting Deformations of Degenerations of Complex Curves (2006)

Vol. 1887: K. Habermann, L. Habermann, Introduction to Symplectic Dirac Operators (2006)

Vol. 1888: J. van der Hoeven, Transseries and Real Differential Algebra (2006)

Vol. 1889: G. Osipenko, Dynamical Systems, Graphs, and Algorithms (2006)

Vol. 1890: M. Bunge, J. Funk, Singular Coverings of Toposes (2006)

Vol. 1891: J.B. Friedlander, D.R. Heath-Brown, H. Iwaniec, J. Kaczorowski, Analytic Number Theory, Cetraro, Italy, 2002. Editors: A. Perelli, C. Viola (2006)

Vol. 1892: A. Baddeley, I. Bárány, R. Schneider, W. Weil, Stochastic Geometry, Martina Franca, Italy, 2004. Editor: W. Weil (2007)

Vol. 1893: H. Hanßmann, Local and Semi-Local Bifurcations in Hamiltonian Dynamical Systems, Results and Examples (2007)

Vol. 1894: C.W. Groetsch, Stable Approximate Evaluation of Unbounded Operators (2007)

Vol. 1895: L. Molnár, Selected Preserver Problems on Algebraic Structures of Linear Operators and on Function Spaces (2007)

Vol. 1896: P. Massart, Concentration Inequalities and Model Selection, Ecole d'Été de Probabilités de Saint-Flour XXXIII-2003. Editor: J. Picard (2007)

Vol. 1897: R. Doney, Fluctuation Theory for Lévy Processes, Ecole d'Été de Probabilités de Saint-Flour XXXV-2005. Editor: J. Picard (2007)

Vol. 1898: H.R. Beyer, Beyond Partial Differential Equations, On linear and Quasi-Linear Abstract Hyperbolic Evolution Equations (2007)

Vol. 1899: Séminaire de Probabilités XL. Editors: C. Donati-Martin, M. Émery, A. Rouault, C. Stricker (2007)

Vol. 1900: E. Bolthausen, A. Bovier (Eds.), Spin Glasses (2007)

Vol. 1901: O. Wittenberg, Intersections de deux quadriques et pinceaux de courbes de genre 1, Intersections of Two Quadrics and Pencils of Curves of Genus 1 (2007)

Vol. 1902: A. Isaev, Lectures on the Automorphism Groups of Kobayashi-Hyperbolic Manifolds (2007)

Vol. 1903: G. Kresin, V. Maz'ya, Sharp Real-Part Theorems (2007)

Vol. 1904: P. Giesl, Construction of Global Lyapunov Functions Using Radial Basis Functions (2007)

Vol. 1905: C. Prévôt, M. Röckner, A Concise Course on Stochastic Partial Differential Equations (2007)

Vol. 1906: T. Schuster, The Method of Approximate Inverse: Theory and Applications (2007)

Vol. 1907: M. Rasmussen, Attractivity and Bifurcation for Nonautonomous Dynamical Systems (2007)

Vol. 1908: T.J. Lyons, M. Caruana, T. Lévy, Differential Equations Driven by Rough Paths, Ecole d'Été de Probabilités de Saint-Flour XXXIV-2004 (2007)

Vol. 1909: H. Akiyoshi, M. Sakuma, M. Wada, Y. Yamashita, Punctured Torus Groups and 2-Bridge Knot Groups (I) (2007)

Vol. 1910: V.D. Milman, G. Schechtman (Eds.), Geometric Aspects of Functional Analysis. Israel Seminar 2004-2005 (2007)

Vol. 1911: A. Bressan, D. Serre, M. Williams, K. Zumbrun, Hyperbolic Systems of Balance Laws. Cetraro, Italy 2003. Editor: P. Marcati (2007)

Vol. 1912: V. Berinde, Iterative Approximation of Fixed Points (2007)

Vol. 1913: J.E. Marsden, G. Misiołek, J.-P. Ortega, M. Perlmutter, T.S. Ratiu, Hamiltonian Reduction by Stages (2007)

Vol. 1914: G. Kutyniok, Affine Density in Wavelet Analysis (2007)

Vol. 1915: T. Bıyıkoğlu, J. Leydold, P.F. Stadler, Laplacian Eigenvectors of Graphs. Perron-Frobenius and Faber-Krahn Type Theorems (2007)

Vol. 1916: C. Villani, F. Rezakhanlou, Entropy Methods for the Boltzmann Equation. Editors: F. Golse, S. Olla (2008)

Vol. 1917: I. Veselić, Existence and Regularity Properties of the Integrated Density of States of Random Schrödinger (2008)

Vol. 1918: B. Roberts, R. Schmidt, Local Newforms for GSp(4) (2007)

Vol. 1919: R.A. Carmona, I. Ekeland, A. Kohatsu-Higa, J.-M. Lasry, P.-L. Lions, H. Pham, E. Taflin, Paris-Princeton Lectures on Mathematical Finance 2004. Editors: R.A. Carmona, E. Çinlar, I. Ekeland, E. Jouini, J.A. Scheinkman, N. Touzi (2007)

Vol. 1920: S.N. Evans, Probability and Real Trees. Ecole d'Été de Probabilités de Saint-Flour XXXV-2005 (2008)

Vol. 1921: J.P. Tian, Evolution Algebras and their Applications (2008)

Vol. 1922: A. Friedman (Ed.), Tutorials in Mathematical BioSciences IV. Evolution and Ecology (2008)

Vol. 1923: J.P.N. Bishwal, Parameter Estimation in Stochastic Differential Equations (2008)

Vol. 1924: M. Wilson, Littlewood-Paley Theory and Exponential-Square Integrability (2008)

Vol. 1925: M. du Sautoy, L. Woodward, Zeta Functions of Groups and Rings (2008)

Vol. 1926: L. Barreira, V. Claudia, Stability of Nonautonomous Differential Equations (2008)

Vol. 1927: L. Ambrosio, L. Caffarelli, M.G. Crandall, L.C. Evans, N. Fusco, Calculus of Variations and Non-Linear Partial Differential Equations. Cetraro, Italy 2005. Editors: B. Dacorogna, P. Marcellini (2008)

Vol. 1928: J. Jonsson, Simplicial Complexes of Graphs (2008)

Vol. 1929: Y. Mishura, Stochastic Calculus for Fractional Brownian Motion and Related Processes (2008)

Vol. 1930: J.M. Urbano, The Method of Intrinsic Scaling. A Systematic Approach to Regularity for Degenerate and Singular PDEs (2008)

Vol. 1931: M. Cowling, E. Frenkel, M. Kashiwara, A. Valette, D.A. Vogan, Jr., N.R. Wallach, Representation Theory and Complex Analysis. Venice, Italy 2004. Editors: E.C. Tarabusi, A. D'Agnolo, M. Picardello (2008)

Vol. 1932: A.A. Agrachev, A.S. Morse, E.D. Sontag, H.J. Sussmann, V.I. Utkin, Nonlinear and Optimal Control Theory. Cetraro, Italy 2004. Editors: P. Nistri, G. Stefani (2008)

Vol. 1933: M. Petkovic, Point Estimation of Root Finding Methods (2008)

Vol. 1934: C. Donati-Martin, M. Émery, A. Rouault, C. Stricker (Eds.), Séminaire de Probabilités XLI (2008)

Vol. 1935: A. Unterberger, Alternative Pseudodifferential Analysis (2008)

Vol. 1936: P. Magal, S. Ruan (Eds.), Structured Population Models in Biology and Epidemiology (2008)

Recent Reprints and New Editions

LECTURE NOTES IN MATHEMATICS

 Springer

Edited by J.-M. Morel, F. Takens, B. Teissier, P.K. Maini

Editorial Policy (for Multi-Author Publications: Summer Schools/Intensive Courses)

1. Lecture Notes aim to report new developments in all areas of mathematics and their applications - quickly, informally and at a high level. Mathematical texts analysing new developments in modelling and numerical simulation are welcome. Manuscripts should be reasonably self-contained and rounded off. Thus they may, and often will, present not only results of the author but also related work by other people. They should provide sufficient motivation, examples and applications. There should also be an introduction making the text comprehensible to a wider audience. This clearly distinguishes Lecture Notes from journal articles or technical reports which normally are very concise. Articles intended for a journal but too long to be accepted by most journals, usually do not have this "lecture notes" character.

2. In general SUMMER SCHOOLS and other similar INTENSIVE COURSES are held to present mathematical topics that are close to the frontiers of recent research to an audience at the beginning or intermediate graduate level, who may want to continue with this area of work, for a thesis or later. This makes demands on the didactic aspects of the presentation. Because the subjects of such schools are advanced, there often exists no textbook, and so ideally, the publication resulting from such a school could be a first approximation to such a textbook. Usually several authors are involved in the writing, so it is not always simple to obtain a unified approach to the presentation.

   For prospective publication in LNM, the resulting manuscript should not be just a collection of course notes, each of which has been developed by an individual author with little or no co-ordination with the others, and with little or no common concept. The subject matter should dictate the structure of the book, and the authorship of each part or chapter should take secondary importance. Of course the choice of authors is crucial to the quality of the material at the school and in the book, and the intention here is not to belittle their impact, but simply to say that the book should be planned to be written by these authors jointly, and not just assembled as a result of what these authors happen to submit.

   This represents considerable preparatory work (as it is imperative to ensure that the authors know these criteria before they invest work on a manuscript), and also considerable editing work afterwards, to get the book into final shape. Still it is the form that holds the most promise of a successful book that will be used by its intended audience, rather than yet another volume of proceedings for the library shelf.

3. Manuscripts should be submitted either online at www.editorialmanager.com/lnm/ to Springer's mathematics editorial, or to one of the series editors. Volume editors are expected to arrange for the refereeing, to the usual scientific standards, of the individual contributions. If the resulting reports can be forwarded to us (series editors or Springer) this is very helpful. If no reports are forwarded or if other questions remain unclear in respect of homogeneity etc, the series editors may wish to consult external referees for an overall evaluation of the volume. A final decision to publish can be made only on the basis of the complete manuscript; however a preliminary decision can be based on a pre-final or incomplete manuscript. The strict minimum amount of material that will be considered should include a detailed outline describing the planned contents of each chapter.

   Volume editors and authors should be aware that incomplete or insufficiently close to final manuscripts almost always result in longer evaluation times. They should also be aware that parallel submission of their manuscript to another publisher while under consideration for LNM will in general lead to immediate rejection.

4. Manuscripts should in general be submitted in English. Final manuscripts should contain at least 100 pages of mathematical text and should always include

   - a general table of contents;
   - an informative introduction, with adequate motivation and perhaps some historical remarks: it should be accessible to a reader not intimately familiar with the topic treated;
   - a global subject index: as a rule this is genuinely helpful for the reader.

   Lecture Notes volumes are, as a rule, printed digitally from the authors' files. We strongly recommend that all contributions in a volume be written in the same LaTeX version, preferably LaTeX2e. To ensure best results, authors are asked to use the LaTeX2e style files available from Springer's web-server at

   ftp://ftp.springer.de/pub/tex/latex/svmonot1/ (for monographs) and
   ftp://ftp.springer.de/pub/tex/latex/svmultt1/ (for summer schools/tutorials).

   Additional technical instructions are available on request from: lnm@springer.com.

5. Careful preparation of the manuscripts will help keep production time short besides ensuring satisfactory appearance of the finished book in print and online. After acceptance of the manuscript authors will be asked to prepare the final LaTeX source files and also the corresponding dvi-, pdf- or zipped ps-file. The LaTeX source files are essential for producing the full-text online version of the book. For the existing online volumes of LNM see: http://www.springerlink.com/openurl.asp?genre=journal&issn=0075-8434.

   The actual production of a Lecture Notes volume takes approximately 12 weeks.

6. Volume editors receive a total of 50 free copies of their volume to be shared with the authors, but no royalties. They and the authors are entitled to a discount of 33.3% on the price of Springer books purchased for their personal use, if ordering directly from Springer.

7. Commitment to publish is made by letter of intent rather than by signing a formal contract. Springer-Verlag secures the copyright for each volume. Authors are free to reuse material contained in their LNM volumes in later publications: a brief written (or e-mail) request for formal permission is sufficient.

Addresses:

Professor J.-M. Morel, CMLA,
École Normale Supérieure de Cachan,
61 Avenue du Président Wilson,
94235 Cachan Cedex, France
E-mail: Jean-Michel.Morel@cmla.ens-cachan.fr

Professor F. Takens, Mathematisch Instituut,
Rijksuniversiteit Groningen, Postbus 800,
9700 AV Groningen, The Netherlands
E-mail: F.Takens@rug.nl

Professor B. Teissier,
Institut Mathématique de Jussieu,
UMR 7586 du CNRS,
Équipe "Géométrie et Dynamique",
175 rue du Chevaleret,
75013 Paris, France
E-mail: teissier@math.jussieu.fr

For the "Mathematical Biosciences Subseries" of LNM:

Professor P.K. Maini, Center for Mathematical Biology,
Mathematical Institute, 24-29 St Giles,
Oxford OX1 3LP, UK
E-mail: maini@maths.ox.ac.uk

Springer, Mathematics Editorial I, Tiergartenstr. 17,
69121 Heidelberg, Germany,
Tel.: +49 (6221) 487-8259
Fax: +49 (6221) 4876-8259
E-mail: lnm@springer.com